OCEAN CIRCULATION

PREPARED BY ANGELA COLLING FOR THE COURSE TEAM

BUTTERWORTH HEINEMANN

in association with

THE OPEN UNIVERSITY, WALTON HALL, MILTON KEYNES, MK7 6AA, ENGLAND

Butterworth–Heinemann
Linacre House, Jordan Hill, Oxford OX2 8DP
A division of Reed Educational and Professional Publishing Ltd

A member of the Reed Elsevier plc group

BOSTON   JOHANNESBURG
MELBOURNE   NEW DELHI   OXFORD

British Library Cataloguing in Publication Data
A catalogue record for this book is available from the British Library

ISBN 0 7506 5278 0

Library of Congress Cataloguing in Publication Data
A catalogue record for this book is available from the Library of Congress

Jointly published by the Open University, Walton Hall, Milton Keynes
MK7 6AA and Butterworth–Heinemann

FOR EVERY TITLE THAT WE PUBLISH, BUTTERWORTH-HEINEMANN
WILL PAY FOR BTCV TO PLANT AND CARE FOR A TREE.

Edited, typeset, illustrated and designed by The Open University
Printed in Singapore by Kyodo under the supervision of MRM
Graphics Ltd., UK

s330v3i2.1

CONTENTS

ABOUT THIS VOLUME

This is one of a Series of Volumes on Oceanography. It is designed so that it can be read on its own, like any other textbook, or studied as part of S330 *Oceanography*, a third level course for Open University students. The science of oceanography as a whole is multidisciplinary. However, different aspects fall naturally within the scope of one or other of the major 'traditional' disciplines. Thus, you will get the most out of this Volume if you have some previous experience of studying physics. Other Volumes in this Series lie more within the fields of geology, chemistry and biology (and their associated sub-branches).

Chapter 1 establishes the essential causes of the circulation patterns of air and water in the atmosphere and the oceans. These are: first, the need for heat to be redistributed over the surface of the Earth; and secondly, that the Earth is both rotating and spherical. Chapter 2 follows on from this by describing the global wind system, and showing how the atmospheric circulation transports heat from low latitudes to high latitudes. It emphasizes that the atmosphere and oceans are continuously interacting, effectively forming one system, and illustrates this idea through the spectacular example of tropical cyclones.

Chapter 3 explains the fundamental principles relating to wind-driven surface currents. The theoretical basis is established with the minimum use of mathematical equations, which are explained and applied in context to enable you to understand their relevance. The Chapter ends with a discussion of the energy of the ocean, and emphasizes the importance of mesoscale eddies which, although they are believed to contain most of the energy of the ocean circulation, were only discovered in the 1970s.

Chapter 4 uses the example of the North Atlantic subtropical gyre, particularly the Gulf Stream, to explore how ideas about ocean circulation have evolved from the sixteenth century up to the present day. Some of the theoretical models of the ocean, devised by oceanographers to help them understand why the subtropical gyres have the form they do, are discussed with a minimum of mathematics. The equations of motion, which are the basis for all dynamic models, are introduced in a qualitative way, and some recent computer models are briefly discussed, so as to bring out the fundamental purpose of modelling. The last Section introduces the idea of 'teleconnections', with discussion of the North Atlantic Oscillation – a continual climatic 'sea-saw', which is the most important factor affecting winter conditions over the northern Atlantic Ocean and Nordic Seas.

Chapter 5 considers the major current systems outside the subtropical gyres, in both high and low latitudes. The characteristics of the current systems of the tropical regions are described and explained in terms of the direct and indirect effects of the wind – the Trade Winds in the Atlantic and Pacific, and the seasonally reversing monsoonal winds in the Indian Ocean. The importance of the Equator as a 'wave guide', which channels disturbances in the upper ocean, is briefly discussed, particularly in the context of the climatic oscillation known as El Niño. The Chapter ends with a description of the Antarctic Circumpolar Wave, a large-scale wave-like disturbance involving the atmosphere, ocean and ice-cover, which travels around the Antarctic continent once every 8–10 years.

The last Chapter takes up the theme of the first, and revises many of the topics introduced in earlier Chapters. It considers the transport of heat and water through the ocean–atmosphere system and begins with a discussion of processes occurring at the air–sea interface, where all oceanic water masses acquire their characteristics. Changes in the rates of formation of the cold, dense water masses which flow through the deep basins of the ocean are closely related to climatic change, and this topic is discussed in relation to climatic fluctuations in the North Atlantic. Finally, there is a discussion of the importance of the World Ocean Circulation Experiment (initiated at the end of the 1980s) and a brief look at future developments in oceanography.

You will find questions designed to help you to develop arguments and/or test your own understanding as you read, with answers provided at the back of this Volume. Important technical terms are printed in **bold** type where they are first introduced or defined.

ABOUT THIS SERIES

The Volumes in this Series are all presented in the same style and format, and together provide a comprehensive introduction to marine science. *Ocean Basins* deals with structure and formation of oceanic crust, hydrothermal circulation, and factors affecting sea-level. *Seawater* considers the seawater solution and leads naturally into *Ocean Circulation*, which is the 'core' of the Series. *Waves, Tides and Shallow-Water Processes* introduces the physical processes which control water movement and sediment transport in the nearshore environment (beaches, estuaries, deltas, shelves). *Ocean Chemistry and Deep-Sea Sediments* is concerned with biogeochemical cycling of elements within the seawater solution and with water–sediment interaction on the ocean floor. *Case Studies in Oceanography and Marine Affairs* examines the effect of human intervention in the marine environment and introduces the essentials of Law of the Sea. The two case studies respectively review marine affairs in the Arctic from an historical standpoint, and outline the causes and effects of the tropical climatic phenomenon known as El Niño.

Biological Oceanography: An Introduction (by C. M. Lalli and T. R. Parsons) is a companion Volume to the Series, and is also in the same style and format. It describes and explains interactions between marine plants and animals in relation to the physical/chemical properties and dynamic behaviour of the seawater in which they live.

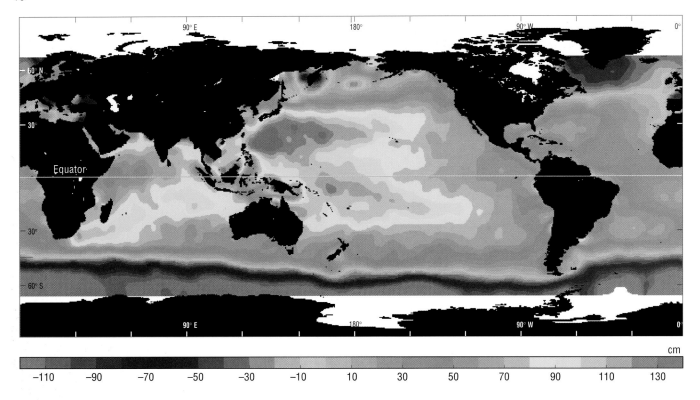

Frontispiece: The sea-surface topography in December 2000, as measured by altimeters aboard the *TOPEX–Poseidon* satellite.

CHAPTER 1 | INTRODUCTION

The waters of the ocean are continually moving – in powerful currents like the Gulf Stream, in large gyres, in features visible from space (Figure 1.1), and in smaller swirls and eddies ranging in size down to a centimetre across or less.

Figure 1.1   These spiral eddies in the central Mediterranean Sea (made visible through the phenomenon of 'sun glint') were photographed from the Space Shuttle *Challenger*. Measuring some 12–15 km across, they are only one example of the wide range of gyral motions occurring in the oceans.

What drives all this motion?

The short answer is: energy from the Sun, and the rotation of the Earth.

The most obvious way in which the Sun drives the oceanic circulation is through the circulation of the atmosphere – that is, winds. Energy is transferred from winds to the upper layers of the ocean through frictional coupling between the ocean and the atmosphere at the sea-surface.

The Sun also drives ocean circulation by causing variations in the temperature and salinity of seawater which in turn control its density. Changes in temperature are caused by fluxes of heat across the air–sea boundary; changes in salinity are brought about by addition or removal of freshwater, mainly through evaporation and precipitation, but also, in polar regions, by the freezing and melting of ice. All of these processes are linked directly or indirectly to the effect of solar radiation.

If surface water becomes denser than the underlying water, the situation is unstable and the denser surface water sinks. Vertical, density-driven circulation that results from cooling and/or increase in salinity – i.e. changes in the content of heat and/or salt – is known as **thermohaline circulation**. The large-scale thermohaline circulation of the ocean will be discussed in Chapter 6.

How does the rotation of the Earth contribute to ocean circulation patterns?

Except for a relatively thin layer close to the solid Earth, frictional coupling between moving water and the Earth is weak, and the same is true for air masses. In the extreme case of a projectile moving above the surface of the Earth, the frictional coupling is effectively zero. Consider, for instance, a missile fired northwards from a rocket launcher positioned on the Equator (Figure 1.2(a)). As it leaves the launcher, the missile is moving eastwards at the same velocity as the Earth's surface as well as moving northwards at its firing velocity. As the missile travels north, the Earth is turning eastwards beneath it. Initially, because it has the same eastward velocity as the surface of the Earth, the missile appears to travel in a straight line. However, the eastward velocity at the surface of the Earth is greatest at the Equator and decreases towards the poles, so as the missile travels progressively northwards, the eastward velocity of the Earth beneath becomes less and less. As a result, *in relation to the Earth*, the missile is moving not only northwards but also *eastwards*, at a progressively greater rate (Figure 1.2(b)). This apparent deflection of objects that are moving over the surface of the Earth without being frictionally bound to it – be they missiles, parcels of water or parcels of air – is explained in terms of an apparent force known as the **Coriolis force**.

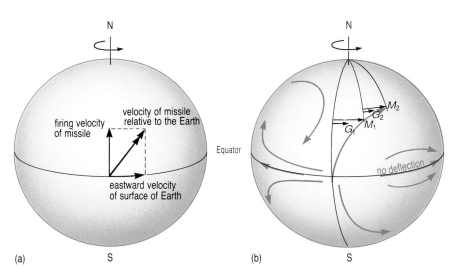

Figure 1.2 (a) A missile launched from the Equator has not only its northward firing velocity but also the same eastward velocity as the surface of the Earth at the Equator. The resultant velocity of the missile is therefore a combination of these two, as shown by the double arrow.

(b) The path taken by the missile in relation to the surface of the Earth. In time interval T_1, the missile has moved eastwards to M_1, and the Earth to G_1; in the time interval T_2, the missile has moved to M_2, and the Earth to G_2. Note that the apparent deflection attributed to the Coriolis force (the difference between M_1 and G_1 and M_2 and G_2) increases with increasing latitude. The other blue curves show possible paths for missiles or any other bodies moving over the surface of the Earth without being strongly bound to it by friction.

N

axis of rotation

Figure 1.3   Diagram of a hypothetical cylindrical Earth, for use with Question 1.1.

QUESTION 1.1   Bearing in mind what you have just read, especially in connection with Figure 1.2, what can you say about the Coriolis force acting on a body moving above the curved surface of a hypothetical *cylindrical* Earth rotating about its axis, as shown in Figure 1.3?

The example given above, of a missile fired northwards from the Equator, was chosen because of its simplicity. In fact, the rotation of the (spherical) Earth about its axis causes deflection of currents, winds and projectiles, irrespective of their initial direction (Figure 1.2(b)). Why this occurs will be explained in Chapter 4, but for now you need only be aware of the following important points:

1   The magnitude of the Coriolis force increases from zero at the Equator to a maximum at the poles.

2   The Coriolis force acts at right angles to the direction of motion, so as to cause deflection to the *right* in the Northern Hemisphere and to the *left* in the Southern Hemisphere.

How these factors affect the direction of current flow in the oceans, and of winds in the atmosphere, is illustrated by the blue arrows in Figure 1.2(b).

When missile trajectories are determined, the effect of the Coriolis force is included in the calculations, but the allowance that has to be made for it is relatively small. This is because a missile travels at high speed and the amount that the Earth has 'turned beneath' it during its short period of travel is small. Winds and ocean currents, on the other hand, are relatively slow moving, and so are significantly affected by the Coriolis force. Consider, for example, a current flowing with a speed of $0.5\,\mathrm{m\,s^{-1}}$ (~ 1 knot) at about 45° of latitude. Water in the current will travel approximately 1800 metres in an hour, and during that hour the Coriolis force will have deflected it about 300 m from its original path (assuming that no other forces are acting to oppose it).

Deflection by the Coriolis force is sometimes said to be *cum sole* (pronounced 'cum so-lay'), or 'with the Sun'. This is because deflection occurs in the same direction as that in which the Sun appears to move across the sky – towards the right in the Northern Hemisphere and towards the left in the Southern Hemisphere.

The Coriolis force thus has the visible effect of deflecting ocean currents. It must also be considered in any study of ocean circulation for another, less obvious, reason. Although it is not a real force in the fixed framework of space, it *is* real enough from the point of view of anything moving in relation to the Earth. We can study, and make predictions about, currents and winds within which the Coriolis force is balanced by horizontal forces resulting from pressure gradients, and for which there is *no* deflection. Such flows are described as *geostrophic* (meaning 'turned by the Earth') and will be discussed in Chapters 2 and 3.

Because solar heating, directly and indirectly, is the fundamental cause of atmospheric and oceanic circulation, the second half of this introductory Chapter will be devoted to the radiation balance of planet Earth.

1.1 THE RADIATION BALANCE OF THE EARTH–OCEAN–ATMOSPHERE SYSTEM

The solid red curves in Figure 1.4(a)(i) and (ii) show the incoming solar radiation reaching the Earth and atmosphere, as a function of latitude, for the northern summer and the southern summer, respectively. The intensity of incoming solar radiation is greatest for mid-latitudes in the hemisphere experiencing summer, while for high latitudes in the winter hemisphere, the oblique angle of the Sun's rays, combined with the long periods of winter darkness, results in amounts of radiation received being low.

However, the Earth not only receives short-wave radiation from the Sun, it also *re*-emits radiation, of a longer wavelength. Little of this long-wave radiation is radiated directly into space; most of it is absorbed by the

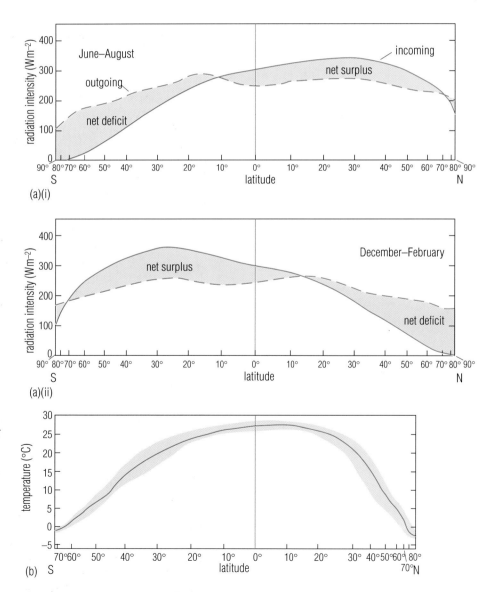

Figure 1.4 (a) The radiation balance at the top of the atmosphere plotted against latitude (scaled according to the Earth's surface area) for (i) the northern summer and (ii) the southern summer. The red solid line is the intensity of incoming solar radiation and the red dashed line is the intensity of radiation lost to space (both determined using satellite-borne radiometer).

(b) The average temperature of surface waters at different latitudes. At a given latitude, there will be surface waters whose annual mean temperatures are higher or lower than shown by the curve; this range is represented by the thickness of the pink envelope.

atmosphere, particularly by carbon dioxide, water vapour and cloud droplets. Thus, the atmosphere is heated from beneath by the Earth and itself re-emits long-wave radiation into space. This generally occurs from the top of the cloud cover where temperatures are surprisingly similar at all latitudes. The intensity of the radiation emitted into space therefore does not vary greatly with latitude, although as can be seen from the dashed curves in Figure 1.4(a), it is highest for the subtropics and lowest for high latitudes in the hemisphere experiencing winter.

QUESTION 1.2

(a) According to Figure 1.4(a), for what latitude band is there always a net surplus of radiation? At what latitudes is there always a deficit?

(b) (i) Approximately how much warmer than surface waters poleward of 70° are surface ocean waters within 10° of the Equator?

(ii) 'Sea-surface temperatures are highest in the vicinity of the Equator because of the long day-length there.' True or false?

In Figure 1.4(a), the two areas labelled 'net surplus' are together about equal to those labelled 'net loss'. This suggests that the radiation budget for the Earth–ocean–atmosphere system as a whole is in balance, i.e. that over the course of a year the system is not gaining more radiation energy than it is losing (or *vice versa*). Note that this use of areas under curves only works here because the horizontal axis is increasingly compressed towards higher latitudes to compensate for the decreasing area of the globe within given latitude bands. Note also that the temperature at the top of the atmosphere is not directly related to the global temperature in general so plots such as those in Figure 1.4(a) do not allow us to determine whether the 'enhanced greenhouse effect' resulting from increasing concentrations of atmospheric CO_2 is leading to global warming.

The positive radiation balance at low latitudes, and the negative one at high latitudes, results in a net transfer of heat energy from low to high latitudes, by means of wind systems in the atmosphere and current systems in the ocean. There has been much debate about the relative importance of the atmosphere and ocean in the poleward transport of heat. Figure 1.5 shows estimates of poleward heat transport, based on satellite observations of the upper atmosphere, and measurements within the atmosphere, published in 1985. Positive values correspond to northward transport and negative values to southward transport.

QUESTION 1.3

(a) How does the total poleward transport of heat in the ocean compare with that in the atmosphere, according to Figure 1.5?

(b) Over which parts of the globe does the ocean contribute significantly more to poleward heat transport, and over which parts does the atmosphere contribute significantly more?

Wind systems redistribute heat partly by the **advection** of warm air masses into cooler regions (and *vice versa*), and partly by the transfer of latent heat which is taken up when water is converted into water vapour, and released when the water vapour condenses in a cooler environment. The tropical storms known as cyclones or hurricanes are dramatic manifestations of the transfer of energy from ocean to atmosphere in the form of latent heat. The generation of cyclones, and their role in carrying heat away from the tropical oceans, will be described in Chapter 2.

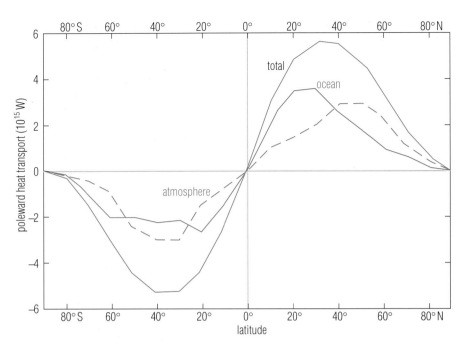

Figure 1.5 Estimates of the contributions to poleward heat transport (red curve) by the ocean (solid blue curve) and the atmosphere (dashed blue curve). Positive values are northward heat transport, negative values are southward heat transport. The total heat transport was derived from satellite measurements at the top of the atmosphere, heat transported by the atmosphere was estimated using measurements of atmospheric heat fluxes, and the heat transported by the ocean was calculated as the difference between the two.

1.2 SUMMARY OF CHAPTER 1

1   Circulation in both the oceans and the atmosphere is driven by energy from the Sun and modified by the Earth's rotation.

2   The radiation balance of the Earth–ocean–atmosphere system is positive at low latitudes and negative at high latitudes. Heat is redistributed from low to high latitudes by means of wind systems in the atmosphere and current systems in the ocean. There are two principal components of the ocean circulation: wind-driven surface currents and the density-driven (thermohaline) deep circulation.

3   Air and water masses moving over the surface of the Earth are only weakly bound to it by friction and so are subject to the Coriolis force. The Coriolis force acts at right angles to the direction of motion, so as to deflect winds and currents to the right in the Northern Hemisphere and to the left in the Southern Hemisphere; the deflections are significant because winds and currents travel relatively slowly. The Coriolis force is zero at the Equator and increases to a maximum at the poles.

Now try the following questions to consolidate your understanding of this Chapter.

QUESTION 1.4
(a) A missile is fired southwards from the Equator. Explain what will happen to the direction of its path in relation to the Earth.

(b) In which direction will a current be deflected by the Coriolis force if it is initially flowing (i) eastwards at 45° N, (ii) westwards on the Equator?

QUESTION 1.5   In Figure 1.4(a), the horizontal axis is scaled according to the surface area of the Earth in different latitude bands. Why do you think the horizontal axis of Figure 1.4(b), showing the annual mean temperature of surface ocean waters, is more compressed at northern than southern high latitudes?

CHAPTER 2   THE ATMOSPHERE AND THE OCEAN

Anyone who has seen images of the Earth from space, like that in Figure 2.1, will have been struck by how much of our planet is ocean, and will have wondered about the swirling cloud patterns. In fact, the atmosphere and the ocean form one system and, if either is to be understood properly, must be considered together. What occurs in one affects the other, and the two are linked by complex feedback loops.

Figure 2.1   The Earth as seen from a geostationary satellite positioned over the Equator. The height of the satellite (about 35 800 km) is such that almost half of the Earth's surface may be seen at once. The outermost part of the image is extremely foreshortened, as can be seen from the apparent position of the British Isles. Colours have been constructed using digital image-processing to simulate natural colours.

The underlying theme of this Chapter is the redistribution of heat by, and within, the atmosphere. We first consider the large-scale atmospheric circulation of the atmosphere, and then move on to consider the smaller-scale phenomena that characterize the moist atmosphere over the oceans.

2.1 THE GLOBAL WIND SYSTEM

Figure 2.2(a) shows what the global wind system would be if the Earth were completely covered with water. As you will see later, the existence of large land masses significantly disturbs this theoretical pattern, but the features shown may all be found in the real atmosphere to greater or lesser extents.

In the lower atmosphere, pressure is low along the Equator, and warm air converges here, rises, and moves polewards. Because the Earth is spherical, air moving polewards in the upper part of the troposphere is forced to converge, as lines of longitude converge. As a result, at about 30° N and S there is a 'piling up' of air aloft, which results in raised atmospheric pressure at the Earth's surface below.* There is therefore a pressure gradient from the subtropical highs (where air is subsiding) towards the equatorial low (Figure 2.2(a)) and, as winds blow from areas of high pressure to areas of low pressure, equatorward winds result. These are the Trade Winds.

As Figure 2.2(a) and (c) show, the Trade Winds blow from the north-east and the south-east, and *do not* blow directly from the north and south. Why is this?

The answer, of course, is because of the Coriolis force. Away from the Equator, the Coriolis force acts to deflect winds and currents to the right in the Northern Hemisphere and to the left in the Southern Hemisphere.

Note that the Trade Winds are named the South-East and North-East Trades because they come *from* the south-east and north-east. However, whereas winds are always described in terms of where they are blowing *from*, currents are described in terms of where they are flowing *towards*. Thus, a southerly current flows *towards* the south and a southerly wind blows *from* the south. To avoid confusion, we will generally use south*ward* rather than south*erly* (for example) when describing current direction.

The Trade Winds form part of the atmospheric circulation known as the Hadley circulation, or **Hadley cells**, which can be seen in Figure 2.2(a) and (b). Strictly speaking, the term 'Hadley cell' refers only to the north–south component of the circulation. Because the flow is deflected by the Coriolis force, in three dimensions the circulation follows an approximately spiral pattern (Figure 2.2(c)).

Figure 2.3(a) and (b) (p. 20) show the prevailing winds at the Earth's surface for July and January, along with the average positions of the main regions of high and low pressure for these months of the year. Figure 2.3(c) shows the surface winds over the Pacific and Atlantic for one particular day in 1999.

How closely do the actual winds and pressure systems over the Earth (Figure 2.3(a),(b)) correspond with the hypothetical wind system and pressure pattern shown in Figure 2.2(a)?

---

*Generally, atmospheric pressure at sea-level is determined by the weight of the column of air above, even if that air is gently rising or subsiding in regions of low or high pressure. Only in intense thunderstorms, cyclones and tornadoes, where the air has an appreciable upward acceleration, is atmospheric pressure directly affected by the vertical motion of the air.

Figure 2.2 (a) Wind system for a hypothetical water-covered Earth, showing major winds and zones of low and high pressure systems. Vertical air movements and circulation cells are shown in exaggerated profile either side, with characteristic surface conditions given on the right. Note that convergence of air at low levels and divergence at high levels results in air rising, while the converse results in air sinking. The two north–south cells on either side of the Equator make up the Hadley circulation. ITCZ = Intertropical Convergence Zone, the zone along which the wind systems of the Northern and Southern Hemispheres meet. (This and other details are discussed further in the text.)

(b) Section through the atmosphere, from polar regions to the Equator, showing the general circulation, the relationship of the polar jet stream to the polar front, and regions of tropical cloud formation. Note that much of the poleward return flow takes place in the upper part of the troposphere (the part of the atmosphere in which the temperature decreases with distance above the Earth); the tropopause is the top of the troposphere and the base of the stratosphere.

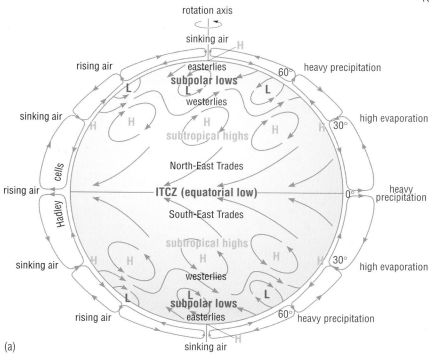

(a)

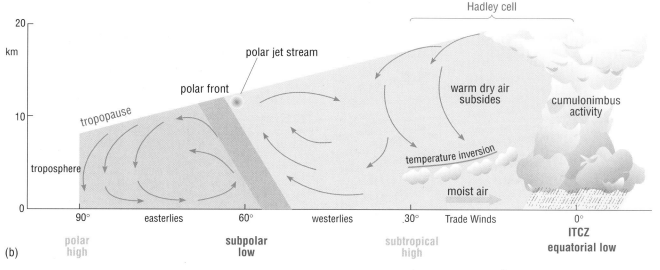

(b)

(c) Schematic diagram to show the spiral circulation patterns of which the Trade Winds form the surface expression; the north–south component of this spiral circulation (see right-hand side) is known as the Hadley circulation or 'Hadley cells' (also shown on Figure 2.2(a)).

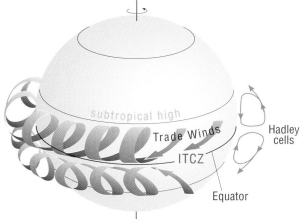

(c)

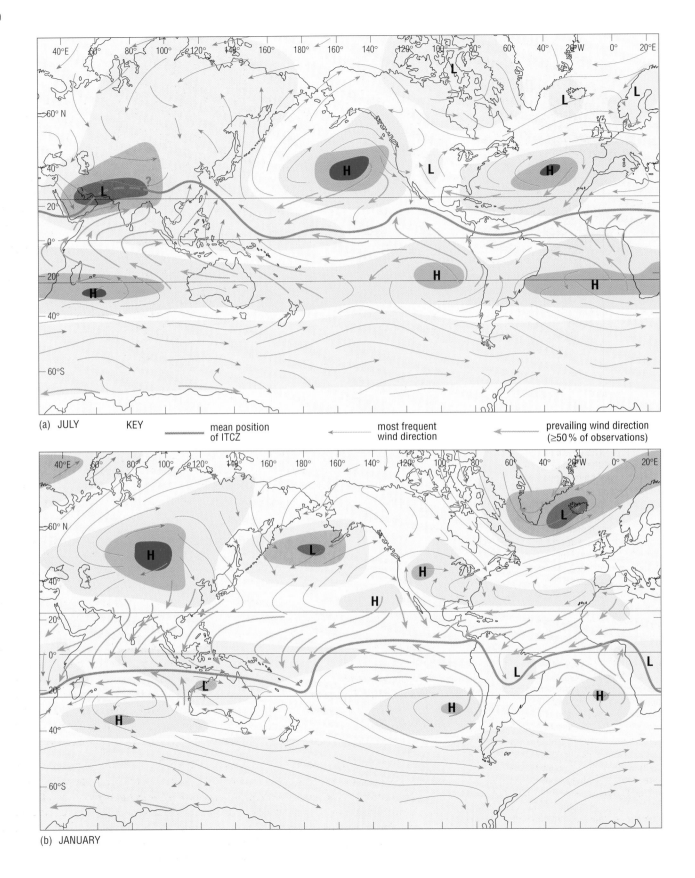

(a) JULY   KEY   ——— mean position of ITCZ   ⟵ most frequent wind direction   ⟵ prevailing wind direction (≥50 % of observations)

(b) JANUARY

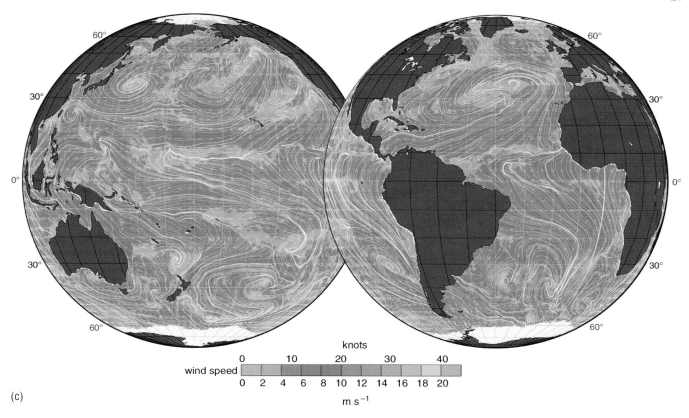

knots

wind speed

| 0 | 10 | 20 | 30 | 40 |

0 2 4 6 8 10 12 14 16 18 20

m s⁻¹

(c)

Figure 2.3 The prevailing winds at the Earth's surface, and the average position of the Intertropical Convergence Zone for (a) July (northern summer/southern winter) and (b) January (southern summer/northern winter). Also shown are the positions of the main regions of high and low atmospheric pressure (red/pink for high pressure, blue for low). Note that as these maps represent conditions averaged over a long period, they show simpler patterns of highs and lows, and of wind flow, than would be observed on any one day (cf. the more complicated patterns in (c)). Extreme conditions are not represented, as they have been 'averaged out' (which is not the case in (c), above).

(c) Surface winds over the Pacific and Atlantic Oceans for one day in 1999 (to be used later, in Question 2.6). Green/pale blue = lowest wind speeds, yellow = highest wind speeds. The maps have been obtained using a satellite-borne 'scatterometer' which measures microwave radar back-scattered from the sea-surface (wind speed is calculated from the estimated roughness of the sea-surface, and wind direction determined from the inferred orientation of wave crests). Note the complexity of the flow pattern, and localized extreme conditions, which do not show up in the maps of averaged data in Figure 2.3(a) and (b).

In general, not that closely, although the actual and hypothetical winds *are* very similar over large areas of ocean, away from the land. You may also have noticed that the simple arrangement of high pressure systems in the subtropics and low pressure in subpolar regions is more clearly seen in the Southern Hemisphere, which is largely ocean.

If you compare Figure 2.3(a) and (b), you will see that the greatest seasonal change occurs in the region of the Eurasian land mass. During the northern winter, the direction of prevailing winds is outwards from the Eurasian land mass; by the summer, the winds have reversed and are generally blowing in towards the land mass. This is because continental masses cool down and heat up faster than the oceans (their thermal capacity is lower than that of the oceans) and so in winter they are colder than the oceans, and in summer they are warmer. Thus, in winter the air above the Eurasian land mass is cooled and becomes denser, so that a large shallow high pressure area develops, from which winds blow out towards regions of lower pressure. In the summer, the situation is reversed: air over the Eurasian land mass heats up, and becomes less dense. There is a region of warm rising air and low pressure which winds blow *towards*. The oceanic regions most affected by these seasonal changes are the Indian Ocean and the western tropical Pacific, where the seasonally reversing winds are known as the monsoons.

The distribution of ocean and continent also influences the position of the zone along which the wind systems of the two hemispheres converge. The zone of convergence – known as the **Intertropical Convergence Zone** or **ITCZ** – is generally associated with the zone of highest surface temperature. Because the continental masses heat up faster than the ocean in summer and cool faster in winter, the ITCZ tends to be distorted southwards over land in the southern summer and northwards over land in the northern summer (Figure 2.3).

2.2 POLEWARD TRANSPORT OF HEAT BY THE ATMOSPHERE

Heat is transported to higher latitudes by the atmosphere both directly and indirectly. If you look at Figure 2.2(a) and (b), you will see that motion in the upper troposphere is generally polewards. Air moving equatorwards over the surface of the Earth takes up heat from the oceans and continents. When it later moves polewards, after rising at regions of low atmospheric pressure such as the Equator, heat is also transported polewards. Thus, any mechanism that transfers heat from the surface of the Earth to the atmosphere also contributes to the poleward transport of heat. The most spectacular example of heat transfer from ocean to atmosphere is the generation of tropical cyclones, which will be discussed in Section 2.3.1.

The Hadley cells, of which the Trade Winds are the surface expressions, may be seen as simple convection cells, in the upper limbs of which heat is transported polewards. How heat is carried polewards at higher latitudes is not quite so obvious.

2.2.1 ATMOSPHERIC CIRCULATION IN MID-LATITUDES

Like the Hadley cells, the low and high pressure centres characteristic of mid-latitudes are a manifestation of the need for heat to be moved polewards, to compensate for the radiation imbalance between low and high latitudes (Figure 1.4). Though less spectacular than tropical cyclones, mid-latitude weather systems transfer enormous amounts of heat – a single travelling depression may be transferring 10–100 times the amount of heat transported by a tropical cyclone.

If, in the long term, no given latitude zone is to heat up, heat must be transported polewards at *all* latitudes. On an idealized, non-rotating Earth, this could be achieved by a simple atmospheric circulation pattern in which surface winds blew from the polar high to the equatorial low, and the warmed air rose at the Equator and returned to the poles at the top of the troposphere to complete the convection cell (see Figure 2.4).

However, because the Earth is rotating, and moving fluids tend to form eddies, this simple system cannot operate. The Coriolis force causes equatorward winds to be deflected to the right in the Northern Hemisphere and to the left in the Southern Hemisphere, so acquiring an easterly component in both hemispheres – as we have seen, the Trade Winds blow towards the Equator from the south-east and north-east. But the Coriolis force increases with latitude, from zero at the Equator to a maximum at the poles, so while winds are deflected relatively little at low latitudes, at higher latitudes the degree of deflection is much greater. The Coriolis force and the inherent instability of air flow together lead to the generation of atmospheric vortices. These are the depressions and anticyclones so familiar to those who live in temperate regions. Their predominantly near-horizontal, or slantwise, circulatory patterns contrast with the near-vertical Hadley circulation of low latitudes.

Before proceeding further, we should briefly summarize the conventions used in describing atmospheric vortices. Circulations around *low* pressure centres, whether in the Northern Hemisphere (where they are anticlockwise) or in the Southern Hemisphere (where they are clockwise), are known as **cyclones** (or lows, or depressions). The way in which air spirals in and up in cyclonic circulation is shown schematically in Figure 2.5(a). Circulations around *high* pressure centres (clockwise in the Northern Hemisphere,

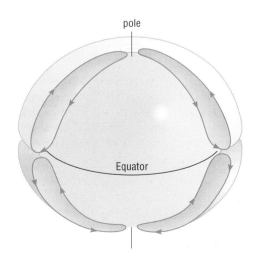

Figure 2.4 Simple atmospheric convection system for a hypothetical non-rotating Earth.

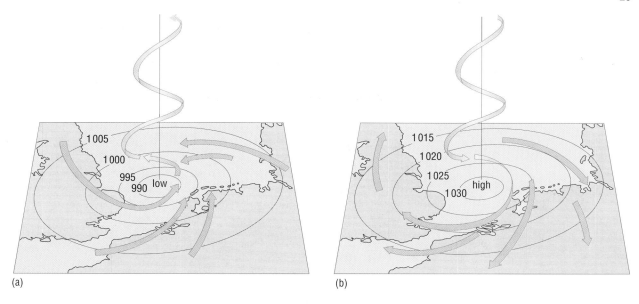

(a)　　　　　　　　　　　　　　　　　　　　　　(b)

Figure 2.5　Schematic diagrams to illustrate how (a) air spirals inwards and upwards in low pressure areas (cyclones) and (b) air spirals downwards and outwards in high pressure areas (anticyclones) (drawn for Northern Hemisphere situations). The contours are **isobars** (lines connecting points of equal atmospheric pressure) and the numbers give typical pressures at ground level, in millibars. Note that here the angle at which the air flow crosses the isobars is somewhat exaggerated. Within ~ 1 km of the Earth's surface, the air flow would cross the isobars at small angles, while above that it would follow the isobars quite closely.

anticlockwise in the Southern Hemisphere), in which air spirals downwards and outwards, are known as **anticyclones** (Figure 2.5(b)).

If you study Figure 2.5(a), you will see that the direction of air flow is such that the pressure gradient (acting from higher to lower pressure, at right angles to the isobars) is to the left of the flow direction, in the opposite direction to the Coriolis force, which (as this is the Northern Hemisphere) is to the right of the flow direction. The same is true in the case of the anticyclone in Figure 2.5(b), although here the pressure gradient is acting outwards and the Coriolis force inwards. If the balance between pressure gradient force and Coriolis force were exact, the winds would blow *along the isobars*. As mentioned in Chapter 1, this type of fluid flow is known as *geostrophic*. If you study weather charts, you will see that, away from the centres of cyclones and anticyclones, air flow is often geostrophic, or almost so, with the wind arrows either parallel to the isobars or crossing them at small angles. Geostrophic flow is also important in the oceans, as will become clear in Chapter 3.

In the atmosphere, how can mid-latitude cyclones and anticyclones contribute to the poleward transport of heat?

Moving air masses mix with adjacent air masses and heat is exchanged between them. For example, air moving northwards around a Northern Hemisphere cyclone or anticyclone will be transporting relatively warm air polewards, while the air that returns equatorwards will have been cooled. This may be likened to the stirring of bath water to encourage the effect of hot water from the tap to reach the far end of the bath, and is shown schematically in Figure 2.6.

Mixing occurs on a variety of scales. Between the warm westerlies and the polar easterlies, there is a more or less permanent boundary region known as the Polar Front (Figure 2.2(b)). Flowing at high level along the polar limit of the westerlies is the fast air current known as the polar jet stream. For days or weeks at a time, this may simply travel eastwards around the globe, but it tends to develop large-scale undulations. As a result, tongues of warm air may be allowed to penetrate to high latitudes, and cold polar air masses may be isolated at relatively low latitudes (Figure 2.7(a)).

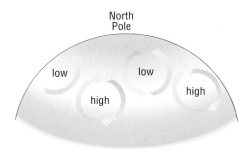

Figure 2.6　Highly schematic diagram to show the poleward transport of heat in the atmosphere by means of mid-latitude vortices (cyclones and anticyclones).

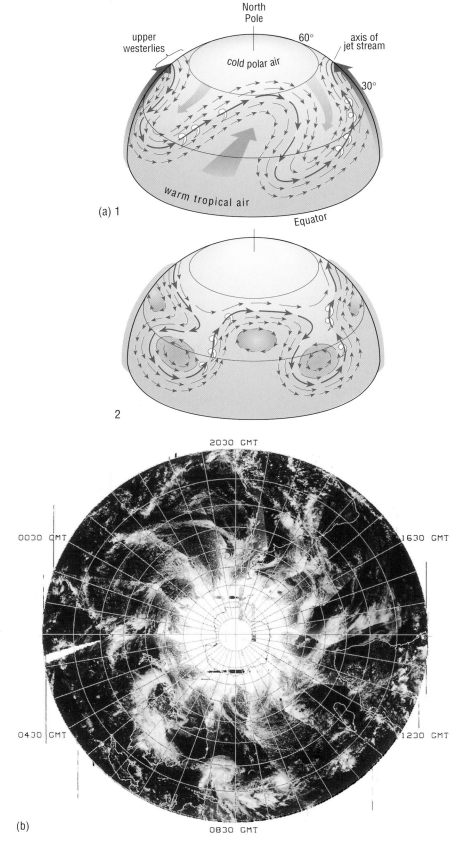

Figure 2.7 (a)   Schematic diagram showing later stages (1, 2) in the development of undulations in the northern polar jet stream, which flows near the tropopause at heights of 10 km or more, cf. Figure 2.2(b)). Meanwhile, cyclones (or depressions, shown here as swirls of cloud) form beneath the poleward-flowing parts of the jet stream, and anticyclones (not shown) form below the equatorward-flowing parts. For clarity, we have omitted the subtropical high pressure regions in mid-latitudes and the Hadley circulation at low latitudes.

(b)   Satellite photomosaic of the Southern Hemisphere showing cloud cover for one day in April 1983 (i.e. during the southern winter). The cyclonic storms that develop below the poleward-trending sections of the jet stream mean that its undulating path is shown up by the cloud pattern. In comparing the cloud swirls with the path of the polar jet stream shown in (a), remember that in the Southern Hemisphere air moves *clockwise* around subpolar lows.

The undulations that develop in the polar jet stream are known as **Rossby waves**, and they always travel *westwards* around the globe. For reasons that will become clear later (Chapter 5), small-scale waves in the polar front tend to develop into anticyclones below equatorward-flowing parts of the jet stream, and into cyclones (indicated by the cloud spirals in Figure 2.7(a)) below poleward-trending parts of the stream. Often referred to as depressions, these low pressure systems are common features of weather charts of the north-east Atlantic region. The overall effect is that heat is transported polewards (Figure 2.8).

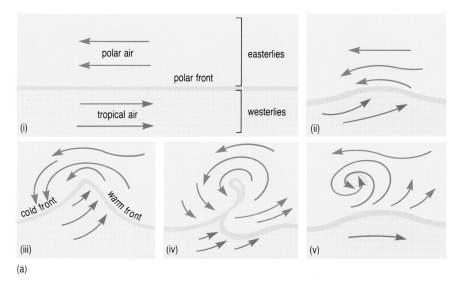

Figure 2.8 (a) Stages in the development of a mid-latitude cyclone (depression) in the Northern Hemisphere, showing how it contributes to the poleward transport of heat. The pale purple band represents the sloping frontal boundary, where warm air overlies a wedge of cold air (as in Figure 2.2(b)). Frontal boundaries can be anything from 50 km to more than 200 km across. That part of the front where cold air is advancing is known as a cold front, and that part where warm air is advancing, as a warm front. Where they combine, an occluded front results.
(b) Typical weather chart for the northern Atlantic/north-west Europe (for 31 October 1999), showing waves in the polar front developing into low pressure systems. The heavy lines are where the warm and cold air masses meet, and the symbols on these lines indicate whether the front is warm or cold (cf. (a)), or a combination (i.e. occluded).

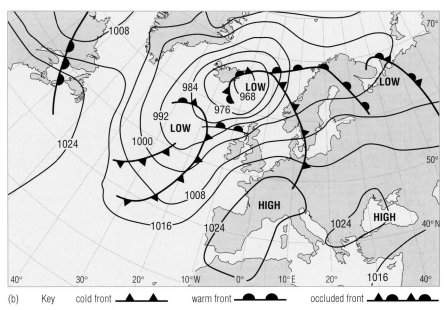

We now move on to consider how heat is redistributed *within* the atmosphere, by means of predominantly vertical motions.

2.2.2 VERTICAL CONVECTION IN THE ATMOSPHERE

Processes occurring at the air–sea interface are greatly affected by the degree of turbulent convection that can occur in the atmosphere above the sea-surface. This in turn is dependent on the degree of *stability* of the air, i.e. on the extent to which, once displaced upwards, it tends to continue rising.

Two ways in which density may vary with height in a fluid are illustrated in Figure 2.9. Situation (a), in which density increases with height, is unstable, and fluid higher up will tend to sink and fluid lower down will tend to rise. Situation (b), in which density decreases with height, is stable: a parcel of fluid (at, say, position O) that is displaced upwards will be denser than its surroundings and will sink back to its original position.

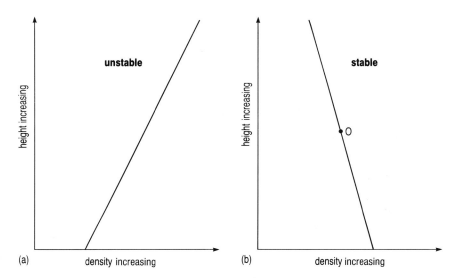

Figure 2.9 Possible variations of fluid density with height, leading to (a) unstable and (b) stable conditions.

The density of air depends on its pressure and its temperature. It also depends on the amount of water vapour it contains – water vapour is less dense than air – but for most practical purposes water vapour content has a negligible effect on density. Thus, the variation of density with height in a column of air is determined by the variation in *temperature* with height.

The variation in temperature with height in the atmosphere is complicated. For one thing, air, like all fluids, is compressible. When a fluid is compressed, the internal energy it possesses per unit volume by virtue of the motions of its constituent atoms, *and which determines its temperature*, is increased. Conversely, when a fluid expands, its internal energy decreases. Thus, a fluid heats up when compressed (a well-known example of this is the air in a bicycle pump), and cools when it expands. Changes in temperature that result from changes in volume/density and internal energy, and *not* because of gain or loss of heat from or to the surroundings, are described as **adiabatic**. Adiabatic temperature changes have a much greater effect on the behaviour of air masses than do other mechanisms for gaining or losing energy (absorbing or emitting radiation, or mixing with other air masses).

Imagine a parcel of air above a warm sea-surface moving upwards in random, turbulent eddying motions. As it rises, it is subjected to decreasing atmospheric pressure and so expands and becomes less dense; this results in an adiabatic decrease in temperature, which for dry air is 9.8 °C per kilometre increase in altitude (black line on Figure 2.10). If the adiabatic decrease in temperature of a rising parcel of air is *less* than the local decrease of temperature with height in the atmosphere, the rising parcel of air will be warmer than the surrounding air and will continue to rise. In other words, this is an unstable situation, conducive to upward convection of air. If, on the other hand, the adiabatic cooling of the rising parcel of air is sufficient to reduce its temperature to a value below that of the surrounding air, it will sink back to its original level – i.e. conditions will be stable.

However, the situation is further complicated by the presence of water vapour in the air. Water vapour has a high latent heat content, with the result that the constant 'dry' adiabatic lapse rate of 9.8 °C per km is of limited relevance. If rising air is saturated with water vapour – or becomes saturated as a result of adiabatic cooling – its continued rise and associated adiabatic cooling results in the condensation of water vapour (onto atmospheric nuclei, such as salt or dust particles), to form water droplets. Condensation releases latent heat of evaporation, which partly offsets the adiabatic cooling, so the rate at which air containing water vapour cools on rising (blue 'saturated' lapse rate in Figure 2.10(a)) is less than the rate for dry air.

Figure 2.10 (a)   Graph to illustrate that the rate at which dry air cools adiabatically (black line) is greater than the rate at which saturated air cools adiabatically (blue curve). Unlike the 'dry' lapse rate, the reduced adiabatic lapse rate for air saturated with water vapour varies with temperature; this is because a small decrease in temperature at high temperatures will result in more condensation than a similar decrease at low temperatures. In the example shown, the air has a temperature of 10 °C at ground level. (As discussed in the text, in practice, it is atmospheric pressure rather than height as such that is important.)
(b)   The implications of the 'dry' and 'saturated' adiabatic lapse rates for the stability of columns of air with various different temperature profiles (dashed red lines).

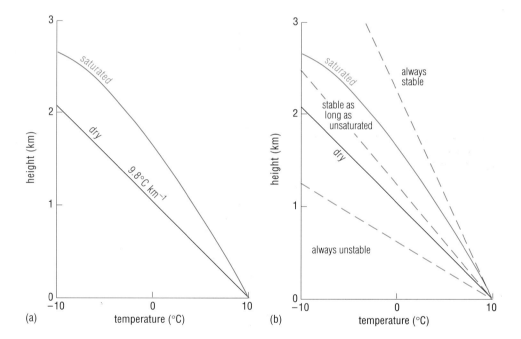

The implications of the 'dry' and 'saturated' adiabatic lapse rate for the stability of various columns of air with various different temperature–height profiles is shown in Figure 2.10(b). Over most of the oceans, particularly in winter when the sea-surface is warmer than the overlying air, the variation of temperature with height in the atmosphere, and the water content of the air, are such that conditions are unstable, air rises, and convection occurs. Convection is further promoted by turbulence resulting from strong winds blowing over the sea-surface. When turbulence is a more effective cause of upward movement of air than the buoyancy forces causing instability, convection is said to be *forced*. The cumulus clouds characteristic of oceanic regions within the Trade Wind belts are a result of such forced convection (Figure 2.11(a)).

(a)

(b)

Figure 2.11 (a)   Cumulus clouds over the ocean in the Trade Wind belt.
(b)   Cumulonimbus and, at lower levels, cumulus clouds, in the Intertropical Convergence Zone over the Java Sea. Like cumulus, cumulonimbus form where moist air rises and cools, so that the water vapour it contains condenses to form droplets. However, cumulonimbus generally extend to much greater heights than cumulus, and their upper parts consist of ice crystals.

As illustrated in Figure 2.2(b), upward development of the Trade Wind cumulus clouds is inhibited by the subsidence of warm air from above. This leads to an *increase* of temperature with height, or a temperature *inversion* (Figure 2.12). Rising air encountering a temperature inversion is no longer warmer than its surroundings and ceases its ascent. The warmer air therefore acts as a 'ceiling' as far as upward convection is concerned.

Along the Intertropical Convergence Zone, the 'ceiling' is the tropopause, so the vigorous convection in the ITCZ can extend much higher than that associated with cumulus formation (Figure 2.2(b)). The towering cumulonimbus that result (Figure 2.11(b)) allow the ITCZ to be easily seen on images obtained via satellites (Figure 2.13). Convection in the ITCZ is the main way that heat becomes distributed throughout the troposphere in low latitudes.

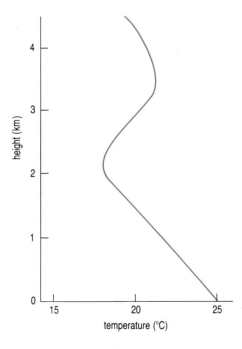

Figure 2.12   Typical variation of temperature with height in the Trade Wind zone, at about 5° of latitude, showing the temperature inversion (here at an altitude of ~2 km).

Figure 2.13   The ITCZ over the tropical Pacific, clearly marked out by cumulonimbus clouds. (This is an infra-red image, obtained from the *GEOS* satellite. The area shown is between ~40° N and ~40° S.)

QUESTION 2.1   The labels on the right-hand side of Figure 2.2(a) indicate characteristic conditions at the Earth's surface. By reference to Figures 2.5 and 2.10, explain briefly why (1) subpolar lows are associated with clouds and rain, and (2) the semi-permanent high pressure regions of the subtropics are associated with dry/arid conditions.

2.3 ATMOSPHERE–OCEAN INTERACTION

As discussed in the previous Section, the ocean influences the atmosphere by affecting its moisture content and hence its stability. Intimately related to this is the effect of sea-surface temperature on the atmospheric circulation. For example, sea-surface temperatures influence the intensity of the Hadley circulation and (as mentioned earlier) the position of the ITCZ generally corresponds to the zone where sea-surface temperatures are highest.

Why might atmospheric circulatory patterns be affected by sea-surface temperature in this way?

The surface of the sea is usually warmer than the overlying air, and the higher the temperature of the sea-surface, the more heat may be transferred from the upper ocean to the lower atmosphere. Warmer air is less dense and rises, allowing more air to flow in laterally to take its place. Thus, an area of exceptionally high surface temperature in the vicinity of the Equator could lead to an increase in the intensity of the Trade Winds and the Hadley circulation. The position of the ITCZ will also be related to the vigorous convection and low pressure zones associated with high sea-surface temperatures; and the warmer the sea-surface, the more buoyancy is supplied to the lower atmosphere, and the more vigorous the vertical convection. As discussed in the previous Section, this effect is enhanced by the fact that the atmosphere over the ocean contains a large amount of water vapour.

In the next Section, you will see a striking example of the influence of sea-surface temperature on the atmosphere.

2.3.1 EASTERLY WAVES AND TROPICAL CYCLONES

A large amount of heat is transported away from low latitudes by strong tropical cyclones – also known as hurricanes and typhoons. Tropical cyclones only develop over oceans and so it is difficult to study the atmospheric conditions associated with their formation. It is known, however, that they are triggered by small low pressure centres, as may occur in small vortices associated with the ITCZ, or in small depressions that have moved in from higher latitudes. Cyclones may also be triggered by linear low pressure areas that form at right angles to the direction of the Trade Winds and travel with them (Figure 2.14). These linear low pressure regions ('troughs') produce wave-like disturbances in the isobaric patterns, with wavelengths of the order of 2500 km; because they move with the easterly Trade Winds, they are known as *easterly waves*.

Easterly waves usually develop in the western parts of the large ocean basins, between about 5° and 20° N. They occur most frequently when the Trade Wind temperature inversion (Figures 2.2(b) and 2.12) is weakest, i.e. during the late summer, when the sea-surface, and hence the lower atmosphere in the Trade Wind belt, are at their warmest.

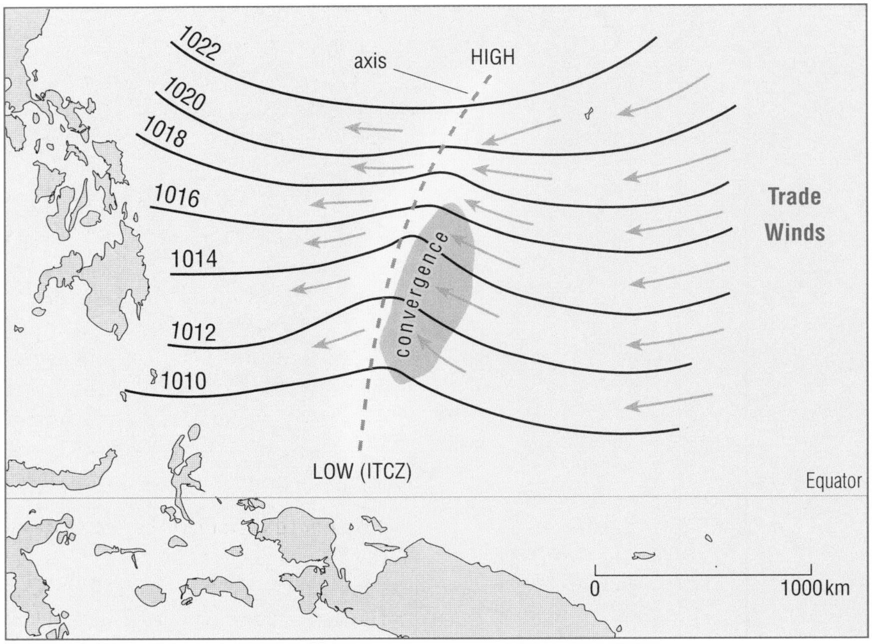

Figure 2.14   Schematic diagram of an easterly wave in the Northern Hemisphere. The black lines are isobars, with sea-level atmospheric pressure given in millibars. Wind direction is shown by the blue arrows. As the wave itself moves slowly westward, lower level air ahead of the 'axis' diverges, and dry air sinking from above to replace it leads to particularly fine weather. Behind the axis, moist air converges and rises, generating showers and thunderstorms (the main rainfall area is shown in greyish-blue).

Ahead of the easterly wave's 'axis', the air flow at low level diverges, so air above sinks, strengthening and lowering the Trade Wind inversion and leading to particularly fine weather. Behind the axis (the 'trough' in atmospheric pressure), moist air converges and rises, temporarily destroying the Trade Wind inversion and generating showers and thunderstorms. Although only a small proportion of easterly waves give rise to cyclones, they are important because they bring large amounts of rainfall to areas that remain generally dry as long as air flow in the Trade Winds is unperturbed.

Figure 2.15   Satellite image of Hurricane 'Andrew' moving across the Gulf of Mexico in August 1992.

Intense tropical cyclones are not generated within about 5° of the Equator. Cyclonic (and anticyclonic) weather systems can only form where flow patterns result from an approximate balance between a horizontal pressure gradient force and the Coriolis force (Section 2.2.1) and, as mentioned earlier, the Coriolis force is zero at the Equator and very small within about 5° of it. Poleward of about 5° of latitude, positive feedback between the ocean and the atmosphere can transform a low pressure area into a powerful tropical cyclone, characterized by closely packed isobars encircling a centre of very low pressure (typically about 950 mbar). Large pressure gradients near the centre of the cyclone cause air to spiral rapidly in towards the low pressure region (anticlockwise in the Northern Hemisphere, clockwise in the Southern Hemisphere), and wind speeds commonly reach 100–200 km hr^{-1} (~30–60 m s^{-1}, see Table 2.2). Violent upward convection of warm humid air results in bands of cumulonimbus and thunderstorms, which spiral in to the core or 'eye' of the cyclone – an area of light winds and little cloud, surrounded by a wall of towering cumulonimbus (see Figures 2.15 and 2.16).

The energy that drives the cyclone comes from the release of latent heat as the water vapour in the rising air condenses into clouds and rain; the resultant warming of the air around the central region of the cyclone causes it to become less dense and to rise yet more, intensifying the divergent anticyclonic flow of air in the upper troposphere that is necessary for the cyclone to be maintained (Figure 2.16). Given their source of energy, it is not surprising that tropical cyclones occur only over relatively large areas of ocean where the surface water temperature is high.

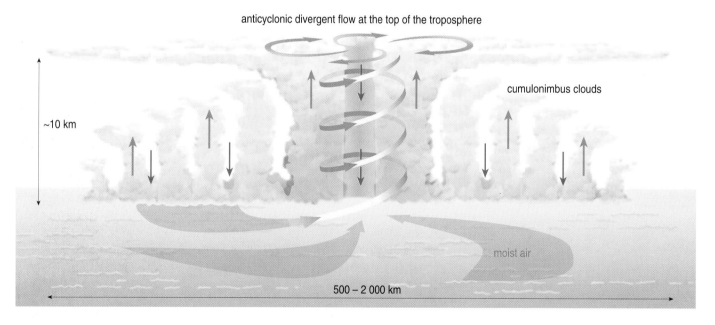

Figure 2.16   Schematic diagram of a tropical cyclone. Moist air spirals inwards and upwards at high speed, forming cumulonimbus clouds arranged in spiral bands which encircle the 'eye'. Meanwhile, in the eye itself (and between the bands), air subsides, warming adiabatically (red arrows), and there are light winds and little cloud.

In practice, the critical sea-surface temperature needed to generate the increased vertical convection which leads to extensive cumulonimbus development and rain and/or cyclone formation is about 27–29 °C. Why should this value be critical? One factor seems to be that the higher the temperature of an air mass the more moisture it can hold, and the greater the upward transfer of latent heat that can occur. Given the positive feedback of the system, a rise in temperature from 27 to 29 °C has a much greater effect on the overlying atmosphere than a rise from, say, 19 to 21 °C. The full answer to this question is, however, as yet unknown. Furthermore, because the local sea-surface temperature – strictly, the heat content of the uppermost 60 m of ocean – has such a marked effect on the rate of development of a tropical cyclone, predicting the intensity of such a system as it moves across expanses of ocean towards land is extremely difficult.

QUESTION 2.2 With the help of Figure 2.3 and bearing in mind what you have been reading about the conditions that favour the initiation and development of easterly waves and tropical cyclones, can you explain why they occur more often in the Northern Hemisphere than the Southern Hemisphere?

Tropical cyclones also occur more often in the western than the eastern parts of the Atlantic and Pacific Oceans. This is because sea-surface temperatures are higher there, for reasons that will become clear in later Chapters.

The violent winds of tropical cyclones generate very large waves on the sea-surface. These waves travel outwards from the central region and as the cyclone progresses the sea becomes very rough and confused. The region where the winds are blowing in the *same* direction as that in which the cyclone is travelling is particularly dangerous because here the waves have effectively been blowing over a greater distance (i.e. they have a greater **fetch**). Furthermore, ships in this region are in danger of being blown into the cyclone's path.

Cyclones also affect the deeper structure of the ocean over which they pass. For reasons that will be explained in Chapter 3, the action of the wind causes the surface waters to diverge so that deeper, cooler water upwells to replace it (Figure 2.17). Thus not only are cyclones *affected* by sea-surface temperature, but they also *modify* it, so that their tracks are marked out by surface water with anomalously low temperatures, perhaps as much as 5 °C below that of the surrounding water.

Characteristic tracks followed by tropical cyclones as they move away from their sites of generation are shown in Figure 2.18. The path of a newly generated cyclone is relatively easy to predict, because it is largely determined by the general air flow in the surrounding atmosphere – compare the paths in Figure 2.18 with the average winds (for summer) in Figure 2.3. The fact that cyclones nearly always move polewards enhances the contribution that they make to the transport of heat from the tropics to higher latitudes.

If they move over extensive areas of land they begin to die away, as the energy conversion system needed to drive them can no longer operate. Their decay over land may be hastened by increased surface friction and the resulting increased variation of wind velocity with height (vertical **wind shear**) which inhibits the maintenance of atmospheric vortices. The lower temperatures of land masses (especially at night) may also play a part in their decay. The average lifespan of a tropical cyclone is about a week.

There is great interannual variation in the frequency of tropical cyclones. Over the past few decades there seems to have been an increase in their occurrence, but it is not clear whether this is a genuine long-term increase or part of a cyclical variation. Information about the number of cyclones occurring annually in the various oceanic regions over a 20-year period is given in Table 2.1.

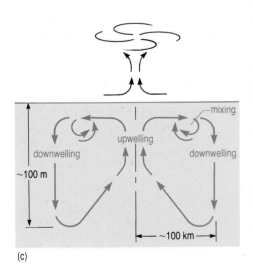

(c)

Figure 2.17   The effect of a cyclone on the surface ocean. Near the centre of the storm – in fact, somewhat to the rear, as it is moving – surface water diverges and upwelling of deeper, cooler water occurs; some distance from the centre, the surface water that has travelled outwards converges with the surrounding ocean surface water and downwelling, or sinking, occurs. In between the zones of upwelling and downwelling, the displaced surface water mixes with cooler subsurface water. (How cyclonic winds lead to upwelling of subsurface water will be explained in Chapter 3.)

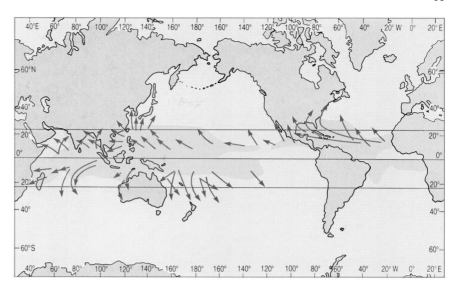

Figure 2.18 Characteristic tracks followed by tropical cyclones, commonly called hurricanes (Atlantic and eastern North Pacific), typhoons (western North Pacific) or cyclones (India and Australia). Initially they move westwards, but then generally curve polewards and eastwards around the subtropical high pressure centres. The pink regions are those areas where the sea-surface temperature exceeds 27 °C in summer (mean values for September and March, in the Northern and Southern Hemispheres, respectively).

Cyclones have a great impact on life in the tropics. They are largely responsible for late summer or autumn rainfall maxima in many tropical areas. The strong winds and large waves associated with them may cause severe damage to natural environments such as reefs; indeed, these catastrophic events play a major role in determining the patterns of distribution of species and life forms within such communities. They may lead to great loss of human life, particularly where there are large populations living near to sea-level, on the flood plains of major rivers or on islands. Quite apart from the high seas caused by the winds, the low atmospheric pressures within a cyclone may cause a **storm surge** – a local rise in sea-level of up to 5 m or more (see Table 2.2, overleaf). Storm surges can lead to widespread flooding of low-lying areas, and deaths from drowning, post-flood disruption and disease greatly exceed those caused by the winds. A storm surge contributed to the havoc caused by tropical cyclones in Orissa, on the north-west of the Bay of Bengal, in October–November 1999, and the cyclonic disaster which devastated the delta region of the Bay in May 1985 was largely the result of a storm surge further amplified by the occurrence of high tides.

Table 2.1 Numbers of tropical cyclones (> 33 m s^{-1} sustained wind speed) occurring annually during a 20-year period (from 1968/69 to 1989/90) in various oceanic regions.

	Maximum	Minimum	Mean
Atlantic, including Caribbean	12	2	5.4
NE Pacific	14	4	8.9
NW Pacific	24	11	16.0
Northern Indian Ocean	6	0	2.5
SW Indian Ocean	10	0	4.4
SE Indian Ocean/NW of Australia	7	0	3.4
SW Pacific/NE of Australia	11	2	4.3
Global total	65	34	44.9

While sea-surface temperature plays a highly significant role in the climate and weather of tropical regions, its effect at higher latitudes cannot be ignored. For example, the 'hurricane' that struck southern England and parts of continental Europe in October 1987 was generated above a patch of anomalously warm water in the Bay of Biscay.

Table 2.2   The Saffir–Simpson scale for potential cyclone damage.

Category	Central pressure (mbar)	Wind speeds (m s^{-1})	(knots)	Height of storm surge (m)	Damage
1	>980	33–42	64–82	~ 1.5	Damage mainly to trees, shrubbery and unanchored mobile homes
2	965–979	42–48	83–95	~ 2.0–2.5	Some trees blown down Major damage to exposed mobile homes Some damage to roofs of buildings
3	945–964	49–58	96–113	~ 2.5–4.0	Foliage removed from trees; large trees blown down Mobile homes destroyed Some structural damage to small buildings
4	920–944	58–69	114–135	~ 4.0–5.5	All signs blown down Extensive damage to roofs, windows and doors Complete destruction of mobile homes Flooding inland as far as 10 km Major damage to lower floors of structures near shore
5	<920	>70	>135	>5.5	Severe damage to windows and doors Extensive damage to roofs of homes and industrial buildings Small buildings overturned and blown away Major damage to lower floors of all structures less than 4.5 m above sea-level within 500 m of the shore

Figure 2.19   A water spout near Lower Malecumbe Key, Florida.

Water spouts

Water spouts are similar to cyclones, in that they are funnel-shaped vortices of air with very low pressures at their centres, so that air and water spiral rapidly inwards and upwards. The funnels extend from the sea-surface to the 'parent' clouds that travel with them (see Figure 2.19). They whip up a certain amount of spray from the sea-surface but are visible mainly because the reduction of pressure within them leads to adiabatic expansion and cooling which causes atmospheric water vapour to condense.

Water spouts are much smaller-scale phenomena than cyclones. They range from a few metres to a few hundred metres in diameter, and they rarely last more than fifteen minutes. Unlike cyclones they are not confined to the tropics, although they occur most frequently there, usually in the spring and early summer. They are particularly common over the Bay of Bengal and the Gulf of Mexico, and also occur frequently in the Mediterranean.

2.3.2   A BRIEF LOOK AHEAD

In the previous Section, we concentrated on phenomena resulting from atmosphere–ocean interaction in the tropics, where the two fluid systems are most closely 'coupled'. The interaction of enormous expanses of warm tropical oceans and a readily convecting atmosphere inevitably has a profound effect on climate, and not just at low latitudes. Because the atmosphere–ocean system is dynamic, it has natural oscillations, and the most famous of these, seated in the tropics, but also affecting higher latitudes, is the phenomenon of El Niño. This and other large-scale atmosphere–ocean oscillations will be discussed in Chapters 4 and 5. A completely different kind of interaction between atmosphere and ocean occurs at high latitudes. As you will see in Chapter 6, this also has a profound effect on the world's climate, over much longer time-scales.

2.4 SUMMARY OF CHAPTER 2

1 The global wind system acts to redistribute heat between low and high latitudes.

2 Winds blow from regions of high pressure to regions of low pressure, but they are also affected by the Coriolis force, to an extent that increases with increasing latitude. Because of the differing thermal capacities of continental masses and oceans, wind patterns are greatly influenced by the geographical distribution of land and sea.

3 In mid-latitudes, the predominant weather systems are cyclones (low pressure centres or depressions) and anticyclones (high pressure centres). At low latitudes, the atmospheric circulation consists essentially of the spiral Hadley cells, of which the Trade Winds form the lowermost limb. The Intertropical Convergence Zone, where the wind systems of the two hemispheres meet, is generally associated with the zone of maximum sea-surface temperature in the vicinity of the Equator.

4 Heat is transported polewards in the atmosphere as a result of warm air moving into cooler latitudes. It is also transported as *latent* heat: heat used to convert water to water vapour is released when the water vapour condenses (e.g. in cloud formation) in a cooler environment. Over the tropical oceans, turbulent atmospheric convection transports large amounts of heat from the sea-surface high into the atmosphere, leading to the formation of cumulus and (especially) cumulonimbus clouds. An extreme expression of this convection of moisture-laden air is the generation of tropical cyclones.

5 Interaction between the atmosphere and the overlying ocean is most intense – i.e. atmosphere–ocean 'coupling' is closest – in the tropics.

Now try the following questions to consolidate your understanding of this Chapter.

QUESTION 2.3 Which of statements (a)–(f) concerned with the global wind system are true and which are false?

(a) Moist air is more easily destabilized (i.e. made to convect) than dry air.

(b) Heavy precipitation is characteristic of regions where air is sinking.

(c) The atmospheric circulation over the North Atlantic is predominantly anticyclonic in both summer and winter.

(d) The atmospheric circulation over Eurasia is predominantly cyclonic in both summer and winter.

(e) In polar regions, the tropopause is about 15 km above the Earth.

(f) Figure 2.7(a) suggests that the variable climate of the British Isles is a result of their position in relation to the fluctuating position of the polar jet stream and polar front.

QUESTION 2.4

(a) In the satellite image of Hurricane 'Andrew' (Figure 2.15), most of the clouds visible are associated with anticyclonic air flow. Can you explain this apparent paradox?

(b) What is the 'fuel' that drives a tropical cyclone?

QUESTION 2.5   Figure 2.20 is a satellite image of the Earth made at the same time as the image in Figure 2.1 but using the electromagnetic wavelengths to which water vapour is opaque. The warmest areas are black and the coldest white. Thus, dark areas represent regions in which the upper troposphere is dry so that radiation from lower, warmer layers is able to reach the satellite; white areas correspond to radiation from the upper troposphere, which is cold. By comparing this image with that in Figure 2.1, use information from this Chapter to identify the following features:

(1) cumulonimbus cloud formation in association with the ITCZ;

(2) the northern and southern polar fronts;

(3) the subtropical high pressure regions.

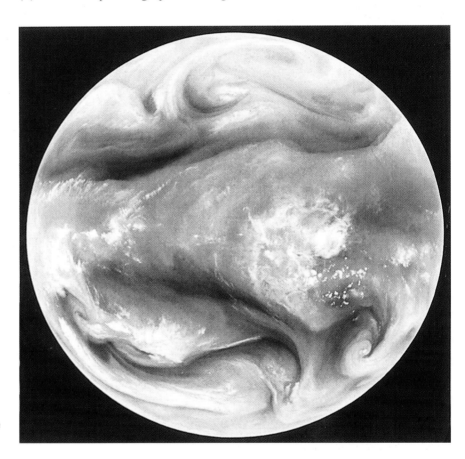

Figure 2.20   Satellite image of the Earth, constructed using the 'water vapour' channel (wavelength 6000 nm); the colour is artificial. This image and that in Figure 2.1 show the same view of the Earth and are for 26 March 1982.

QUESTION 2.6   As stated in the caption, Figure 2.3(c) shows surface winds over the Pacific and Atlantic during one particular day in 1999. On the basis of what you have read in this Chapter, explain whether this was a day during the northern summer or the southern summer. You should be able to cite at least three particular aspects of the wind pattern in support of your answer.

QUESTION 2.7   In Chapter 5, we will be discussing how disturbances may be transmitted from one part of the ocean to another by means of large-scale wave motions. From your reading of Chapter 2, give at least two examples of wave-like *atmospheric* disturbances.

CHAPTER 3   OCEAN CURRENTS

In Chapter 2, we saw how the atmospheric circulation transports heat from low to high latitudes. Figure 3.1 shows how the same is true in the oceans, where surface currents warmed in low latitudes carry heat polewards, while currents cooled at high latitudes flow equatorwards.

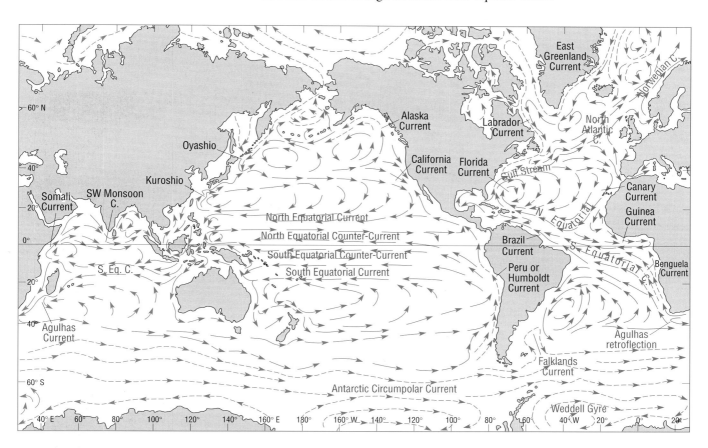

Figure 3.1   Highly generalized representation of the global surface current system. Cool currents are shown by dashed arrows; warm currents are shown by solid arrows. The map shows average conditions for summer months in the Northern Hemisphere. There are local differences in the winter, particularly in regions affected by monsoonal circulations (those parts of the tropics where there are large seasonal changes in the position of the ITCZ, cf. Figure 2.3).

QUESTION 3.1

(a) Which are the main cool surface currents in (i) the subtropical North and South Pacific, and (ii) the subtropical North and South Atlantic?

(b) What do these currents have in common?

As far as the transport of heat is concerned, the 'warm' and 'cool' surface currents shown in Figure 3.1 are only part of the story. In certain high-latitude regions, water that has been subjected to extreme cooling sinks and flows equatorwards in the thermohaline circulation (Chapter 1). In order to know the net poleward heat transport in the oceans at any location, we would need to know the direction and speed of flow of water, and its temperature, at all depths. In fact, the three-dimensional current structure of the oceans is complex and has only relatively recently begun to be investigated successfully.

If you compare Figures 2.3 and 3.1, you will see immediately that the surface wind field and the surface current system have a general similarity. The most obvious difference is that because the flow of ocean currents is constrained by coastal boundaries, the tendency for circular or gyral motion seen in the atmospheric circulation is even more noticeable in the oceans. However, the way in which ocean currents are driven by the atmospheric circulation is not as obvious as it might at first appear, and in this Chapter we consider some of the mechanisms involved.

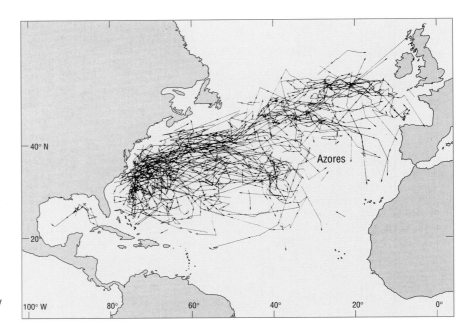

Figure 3.2 The paths of drifting derelict sailing vessels (and a few drifting buoys) over the period 1883–1902. This chart was produced using data from the monthly Pilot Charts of the US Navy Hydrographic Office, which is why the paths are shown as straight lines between points. The paths are extremely convoluted and cross one another, but the general large-scale anticyclonic circulation of the North Atlantic may just be distinguished.

This is perhaps a suitable point to emphasize that maps like Figures 2.3(a) and (b) and 3.1 represent *average* conditions. If you were to observe the wind and current at, for instance, a locality in the region labelled 'Gulf Stream' on Figure 3.1, you might well find that the wind and/or current directions were quite different from those shown by the arrows in Figures 2.3 and 3.1, perhaps even in the opposite direction. Moreover, currents should not be regarded as river-like. Even powerful, relatively well-defined currents like the Gulf Stream continually shift and change their position to greater or lesser extents, forming meanders and eddies, or splitting into filaments. The actual spatial and temporal variations in velocity (speed and direction) are much more complex than could be shown by the most detailed series of current charts. The drift paths in Figure 3.2 convey something of the variability of surface currents in the North Atlantic, while the satellite image on the book cover shows the complexity of the Gulf Stream flow pattern at one moment in time.

Despite the complex and variable nature of the marine environment, certain key theoretical ideas have been developed about the causes and nature of current flow in the ocean, and have been shown to be reasonable approximations to reality, at least under certain circumstances. The most fundamental of these are introduced in Sections 3.1 to 3.4.

direction of wave propagation

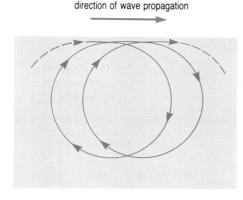

Figure 3.3   In a surface wave, water particles make orbits in the vertical plane. The particles advance slightly further in the crest (the top of the orbit) than they retreat in the trough (bottom of the orbit), so a small net forward motion (known as 'wave drift') results. In deep water, this motion may be of the order of several millimetres to several centimetres per second.

3.1   THE ACTION OF WIND ON SURFACE WATERS

When wind blows over the ocean, energy is transferred from the wind to the surface layers. Some of this energy is expended in the generation of surface gravity waves (which lead to a small net movement of water in the direction of wave propagation; see Figure 3.3) and some is expended in driving currents. The processes whereby energy is transferred between waves and currents are complex; it is not a simple task to discover, for example, how much of the energy of a breaking wave is dissipated and how much is transferred to the surface current.

Nevertheless, it is still possible to make some general statements and predictions about the action of wind on the sea. The greater the speed of the wind, the greater the frictional force acting on the sea-surface, and the stronger the surface current generated. The frictional force acting on the sea-surface as a result of the wind blowing over it is known as the **wind stress**. Wind stress, which is usually given the symbol τ (Greek 'tau'), has been found by experiment to be proportional to the square of the wind speed, W. Thus:

$$\tau = cW^2 \tag{3.1}$$

where the value of c depends on the prevailing atmospheric conditions. The more turbulent the atmosphere overlying the sea-surface (Section 2.2.2), the higher the value of c.

How would you expect the value of c to be affected by the wind speed?

The value of c will increase with increasing wind speed, which not only increases the amount of turbulent convection in the atmosphere over the sea (Section 2.2.2) but also increases the roughness of the sea-surface.

Because of friction with the sea-surface, wind speed decreases with increasing proximity to the sea, and so the value of c to be used also depends critically on the *height* at which the wind speed is measured; this is commonly about 10 m, the height of the deck or bridge of a ship. As a rough guide, we can say that a wind which has a speed of $10\,\mathrm{m\,s^{-1}}$ (nearly 20 knots) 5–10 m above the sea-surface, will give rise to wind stress on the sea-surface of the order of $0.2\,\mathrm{N\,m^{-2}}$ (1 newton = $1\,\mathrm{kg\,m\,s^{-2}}$).

QUESTION 3.2   What value does that imply for c, given the relationship in Equation 3.1? What are its units?

It is important to remember that, for the reasons given above, c is not constant. Nevertheless, a value of c of about 2×10^{-3} gives values of τ that are accurate to within a factor of 2, and often considerably better than that.

Another useful empirical observation is that the surface current speed is typically about 3% of the wind speed, so that a $10\,\mathrm{m\,s^{-1}}$ wind might be expected to give rise to a surface current of about $0.3\,\mathrm{m\,s^{-1}}$. Again, this is only a rough 'rule of thumb', for reasons that should become clear shortly.

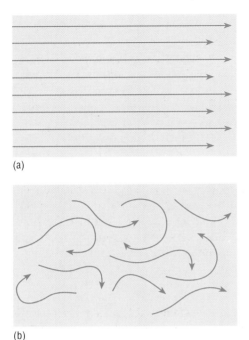

(a)

(b)

Figure 3.4 The difference between (a) laminar flow and (b) turbulent flow. The arrows represent the paths taken by individual parcels of water.

3.1.1 FRICTIONAL COUPLING WITH THE OCEAN

The effect of wind stress on the sea-surface is transmitted downwards as a result of internal friction within the upper ocean. This internal friction results from turbulence and is not simply the viscosity of a fluid moving in a laminar fashion (see Figure 3.4).

Friction in a moving fluid results from the transfer of momentum (mass × velocity) between different parts of the fluid. In a fluid moving in a laminar manner, momentum transfer occurs as a result of the transfer of *molecules* (and their associated masses and velocities) between adjacent layers, as shown schematically in Figure 3.5(a), and should therefore strictly be called **molecular viscosity**. At the sea-surface, as in the rest of the ocean, motion is rarely laminar, but instead turbulent, so that parcels of water, rather than individual molecules, are exchanged between one part of the moving fluid and another (Figure 3.5(b)). The internal friction that results is much greater than that caused by exchange of individual molecules, and is known as **eddy viscosity**.

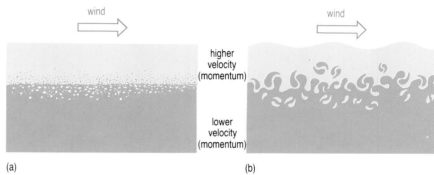

(a)   (b)

Figure 3.5 Schematic diagram to illustrate the difference between (a) molecular viscosity and (b) eddy viscosity. In (a), the momentum transferred between layers is that associated with individual molecules, whereas in (b) it is that associated with parcels of fluid. (For simplicity, we have only shown two layers of differing velocity; in reality, of course, there are an infinite number of layers.)

Turbulent eddies in the upper layer of the ocean act as a 'gearing' mechanism that transmits motion at the surface to deeper levels. The extent to which there is turbulent mixing, and hence the magnitude of the eddy viscosity, depend on how well stratified the water column is. If the water column is well mixed and hence fairly homogeneous, density will vary little with depth and the water column will be easily overturned by turbulent mixing; if the water column is well stratified so that density increases relatively sharply with depth, the situation is stable and turbulent mixing is suppressed.

QUESTION 3.3 Between the warm well-mixed surface layer and the cold waters of the main body of the ocean is the **thermocline**, the zone within which temperature decreases markedly with depth. Explain whether you would expect eddy viscosity to be greater in the thermocline or in the mixed surface layer.

An extreme manifestation of the answer to Question 3.3 is the phenomenon of the **slippery sea**. If the surface layers of the ocean are exceptionally warm or fresh, so that density increases abruptly not far below the sea-surface, there is very little frictional coupling between the thin surface layer and the underlying water. The energy and momentum of the wind are then transmitted only to this thin surface layer, which effectively slides over the water below.

As you might expect from the foregoing discussion, values of eddy viscosity in the ocean vary widely, depending on the degree of turbulence. Eddy viscosity (strictly, the *coefficient* of eddy viscosity) is usually given the symbol A, and we may distinguish between A_z, which is the eddy viscosity resulting from vertical mixing (as discussed above), and A_h, which is the eddy viscosity resulting from horizontal mixing – for example, that caused by the turbulence between two adjacent currents, or between a current and a coastal boundary. Values of A_z typically range from $10^{-5}\,\mathrm{m^2\,s^{-1}}$ in the deep ocean to $10^{-1}\,\mathrm{m^2\,s^{-1}}$ in the surface layers of a stormy sea; those for A_h are generally much greater, ranging from 10 to $10^5\,\mathrm{m^2\,s^{-1}}$. (By contrast, the molecular viscosity of seawater is $\sim 10^{-6}\,\mathrm{m^2\,s^{-1}}$.)

Why are the values of A_h so much higher than values of A_z?

The high values of A_h and the relatively low values of A_z reflect the differing extents to which mixing can occur in the vertical and horizontal directions. The ocean is stably stratified nearly everywhere, most of the time, and stable stratification acts to suppress vertical mixing; motion in the ocean is nearly always in a horizontal or near-horizontal direction. In addition, the oceans are many thousands of times wider than they are deep and so the spatial extent of horizontal eddying motions is much less constrained than is vertical mixing.

The fact that frictional coupling in the oceans occurs through turbulence rather than through molecular viscosity has great significance for those seeking to understand wind-driven currents. Currents developing in response to wind increase to their maximum strength many times faster than would be possible through molecular processes alone. For example, in the absence of turbulence, the effect of a $10\,\mathrm{m\,s^{-1}}$ wind would hardly be discernible 2 m below the surface, even after the wind had been blowing steadily for two days; in reality, wind-driven currents develop very quickly. Similarly, when a wind driving a current ceases to blow, the current is slowed down by friction from turbulence many times faster than would otherwise be possible. Turbulence redistributes and dissipates the kinetic energy of the current; ultimately, it is converted into heat through molecular viscosity.

The types of current motion that are easiest to study, and that will be considered here, are those that result when the surface ocean has had time to adjust to the wind, and ocean and atmosphere have locally reached a state of equilibrium. When a wind starts to blow over a motionless sea-surface, the surface current which is generated takes some time to attain the maximum speed that can result from that particular wind speed; in other words, it first accelerates. The situations that are generally studied are those in which acceleration has ceased and the forces acting on the water are in balance.

The first satisfactory theory for wind-driven currents was published by V.W. Ekman in 1905. It is to his ideas, and their surprising implications, that we now turn.

3.1.2 EKMAN MOTION

In the 1890s, the Norwegian scientist and explorer Fridtjof Nansen led an expedition across the Arctic ice. His specially designed vessel, the *Fram*, was allowed to freeze into the ice and drift with it for over a year. During this period, Nansen observed that ice movements in response to wind were *not* parallel to the wind, but at an angle of 20–40° to the right of it. Ekman developed his theory of wind-driven currents in order to explain this observation.

Ekman considered a steady wind blowing over an ocean that was infinitely deep, infinitely wide, and with no variations in density. He also assumed that the surface of the ocean remained horizontal, so that at any given depth the pressure due to the overlying water was constant. This hypothetical ocean may be considered to consist of an infinite number of horizontal layers, of which the topmost is subjected to friction by the wind (wind stress) at its upper surface and to friction (eddy viscosity) with the next layer down at its lower surface; this second layer is acted upon at its upper surface by friction with the top layer, and by friction with layer three at its lower surface; and so on. In addition, because they are moving in relation to the Earth, all layers are affected by the Coriolis force.

By considering the balance of forces – friction and the Coriolis force – on the infinite number of layers making up the water column, Ekman deduced that the speed of the wind-driven current decreases exponentially with depth. He further found that in his hypothetical ocean the direction of the current deviates 45° *cum sole* from the wind direction at the surface, and that the angle of deviation increases with increasing depth. The current vectors therefore form a spiral pattern (Figure 3.6(a)) and this theoretical current pattern is now known as the **Ekman spiral**. The layer of ocean under the influence of the wind is often referred to as the **Ekman layer**.

We do not need to go into the calculations that Ekman used to deduce the spiral current pattern, but we can consider the forces involved in a little more detail. As discussed in Chapters 1 and 2, the Coriolis force acts 90° *cum sole* of the direction of current flow (i.e. to the right of the flow in the

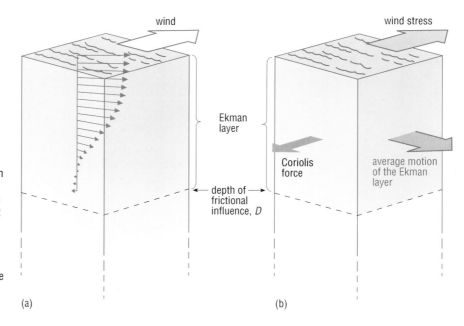

Figure 3.6 (a)   The Ekman spiral current pattern which under ideal conditions would result from the action of wind on surface waters. The lengths and directions of the dark blue arrows represent the speed and direction of the wind-driven current.
(b)   For the Ekman layer as a whole, the force resulting from the wind is balanced by the Coriolis force, which in the Northern Hemisphere is 90° to the right of the average motion of the layer (broad blue arrow).

Northern Hemisphere, to the left in the Southern), and increases with latitude. More specifically, the Coriolis force is proportional to the sine of the latitude, and for a particle of mass m moving with speed u it is given by:

$$\text{Coriolis force} = m \times 2\Omega \sin \phi \times u \qquad (3.2a)$$

where Ω (capital omega) is the angular velocity of the Earth about its axis and ϕ (phi) is the latitude. The term $2\Omega \sin \phi$ is known as the **Coriolis parameter** and is often abbreviated to f, so that the expression given above becomes simply:

$$\text{Coriolis force} = mfu \qquad (3.2b)$$

Thus, the Coriolis force – the force experienced by a moving parcel of water by virtue of it being on the rotating Earth – is at right angles to the direction of motion and *increases* with the speed of the parcel. This is important for the balances of forces that determine motion in the ocean – for example, the balance between frictional forces and the Coriolis force for each of the infinite number of layers of water making up the Ekman spiral.

Ekman deduced that in a homogeneous infinite ocean the speed of the surface current, u_0, is given by

$$u_0 = \frac{\tau}{\rho\sqrt{A_z f}} \qquad (3.3)$$

where τ is the surface wind stress, A_z is the coefficient of eddy viscosity for vertical mixing, ρ (rho) is the density of seawater and f is the Coriolis parameter.

QUESTION 3.4

(a) What is the magnitude of f, the Coriolis parameter, at 40° S, given that $\Omega = 7.29 \times 10^{-5}\,\text{s}^{-1}$?

(b) (i) A westerly wind with a speed of $5\,\text{m s}^{-1}$ blows over the surface of the ocean at 40° S. Assuming that the wind stress that results is $0.1\,\text{N m}^{-2}$, and that $A_z = 0.1\,\text{m}^2\,\text{s}^{-1}$ and $\rho = 10^3\,\text{kg m}^{-3}$, what is the speed of the wind-generated current at the surface, according to Equation 3.3?

(ii) In which direction does it flow?

(c) How well does the value you have calculated for the surface current speed correspond with the 'rule of thumb' mentioned earlier, given that the wind speed is $5\,\text{m s}^{-1}$?

Observations of wind-driven currents in the real oceans, away from coastal boundaries, have shown that surface current speeds are similar to those predicted by Ekman and that the surface current direction does deviate *cum sole* of the wind direction. However, the deviations observed are less than the 45° predicted, and the full spiral pattern has not been recorded. There are several reasons for this. For example, in many regions the ocean is too shallow, so that the full spiral cannot develop, and friction with the sea-bed becomes significant.

Bearing in mind the calculation you made in Question 3.4(b), can you suggest another more fundamental reason?

For the purpose of his calculations, Ekman assumed that the coefficient of eddy viscosity A_z remains constant with depth. As discussed in Section 3.1.1, values of A_z vary over several orders of magnitude and are generally much higher in the surface layers of the ocean than at depth.

The significant prediction of Ekman's theory is not the spiral current pattern but the fact that *the mean motion of the wind-driven (or 'Ekman') layer is at right angles to the wind direction*, to the right in the Northern Hemisphere and to the left in the Southern Hemisphere. This may be understood by imagining the averaged effect of the current spiral (Figure 3.6(a)), or by considering the forces acting on the wind-driven layer as a whole. The only forces acting on the wind-driven layer are wind stress and the Coriolis force (ignoring friction at the bottom of the wind-driven layer), and once an equilibrium situation has been reached, these two forces must be equal and must act in opposite directions. Such a balance is obtained by the average motion, or **depth mean current** of the wind-driven layer, being at right angles to the wind direction; Figure 3.6(b) illustrates this for the Northern Hemisphere.

The magnitude of the depth mean current, $\bar{u}$ (where the bar means 'averaged'), is given by

$$\bar{u} = \frac{\tau}{D\rho f} \tag{3.4}$$

where D is the depth at which the direction of the wind-driven current is directly opposite to its direction at the surface (Figure 3.6(a)). At this depth the current has decreased to 1/23 of its surface value, and the effect of the wind may be regarded as negligible. D, which depends on the eddy viscosity and the latitude, is therefore often equated with the **depth of frictional influence** of the wind. If it is assumed that the eddy viscosity $A_z = 10\,\mathrm{m^2\,s^{-1}}$, D works out to about $40\,\mathrm{m}$ at the poles and about $50\,\mathrm{m}$ for middle latitudes; close to the Equator (where the Coriolis force is zero) it rapidly approaches infinity. As you might expect from Section 3.1.1, in the real ocean the depth of frictional influence of the wind is largely determined by the thickness of the mixed surface layer, as frictional coupling between this and deeper water is often relatively weak. It therefore varies from a few tens of metres to as much as 100–200 m, depending on the time of year, wind speed etc.

The total *volume* of water transported at right angles to the wind direction per second may be calculated by multiplying $\bar{u}$ by the thickness of the wind-driven layer. This volume transport (in $\mathrm{m^3\,s^{-1}}$) is generally known as the **Ekman transport** and, as indicated above, the wind-driven layer (theoretically of depth D) is often known as the Ekman layer. Ekman transports in response to prevailing wind fields contribute significantly to the general ocean circulation, as will become clear in Sections 3.4 and 4.2.2.

3.2 INERTIA CURRENTS

We should now briefly consider what happens when the wind that has been driving a current suddenly ceases to blow. Because of its momentum, the water will not come to rest immediately, and as long as it is in motion, both friction and the Coriolis force will continue to act on it. In the open ocean away from any boundaries, frictional forces may be very small so that the energy imparted to the water by the wind takes some time to be dissipated; meanwhile, the Coriolis force continues to turn the water *cum sole*. The resulting curved motion, under the influence of the Coriolis force, is known as an **inertia current** (Figure 3.7(a)). If the Coriolis force is the *only* force acting in a horizontal direction, and the motion involves only a small change in latitude, the path of the inertia current will be circular (Figure 3.7(b)).

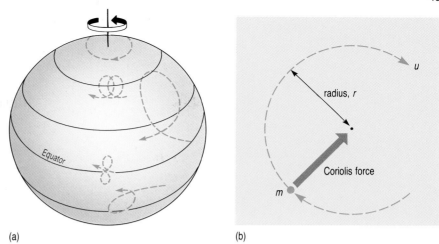

Figure 3.7 (a) Various possible paths for inertia currents.
(b) Plan view showing inertial motion in the Northern Hemisphere. For details, see text.

(a)

(b)

In an inertia current, the Coriolis force is acting as a centripetal force, towards the centre of the circle (Figure 3.7(b)). Now, if a body of mass m moves around a circle of radius r at a speed u, the centripetal force is given by

$$\text{centripetal force} = \frac{mu^2}{r} \tag{3.5}$$

In this case, the centripetal force is the Coriolis force (given by mfu, where f is the Coriolis parameter), so we can write:

centripetal force = Coriolis force,

$$\text{i.e. } \frac{mu^2}{r} = mfu$$

Dividing throughout by mu, this equation simplifies to:

$$\frac{u}{r} = f \tag{3.6}$$

If the motion is small-scale and does not involve any appreciable changes in latitude, f is constant and so the water will follow a circular path of radius r, at a constant speed u. The time, T, taken for a water parcel to complete one circuit, i.e. the *period* of the inertia current, is given by the circumference of the circle divided by the speed:

$$T = \frac{2\pi r}{u}$$

From Equation 3.6, $r/u = 1/f$, so

$$T = \frac{2\pi}{f} \tag{3.7}$$

This equation shows that in the 'ideal' situation, the only variable affecting T is f, i.e. latitude. Substituting appropriate values for $\sin \phi$ would show that at latitude 45°, T is approximately 17 hours, while at the Equator it becomes infinite.

Inertia currents have been identified from current measurements in many parts of the ocean. A classic example, observed in the Baltic Sea, is illustrated in Figure 3.8. Here a wind-driven current flowing to the north-north-west was superimposed on the inertial motion, which had a period of about 14 hours (the theoretical value of T for this latitude is a 14 hours 8 minutes) and died out after some nine or ten rotations.

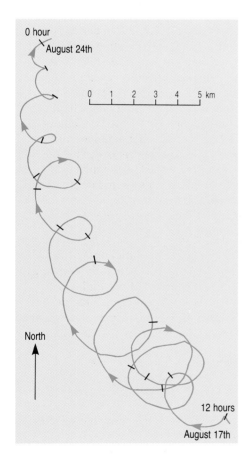

0 hour
August 24th

0 1 2 3 4 5 km

North

12 hours
August 17th

Figure 3.8 Plan view showing inertial motion observed in the Baltic Sea at about 57° N. The diagram shows the path of a parcel of water; if this path was representative of the general flow, the surface water in the region was both rotating and moving north-north-west. The observations were made between 17 and 24 August, 1933, and the tick marks on the path indicate intervals of 12 hours.

In Question 3.4(a) we used the fact that Ω, the angular speed of rotation of the Earth about its axis, is $7.29 \times 10^{-5}\,\text{s}^{-1}$. This may also be written as 7.29×10^{-5} radians* s^{-1}, and is calculated by dividing the angle the Earth turns through each day (2π) by the time it takes to do so (one day). Strictly, we should use one sidereal day, i.e. the time it takes the Earth to complete one revolution relative to the fixed stars. This is 23 hr 56 min or 86 160 seconds.

$$\text{Thus: } \Omega = \frac{2\pi}{86\,160}\,\text{s}^{-1}$$
$$= \frac{2 \times 3.142}{86\,160}\,\text{s}^{-1} = 7.29 \times 10^{-5}\,\text{s}^{-1}$$

For many purposes, however, we can use $\Omega = 2\pi/24\,\text{hr}^{-1}$.

QUESTION 3.5   Bearing this in mind, show that the period of an inertia current at the North Pole would be about 12 hours, while at a latitude of 30° it would be about 24 hours.

3.3   GEOSTROPHIC CURRENTS

So far, we have been assuming that the ocean is infinitely wide, as Ekman did in formulating his theory. When the effect of coastal boundaries is brought into the picture it becomes rather more complicated, because boundaries impede current flow and lead to slopes in the sea-surface – put simply, water tends to 'pile up' against boundaries. This is important, because if the sea-surface is sloping, the hydrostatic pressure acting on horizontal surfaces at depth in the ocean will vary accordingly; in other words there will be *horizontal pressure gradients*. In the same way that winds blow from high to low pressure, so water tends to flow so as to even out lateral differences in pressure. The force that gives rise to this motion is known as the **horizontal pressure gradient force**. If the Coriolis force acting on moving water is balanced by a horizontal pressure gradient force, the current is said to be in geostrophic equilibrium and is described as a **geostrophic current**. (You will remember that we touched on geostrophic *winds* in Section 2.2.1.)

Before we discuss geostrophic currents further, we should consider oceanic pressure gradients in a little more detail.

3.3.1   PRESSURE GRADIENTS IN THE OCEAN

As stated above, lateral variation in pressure at a given depth results in a horizontal pressure gradient force. Pressure at any location in the ocean is affected to some extent by the motion of the water, but because ocean currents are relatively slow, particularly in a vertical direction, for most purposes the pressure at depth may be taken to be the *hydrostatic* pressure – i.e. the pressure resulting from the overlying water, which is assumed to be static.

---

*A radian is the angle subtended at the centre of a circle by an arc of the circle equal in length to the radius, r. As the circumference of a circle is given by $2\pi r$, there are $2\pi r/r = 2\pi$ radians in 360°. Because a radian is one length divided by another, it is dimensionless and therefore not a unit in the usual sense.

Hydrostatic pressure

The hydrostatic pressure at any depth, z, in the ocean is simply the weight of the water acting on unit area (say 1 m^2). This is given by:

hydrostatic pressure at depth $z = p =$ (mass of overlying seawater) $\times g$

where g is the acceleration due to gravity. If the density of seawater is ρ, this can be written as:

$p =$ (volume of overlying seawater) $\times \rho \times g$

which for hydrostatic pressure acting on unit area is simply

$$p = \rho g z \qquad (3.8a)$$

This is known as the **hydrostatic equation**.

It is usually written:

$$p = -\rho g z \qquad (3.8b)$$

because in oceanography the vertical z-axis has its origin at sea-level and is positive upwards and negative downwards.

In the real ocean, the density ρ varies with depth. A column of seawater may be seen as consisting of an infinite number of layers, each of an infinitesimally small thickness dz and contributing an infinitesimally small pressure dp to the total hydrostatic pressure at depth z (Figure 3.9(a)). In this case,

$$dp = -\rho g \, dz \qquad (3.8c)$$

and the total pressure at depth z is the sum of the dp's. However, if we make the assumption that density is independent of depth, or (more realistically) we take an average value for ρ, the hydrostatic pressure at any depth may be calculated using Equation 3.8a or 3.8b (Figure 3.9(b)).

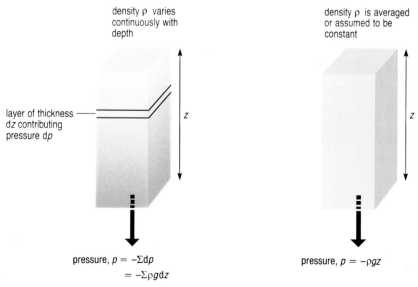

Figure 3.9   Diagrams to illustrate how pressure at depth in the ocean may be calculated.
(a)   The pressure p at depth z is the sum (Σ 'sigma') of all the pressures dp.
(b)   If an average value is taken for the density ρ, or it is assumed to be constant, the hydrostatic pressure at depth z is simply $-\rho g z$.

density ρ varies continuously with depth

density ρ is averaged or assumed to be constant

layer of thickness dz contributing pressure dp

z

z

pressure, $p = -\Sigma dp$
$= -\Sigma \rho g dz$

pressure, $p = -\rho g z$

(a)

(b)

48

Horizontal pressure gradients

Horizontal pressure gradients are easiest to imagine by first considering a highly unrealistic situation. Seawater of constant density ρ occupies an ocean basin (or some other container) and a sea-surface slope is maintained across it without giving rise to any current motion (see Figure 3.10). The hydrostatic pressure acting at depth at site A in Figure 3.10 is given by the hydrostatic equation (3.8a):

$$p_A = \rho g z$$

where z is the height of the overlying column of water (and we are omitting the minus sign). At site B, where the sea-level is higher by an amount Δz (delta z),

$$p_B = \rho g(z + \Delta z)$$

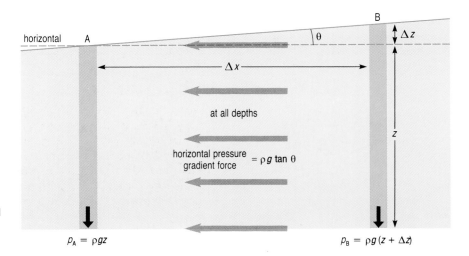

Figure 3.10 Schematic diagram to illustrate how a sloping sea-surface results in a horizontal pressure gradient. If the seawater density is constant, the horizontal pressure gradient force (grey arrows) is ρg tan θ (where tan θ = Δz/Δx) at all depths. (For further information, see text.)

The pressure at depth at site B is therefore greater than that at site A by a small amount, which we will call Δp, so that:

$$\Delta p = p_B - p_A = \rho g(z + \Delta z) - \rho g z = \rho g \, \Delta z$$

If A and B are a distance Δx apart, the horizontal pressure gradient between them is given by $\Delta p / \Delta x$. And since $\Delta p = \rho g \, \Delta z$,

$$\frac{\Delta p}{\Delta x} = \rho g \frac{\Delta z}{\Delta x}$$

and because $\dfrac{\Delta z}{\Delta x} = \tan \theta$,

$$\frac{\Delta p}{\Delta x} = \rho g \, \tan \theta \tag{3.9}$$

If we assume that Δp and Δx are extremely small, we can call them dp and dx, where dp is the incremental increase in pressure over a horizontal distance dx. Equation 3.9 then becomes

$$\frac{dp}{dx} = \rho g \, \tan \theta \tag{3.10}$$

This is the rate of change of pressure, or the horizontal pressure gradient force, in the *x*-direction. More precisely, it is the horizontal pressure gradient force acting on *unit volume* of seawater. In order to obtain the force acting on *unit mass* of seawater, we simply divide by the density, ρ (because density = mass/volume):

$$\text{horizontal pressure gradient force per unit mass} = \frac{1}{\rho}\frac{dp}{dx} = g\tan\theta \quad (3.10a)$$

If the ocean has a uniform density, the horizontal pressure gradient force from B to A is the same at all depths. If there are no other horizontal forces acting, the entire ocean will be uniformly accelerated from the high pressure side to the low pressure side.

3.3.2 BAROTROPIC AND BAROCLINIC CONDITIONS

In the idealized situation discussed in the previous Section, a sea-surface slope had been set up across an ocean consisting of seawater of constant density. Consequently, the resulting horizontal pressure gradient was simply a function of the degree of slope of the surface (tan θ) – the greater the sea-surface slope, the greater the horizontal pressure gradient. In such a situation, the surfaces of equal pressure within the ocean – the **isobaric surfaces** – are parallel to the sea-surface, itself the topmost isobaric surface since the pressure acting all over it is that of the atmosphere.

In real situations where ocean waters are well-mixed and therefore fairly homogeneous, density nevertheless increases with depth because of compression caused by the weight of overlying water. In these circumstances, the isobaric surfaces are parallel not only to the sea-surface but also to the surfaces of constant density or **isopycnic surfaces**. Such conditions are described as **barotropic** (see Figure 3.11(a)).

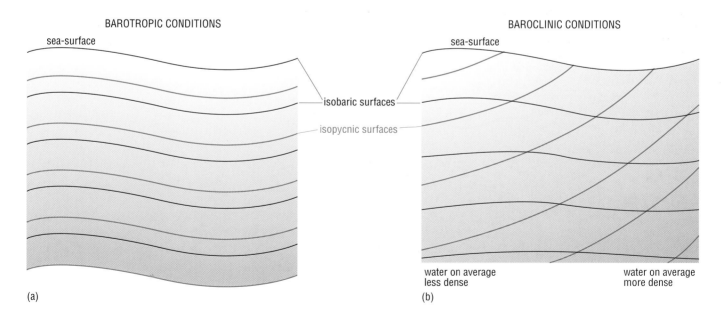

Figure 3.11   The relationship between isobaric and isopycnic surfaces in (a) barotropic conditions and (b) baroclinic conditions. In barotropic conditions, the density distribution (indicated by the intensity of blue shading) does not influence the shape of isobaric surfaces, which follow the slope of the sea-surface at all depths. By contrast, in baroclinic conditions, lateral variations in density *do* affect the shape of isobaric surfaces, so that with increasing depth they follow the sea-surface less and less.

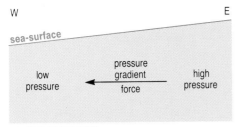

(a) CROSS-SECTION

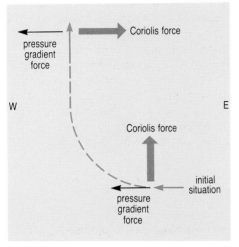

(b) PLAN

Figure 3.12 (a)   A sea-surface slope up towards the east results in a horizontal pressure gradient force towards the west.
(b)   Initially, this causes motion 'down the pressure gradient', but because the Coriolis force acts at right angles to the direction of motion (to the right, if this is in the Northern Hemisphere), the equilibrium situation is one in which the direction of flow (dashed blue line) is at right angles to the pressure gradient.

As discussed in Section 3.3.1, the hydrostatic pressure at any given depth in the ocean is determined by the weight of overlying seawater. In barotropic conditions, the variation of pressure over a horizontal surface at depth is determined *only* by the slope of the sea-surface, which is why isobaric surfaces are *parallel* to the sea-surface. However, any variations in the density of seawater will also affect the weight of overlying seawater, and hence the pressure, acting on a horizontal surface at depth. Therefore, in situations where there are lateral variations in density, isobaric surfaces follow the sea-surface less and less with increasing depth. They *intersect isopycnic surfaces* and the two slope in opposite directions (see Figure 3.11(b)). Because isobaric and isopycnic surfaces are *inclined* with respect to one another, such conditions are known as **baroclinic**.

Geostrophic currents – currents in which the horizontal pressure gradient force is balanced by the Coriolis force – may occur whether conditions in the ocean are barotropic (homogeneous), or baroclinic (with lateral variations in density). At the end of the previous Section we noted that in the hypothetical ocean where the pressure gradient force was the only horizontal force acting, motion would occur in the direction of the pressure gradient. In the real ocean, as soon as water begins to move in the direction of a horizontal pressure gradient, it becomes subject to the Coriolis force. Imagine, for example, a region in a Northern Hemisphere ocean, in which the sea-surface slopes up towards the east, so that there is a horizontal pressure gradient force acting from east to west (Figure 3.12(a)). Water moving westwards under the influence of the horizontal pressure gradient force immediately begins to be deflected towards the north by the Coriolis force; eventually an equilibrium situation may be attained in which the water flows northwards, the Coriolis force acting on it is towards the east, and is balancing the horizontal pressure gradient force towards the west (Figure 3.12(b)). Thus, in a geostrophic current, instead of moving *down* the horizontal pressure gradient, water moves *at right angles* to it.

Moving fluids tend towards situations of equilibrium, and so flow in the ocean is often geostrophic or nearly so. As discussed in Section 2.2.1, this is also true in the atmosphere, and geostrophic winds may be recognized on weather maps by the fact that the wind-direction arrows are parallel to relatively straight isobars. Even when the motion of the air is strongly curved, and centripetal forces are important (as in the centres of cyclones and anticyclones), wind-direction arrows do not cross the isobars at right angles – which would be directly down the pressure gradient – but obliquely (see Figure 2.5 and associated discussion).

So far, we have only been considering forces in a horizontal plane. Figure 3.13 is a partially completed diagram showing the forces acting on a parcel of water of mass *m*, in a region of the ocean where isobaric surfaces make an angle θ (greatly exaggerated) with the horizontal. If equilibrium has been attained so that the current is steady and not accelerating, the forces acting in the horizontal direction must balance one another, as must those acting in a vertical direction. The vertical arrow labelled *mg* represents the weight of the parcel of water, and the two horizontal arrows represent the horizontal pressure gradient force and the Coriolis force.

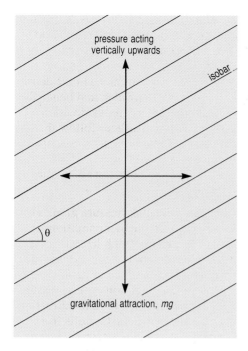

Figure 3.13   Cross-section showing the forces acting on a parcel of water, of mass m and weight mg, in a region of the ocean where the isobars make an angle θ (greatly exaggerated) with the horizontal. (Note that if the parcel of water is not to move vertically downwards under the influence of gravity, there must be an equal force acting upwards; this is supplied by pressure.)

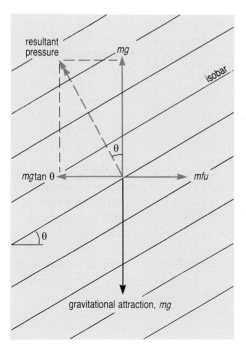

QUESTION 3.6

(a) Given the direction of slope of the isobars on Figure 3.13, which of the horizontal arrows represents the horizontal pressure gradient force and which the Coriolis force?

(b) If the current is flowing '*into* the page', is the situation illustrated in Figure 3.13 in the Northern or Southern Hemisphere?

Important: Ensure that you understand the answer to this question before moving on.

In Section 3.3.1 we considered a somewhat unrealistic ocean in which conditions were barotropic. We deduced that if the sea-surface (and all other isobars down to the bottom) made an angle of θ with the horizontal, the horizontal pressure gradient force acting on *unit mass* of seawater is given by $g \tan \theta$ (Equation 3.10a). The horizontal pressure gradient force acting on a water parcel of mass m is therefore given by $mg \tan \theta$. We also know that the Coriolis force acting on such a water parcel moving with velocity u is mfu, where f is the Coriolis parameter (Equation 3.2). In conditions of geostrophic equilibrium, the horizontal pressure gradient force and the Coriolis force balance one another, and we can therefore write:

$$mg \tan \theta = mfu$$

$$\text{or } \tan \theta = \frac{fu}{g} \tag{3.11}$$

Equation 3.11 is known as the **gradient equation**, and in geostrophic flow *is true for every isobaric surface* (Figure 3.14).

It is worth noting that the Coriolis force and the horizontal pressure gradient force are extremely small: they are generally less than $10^{-4}\,\mathrm{N\,kg^{-1}}$ and therefore several orders of magnitude smaller than the forces acting in a vertical direction. Nevertheless, in much of the ocean, the Coriolis force and the horizontal pressure gradient force are the largest forces acting in a horizontal direction.

As shown in Figure 3.11(a), in barotropic conditions the isobaric surfaces follow the sea-surface, even at depth; in baroclinic conditions, by contrast, the extent to which isobaric surfaces follow the sea-surface *decreases* with increasing depth, as the density distribution has more and more effect. Thus near the surface in Figure 3.11(b), pressure at a given horizontal level is greater on the left-hand side because the sea-surface is higher and there is a longer column of seawater weighing down above the level in question. However, water on the right-hand side is more dense, and with increasing depth the greater density increasingly compensates for the lower sea-surface, and the pressures on the two sides become more and more similar.

Figure 3.14   Diagram to illustrate the dynamic equilibrium embodied in the gradient equation. In geostrophic flow, the horizontal pressure gradient force ($mg \tan \theta$) is balanced by the equal and opposite Coriolis force (mfu). If you wish to verify for yourself trigonometrically that the horizontal pressure gradient force is given by $mg \tan \theta$, you will find it helpful to regard the horizontal pressure gradient force as the horizontal component of the resultant pressure which acts at right angles to the isobars; the vertical component is of magnitude mg, balancing the weight of the parcel of water (cf. Figure 3.13).

At some depth, the isobaric surface may well become horizontal. What implication does this have for the velocity of the geostrophic current, given the relationship between isobaric slope and geostrophic velocity, u, expressed by the gradient equation (Equation 3.11)?

As geostrophic velocity, u, is proportional to tan θ, the smaller the slope of the isobars, the smaller the geostrophic velocity. If tan θ becomes zero – i.e. isobaric surfaces become horizontal – the geostrophic velocity will also be zero. This contrasts with the situation in barotropic conditions, where the geostrophic velocity remains constant with depth.

Figure 3.15 summarizes the differences between barotropic and baroclinic conditions, and illustrates for the two cases how the distribution of density affects the slopes of the isobaric surfaces and how these in turn affect the variation of the geostrophic current velocity with depth.

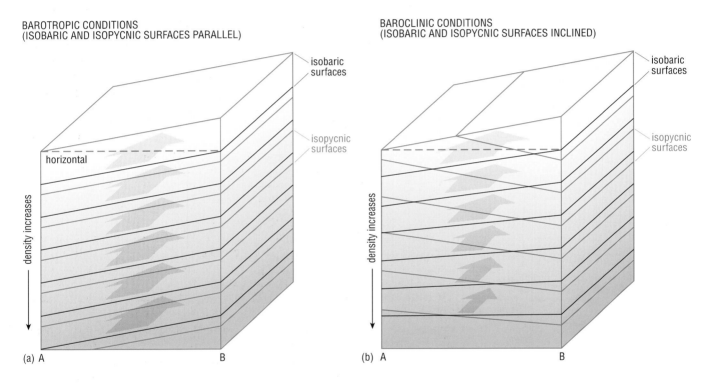

Figure 3.15   Diagrams to summarize the difference between (a) barotropic and (b) baroclinic conditions. The intensity of blue shading corresponds to the density of the water, and the broad arrows indicate the strength of the geostrophic current.
(a)   In barotropic flow, isopycnic surfaces (surfaces of constant density) and isobaric surfaces are parallel and their slopes remain constant with depth, because the average density of columns A and B is the same. As the horizontal pressure gradient from B to A is constant with depth, so is the geostrophic current at right angles to it.
(b)   In baroclinic flow, the isopycnic surfaces intersect (or are *inclined* to) isobaric surfaces. At shallow depths, isobaric surfaces are parallel to the sea-surface, but with increasing depth their slope becomes smaller, because the average density of a column of water at A is more than that of a column of water at B, and with increasing depth this compensates more and more for the effect of the sloping sea-surface. As the isobaric surfaces become increasingly near horizontal, so the horizontal pressure gradient decreases and so does the geostrophic current, until at some depth the isobaric surface is horizontal and the geostrophic current is zero.

Barotropic conditions may be found in the well-mixed surface layers of the ocean, and in shallow shelf seas, particularly where shelf waters are well-mixed by tidal currents. They also characterize the deep ocean, below the permanent thermocline, where density and pressure are generally only a function of depth, so that isopycnic surfaces and isobaric surfaces are parallel. Conditions are most strongly baroclinic (i.e. the angle between isobaric and isopycnic surfaces is greatest) in regions of fast surface current flow.

Currents are described as geostrophic when the Coriolis force acting on the moving water is balanced by the horizontal pressure gradient force. This is true whether the water movement is being maintained by wind stress and the waters of the upper ocean have 'rearranged themselves' so that the density distribution is such that geostrophic equilibrium is attained, or whether the density distribution is itself the cause of water movement. Indeed, it is often impossible to determine whether a horizontal pressure gradient within the ocean is the cause or the result of current flow, and in many cases it may not be appropriate to try to make this distinction.

For there to be exact geostrophic equilibrium, the flow should be steady and the pressure gradient and the Coriolis force should be the only forces acting on the water, other than the attraction due to gravity. In the real oceans, other influences may be important; for example, there may be friction with nearby coastal boundaries or adjacent currents, or with the sea-floor. In addition, there may be local accelerations and fluctuations, both vertical (resulting perhaps from internal waves) and horizontal (as when flow paths are curved). Nevertheless, within many ocean currents, including all the major surface current systems – e.g. the Gulf Stream, the Antarctic Circumpolar Current and the equatorial currents – flow is, to a first approximation, in geostrophic equilibrium.

It is important to remember that the slopes shown in diagrams like Figures 3.10 to 3.15 are greatly exaggerated. Sea-surface slopes associated with geostrophic currents are broad, shallow, topographic irregularities. They may be caused by prevailing winds 'piling up' water against a coastal boundary, by variations in pressure in the overlying atmosphere, or by lateral variations in water density resulting from differing temperature and salinity characteristics (in which case, conditions are baroclinic), or by some combination of these factors. The slopes have gradients of about 1 in 10^5 to 1 in 10^8, i.e. a few metres in 10^2–10^5 km, so they are extremely difficult to detect, let alone measure. However, under baroclinic conditions the isopycnic surfaces may have slopes that are several hundred times greater than this, and these can be determined. How this is done is outlined in Section 3.3.3.

3.3.3 DETERMINATION OF GEOSTROPHIC CURRENT VELOCITIES

Note From now on, we will usually follow common practice and use the terms isobar and isopycnal – lines joining respectively points of equal pressure and points of equal density – instead of the more cumbersome 'isobaric surface' and 'isopycnic surface'.

You have seen that in geostrophic flow the slopes of the sea-surface, isobars and isopycnals are all related to the current velocity. This means that measurement of these slopes can be used to determine the current velocity.

In barotropic conditions, how might the geostrophic current velocity be estimated?

In theory, if the slope of the sea-surface could be measured, the current velocity could be determined using the gradient equation (Equation 3.11). In practice, it is only convenient to do this for flows through straits where the average sea-level on either side may be calculated using tide-gauge data – measuring the sea-surface slope in the open ocean would be extremely difficult, if not impossible, by traditional oceanographic methods.

In practice, it is the density distribution, as reflected by the slopes of the isopycnals, that is used to determine geostrophic current flow in the open oceans.

Why should it be relatively straightforward to determine the density distribution in a volume of ocean?

Because density is a function of temperature and salinity, both of which are routinely and fairly easily measured with the necessary precision. Density is also a function of pressure or, to a first approximation, depth, which is also fairly easy to measure precisely.

Measurement of the density distribution also has the advantage – as observed earlier – that in baroclinic conditions the slopes of the isopycnals are several hundred times those of the sea-surface and other isobars. However, it is important not to lose sight of the fact that it is the slopes of the *isobars* that we are ultimately interested in, because they control the horizontal pressure gradient; determination of the slopes of the isopycnals is, in a sense, only a means to an end.

Figure 3.16(a) and (b) are two-dimensional versions of Figure 3.15(a) and (b). In the situation shown in (a), conditions are barotropic and u, the current velocity at right angles to the cross-section, is constant with depth.

In these circumstances, how may we determine u?

We know that for any isobaric surface making an angle θ with the horizontal, the velocity u along that surface may be calculated using the gradient equation:

$$\tan \theta = \frac{fu}{g} \qquad \text{(Eqn 3.11)}$$

where f is the Coriolis parameter and g is the acceleration due to gravity.

As conditions here are barotropic, the isopycnals are parallel to the isobars and we can therefore determine θ by measuring the slope of the isopycnals (or the slope of the sea-surface). The velocity, u, will then be given by:

$$u = \frac{g}{f} \tan \theta \qquad \text{(3.11a)}$$

at all depths in the water column. The depth-invariant currents that flow in barotropic conditions (isobars parallel to the sea-surface slope) are sometimes described as 'slope currents'; they are often too small to be measured directly.

By contrast, the geostrophic currents that flow in baroclinic conditions *vary* with depth (Figure 3.15(b)). Unfortunately, from the density distribution alone we can only deduce *relative* current velocities; that is, we can only deduce *differences* in current velocity between one depth and another. However, if we know the isobaric slope or the current velocity at some depth, we may use the density distribution to calculate how much greater (or less) the isobaric slope, and hence geostrophic current velocity, will be at other depths. For convenience, it is often assumed that at some fairly deep level the isobars are horizontal (i.e. the horizontal pressure gradient force is zero) and the geostrophic velocity is therefore also zero. Relative current velocities calculated with respect to this level – known as the 'reference level' or 'level of no motion' – may be assumed to be absolute velocities. This is the approach we will take here.

Now look at Figure 3.16(b), which we will again assume represents a cross-section of ocean at *right angles* to the geostrophic current. A and B are two oceanographic stations a distance L apart. At each station, measurements of temperature and salinity have been made at various depths, and used to deduce how *density* varies with depth. However, if we are to find out what the geostrophic velocity is at depth z_1 (say), we really need to know how *pressure* varies with depth at each station, so that we can calculate the slope of the isobars at depth z_1.

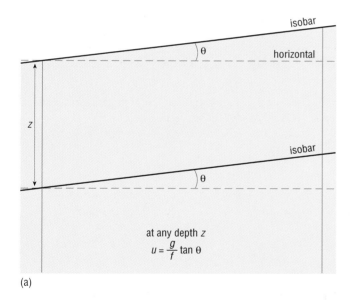

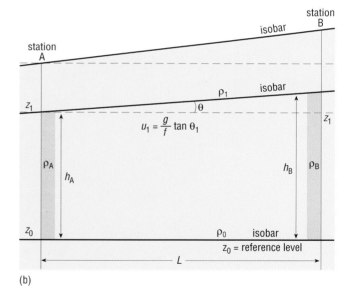

(a)

(b)

Figure 3.16 (a)   In barotropic conditions, the slope of the isobars is tan θ at all depths; the geostrophic current velocity u is therefore (g/f) tan θ (Eqn 3.11a) at all depths.
(b)   In baroclinic conditions, the slope of the isobars varies with depth. At depth z_1 the isobar corresponding to pressure p_1 has a slope of tan θ_1. At depth z_0 (the reference level), the isobar corresponding to pressure p_0 is assumed to be horizontal. (For further details, see text.)

Why do we want to know the slope of the isobars?

So that we can apply the gradient equation (3.11a), and hence obtain a value for u.

Assume that in this theoretical region of ocean the reference level has been chosen to be at depth z_0. From our measurements of temperature and salinity, we know that the average density ρ of a column of water between depths z_1 and z_0 is greater at station A than at station B; i.e. $\rho_A > \rho_B$. The distance between the isobars p_1 and p_0 must therefore be greater at B than at A, because hydrostatic pressure is given by ρgh, where h is the height of the column of water (cf. Equation 3.8). Isobar p_1 must therefore slope up from A to B, making an angle θ_1 with the horizontal, as shown in Figure 3.16(b).

How can $\tan \theta_1$ be expressed in terms of distances shown on Figure 3.16(b)?

It is given by

$$\tan \theta_1 = \frac{h_B - h_A}{L}$$

Substituting for $\tan \theta_1$ in the gradient equation (3.11a), we get:

$$u = \frac{g}{f}\left(\frac{h_B - h_A}{L}\right) \tag{3.12}$$

The difference in hydrostatic pressure between isobars p_1 and p_0 is the same at both A and B, and so

$$\rho_A g h_A = \rho_B g h_B$$

i.e. $h_A = h_B\left(\frac{\rho_B}{\rho_A}\right)$

Therefore, substituting for h_A in Equation 3.12 we get:

$$u = \frac{g}{f}\left(\frac{h_B - h_B\left(\frac{\rho_B}{\rho_A}\right)}{L}\right)$$

i.e. $u = \dfrac{g h_B}{fL}\left(1 - \dfrac{\rho_B}{\rho_A}\right) \tag{3.13}$

We now have an equation that enables us to deduce the geostrophic current velocity u from the density distribution; we will refer to it as the 'geostrophic equation'. Try using the geostrophic equation, in conjunction with Figure 3.16(b), in Question 3.7.

QUESTION 3.7

(a) Hydrographic stations A and B are 100 km apart at 30° of latitude. Temperature and salinity measurements at the two stations indicate that $\rho_A = 1.0265 \times 10^3 \, kg \, m^{-3}$ and $\rho_B = 1.0262 \times 10^3 \, kg \, m^{-3}$. Calculate the geostrophic velocity u at depth $z_1 = 1000 \, m$, taking your reference level (z_0) as 2000 m (in order to apply Equation 3.13 you will have to make the assumption that h_B and $z_0 - z_1$ are equal). The value for g is $9.8 \, m \, s^{-2}$.

(b) (i) Use Equation 3.12 to calculate the difference $h_B - h_A$. How valid was the assumption, made above, that $h_B = z_0 - z_1$? (ii) Use your value for $h_B - h_A$ to calculate $\tan \theta$ (i.e. the slope of the isobar at 1000 m depth), and the angle θ itself.

(c) If stations A and B are in the Southern Hemisphere, with B due east of A, in which direction is the geostrophic current flowing? Given that the conventional symbols for indicating flow direction are $\otimes$ for 'flow into the page' and $\odot$ for 'flow out of the page', draw a simple diagram to illustrate this situation (A on the left, B on the right). Show the sloping sea-surface and the sloping isobar at 1000 m depth, and arrows representing the horizontal pressure gradient force and the Coriolis force either side of the appropriate symbol for flow direction.

Note: Before moving on, you should ensure that you can follow the answer to Question 3.7, especially part (c). The sketch for this is shown overleaf, in Figure 3.17(a).

By setting h_B equal to $z_0 - z_1$ in Question 3.7(a), you were effectively using average densities ρ_A and ρ_B that had been calculated assuming that the columns of water between isobars p_1 and p_0 were the same height at A and B. As you will have found in part (b), the difference in height between the two columns $(h_B - h_A)$ is a very small fraction (0.03%) of $z_0 - z_1$, and the resulting error in ρ_B, and hence in u, is also extremely small.

In Question 3.7(a) you calculated the velocity of the geostrophic current at one depth, z_1. A complete profile of the variation of the geostrophic current velocity with depth may be obtained by choosing a reference level z_0 and then applying Equation 3.13 at successively higher levels, each time calculating the geostrophic current velocity in relation to the level below (see for example Figure 3.18(a)). This is how geostrophic current velocities have been calculated since the beginning of the 20th century; the full version of Equation 3.13 used for this purpose is often referred to as Helland-Hansen's equation, after the Scandinavian oceanographer Bjørn Helland-Hansen who, with Johan W. Sandström, did much pioneering work in this field.

Until relatively recently, oceanographers had to calculate seawater densities from measured values of temperature, salinity and depth, using standard tables. Today, temperature, salinity and depth are routinely recorded electronically, and the necessary calculations (effectively, many iterations of the gradient equation) are done by computer, but it is still important for oceanographers to be aware of the limitations of the geostrophic method and the various assumptions and approximations that are implicit in the calculations.

The first point to note is that we have been implicitly assuming that the geostrophic current we are attempting to quantify is flowing *at right angles* to the section A–B. In reality, there is no way of ensuring that this is the case. If the current direction makes an angle with the section, the geostrophic equation (3.13) will only provide a value for that component of the flow at right angles to it. The calculated geostrophic velocities will therefore be underestimates. To take an extreme example, if the average flow in a region is north-eastwards and the section taken is from the south-west to the north-east (i.e. effectively parallel to the flow), the geostrophic velocity calculated will be only a very small proportion of the actual geostrophic velocity to the north-east. To get over this problem, a second section may be made at right angles to the first, and the total current calculated by combining the two components. In practice, the geostrophic flow in an area is usually determined using data from a grid of stations.

The second important limitation of this method is that, as discussed earlier, it can only be used to determine *relative* velocity. Thus in calculating the geostrophic current in Question 3.7(a), it was assumed that at a depth of 2000 m – the reference level – the current was zero. If in fact current measurements revealed that there was a current of, say, 0.05 m s^{-1} at this depth, that value could be added on to the (calculated) velocity at all depths.

Current that persists below the chosen reference level is sometimes referred to as the 'barotropic' part of the flow (Figure 3.18(b)). As mentioned in Section 3.3.2, in the deep ocean below the permanent thermocline, density and pressure usually vary as a function of depth only, and so even if isobars and isopycnals are not horizontal, they are likely to be parallel to one another.

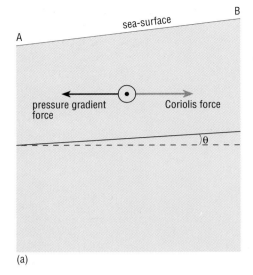

(a)

(b)

Figure 3.17 (a)   Answer to Question 3.7(c).
Note that the pressure gradient arrows are
horizontal. Remember that the sketch relates
to the *Southern Hemisphere*.
(b)   Drawing to illustrate the rationale for
the conventional current-direction symbols,
⊗ and ⊙.

Figure 3.18 (a)   Example of a profile of
geostrophic current velocity, calculated on the
assumption that the horizontal pressure
gradient, and hence the geostrophic current
velocity, are zero at 1000 m depth (the
reference level).
(b)   If direct current measurements reveal
that the current velocity below about 1000 m is
not zero as assumed but some finite value (say
0.05 m s⁻¹), the geostrophic current velocity
profile would look like this. The geostrophic
velocity at any depth may therefore be
regarded as a combination of baroclinic and
barotropic components.

Would the barotropic part of the flow show up in calculations like those you
made in Question 3.7(a)?

No. The geostrophic velocity u obtained using Equation 3.13 is that for flow
resulting from lateral variations in density (i.e. attributable to the difference
between ρ_A and ρ_B). The effect of any horizontal pressure gradient that
remains constant with depth is not included in Equation 3.13. In reality,
'barotropic flow' resulting from a sea-surface slope caused by wind, and
'baroclinic flow' associated with lateral variations in density – or, put
another way, the 'slope current' and the 'relative current' – may not be as
easily separated from one another as Figure 3.18 suggests.

Another point that must be borne in mind is that the geostrophic equation
only provides information about the *average* flow between stations (which
may be many tens of kilometres apart) and gives no information about
details of the flow. However, this is not a problem if the investigator is
interested only in the large-scale mean conditions. Indeed, in some ways it
may even be an advantage because it means that the effects of small-scale
fluctuations are averaged out, along with variations in the flow that take
place during the time the measurements are being made (which may be
from a few days to a few weeks).

We have seen how information about the distribution of density with depth
may be used to determine a detailed profile of geostrophic current velocity
with depth. Although both density and current velocity generally vary
continuously with depth (e.g. Figure 3.19(a) and Figure 3.18), for some
purposes it is convenient to think of the ocean as a number of homogeneous
layers, each with a constant density and velocity. This simplification is most
often applied in considerations of the motion of the mixed surface layer,
which may be assumed to be a homogeneous layer separated from the
deeper, colder waters by an abrupt density discontinuity (Figure 3.19(b)),
rather than by a **pycnocline** – an increase in depth over a finite depth. In
this situation, the slopes of the sea-surface and the interface will be as

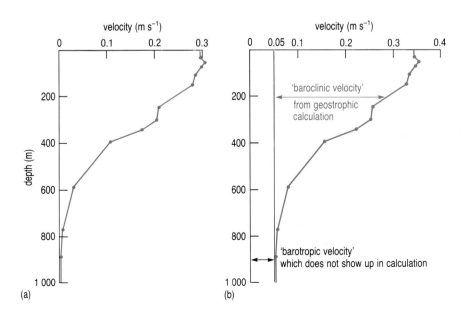

(a)   (b)

shown in Figure 3.20(a) and, for convenience, the geostrophic velocity of the upper layer may be calculated on the assumption that the lower layer is motionless. Figure 3.20(b) is an example of a more complex model, which may also sometimes approximate to reality; here there are three homogeneous layers, with the intermediate layer flowing in the opposite direction to the other two.

In Chapter 5, you will see how such simplifications may help us to interpret the density and temperature structure of the upper ocean in terms of geostrophic current velocity.

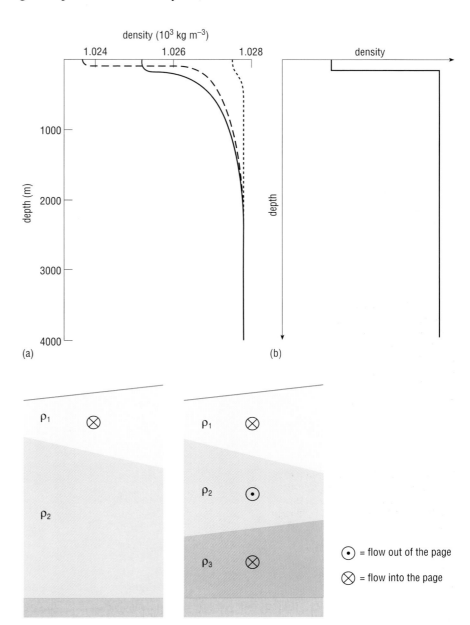

Figure 3.19 (a) Typical density profiles for different latitudes: solid line = tropical latitudes; dashed line = equatorial latitudes; dotted line = high latitudes.
(b) The type of simplified density distribution sometimes assumed in order to estimate geostrophic currents in the mixed surface layer.

Figure 3.20 Diagrams to show how the interfaces between layers slope in the case of (a) two and (b) three homogeneous layers, where $\rho_1 < \rho_2 < \rho_3$. The symbols for flow direction are drawn on the assumption that the locations are in the Northern Hemisphere.

3.3.4 PRESSURE, DENSITY AND DYNAMIC TOPOGRAPHY

We stated earlier that seawater densities used in the calculation of geostrophic velocities are computed from data on temperature, salinity and depth. Of course, it is not depth *per se* that affects the density of a sample of seawater but the pressure it is under, which is related to the depth by the hydrostatic equation (3.8). For the purpose of making geostrophic calculations, it is usual to interconvert hydrostatic pressure and depth by using the following approximate relationship:

$$\text{pressure} \approx 10^4 \times \text{depth} \tag{3.14}$$

where pressure is in newtons per square metre ($N\,m^{-2}$) or pascals ($1\,Pa = 1\,N\,m^{-2}$), and depth is in metres.

Can you see where the factor of 10^4 comes from?

Relationship 3.14 is simply the hydrostatic equation $p = \rho gz$, in which it has been assumed that g is exactly $10\,m\,s^{-2}$ (instead of $9.8\,m\,s^{-2}$) and that the density of seawater is a constant $1 \times 10^3\,kg\,m^{-3}$ (rather than ranging between 1.025 and $1.029 \times 10^3\,kg\,m^{-3}$). This approximation is very useful for 'back-of-envelope' calculations involving pressure at depth in the ocean, as we shall now demonstrate.

One possible source of error in geostrophic calculations is the variable pressure of the atmosphere. An oceanographic section may be of the order of 100 km across and it is quite likely that there will be a difference in atmospheric pressure between the two hydrographic stations.

Would such a difference in atmospheric pressure be taken into account by Equation 3.13?

No, it would not. In deducing Equation 3.13 we used the *hydro*static equation (3.8), which gives pressures resulting from the weight of overlying *water* only. Nevertheless, lateral variations in atmospheric pressure would contribute to any slope in the sea-surface and hence to the slopes of isobars at depth. As the ultimate purpose of Equation 3.13 is to obtain geostrophic current velocities from isobaric slopes, it would be useful to know the extent to which variations in atmospheric pressure affect these slopes.

The pressure resulting from a standard atmosphere is 1 bar, or $10^5\,Pa$. Given relationship 3.14 above, what depth of seawater would give rise to the same pressure?

The answer must be $10^5/10^4 = 10\,m$. If a standard atmosphere is equivalent to 10 m of seawater, a difference in atmospheric pressure of a few millibars – which is what we might expect over distances of the order of 100 km – would be equivalent to a few centimetres of seawater. Thus, variations in atmospheric pressure typically give rise to isobaric slopes of a few centimetres in 100 km or about 1 in 10^7, and hence contribute between about 1 and 10% of the total horizontal pressure gradient (cf. end of Section 3.3.2).

Would geostrophic current flow resulting from spatial variations in atmospheric pressure change with depth?

No, it would not, because it would be a 'slope current' and as such would be in the 'barotropic' component of the flow (cf. Figure 3.18(b)). Thus, although the effects of variations in atmospheric pressure cannot be completely ignored, in theory they could easily be corrected for. However, weather systems may travel several hundred kilometres per day, so differences in atmospheric pressure between two stations are likely to fluctuate over the period in which measurements are made. In practice, therefore, the effect of atmospheric pressure on geostrophic current flow is difficult to take account of, and in most situations is considered sufficiently small to be ignored.

We deduced above that 10 m of seawater is equivalent to a pressure of 1 bar, i.e. that 1 m of seawater is equivalent to 1 decibar (dbar). For many purposes this is a very useful approximation. However, like all fluids, seawater is compressible, and for greater depths quite significant errors result from converting pressures in decibars directly to depths in metres. For example, a pressure of 11 240 dbar has been recorded at the bottom of the Marianas Trench, but soundings of the area indicate that the maximum depth is about 10 880 m.

Dynamic topography

You have seen how, to determine geostrophic current velocities, we need to quantify departures of isobaric surfaces from the horizontal. But what does 'horizontal' really mean? The simplest answer is that a horizontal surface is any surface at right angles to a vertical plumb-line, i.e. a plumb-line hanging so that it is parallel to the direction in which the force of gravity acts. The reason for this apparently perverse approach is that the upper layers of the solid Earth are neither level nor uniformly dense, so that a surface over which gravitational potential energy is constant is not smooth but has a topography, with bumps and dips on a horizontal scale of tens to thousands of kilometres and with a relief of up to 200 m. If there were no currents, the sea-surface would be coincident with an equipotential surface. The particular equipotential surface that corresponds to the sea-surface of a hypothetical motionless ocean is known as the marine **geoid**.

In the context of geostrophic current flow, the important aspect of a 'horizontal' or equipotential surface is that the potential energy of a parcel of water moving over such a surface remains constant. If a parcel of water moves from one equipotential surface to another, it gains or loses potential energy, and the amount of potential energy gained or lost depends on the vertical distance moved and the value of g at the location in question.

Now imagine a situation in which a steady wind has been blowing sufficiently long for the slopes of isobaric and isopycnic surfaces to have adjusted so that there is a situation of geostrophic equilibrium. If the wind speed now increases up to another steady value, more energy is supplied to the upper ocean; the speed of the current increases and the slopes of isobaric and isopycnic surfaces become steeper – i.e. they depart even further from the horizontal, the position of least potential energy. In other words, the ocean gains potential energy as well as kinetic energy. You might like to think of this as being analogous to motorcyclists riding around a 'wall of death'. As the motorcyclists travel faster and faster, so their circuits move higher and higher up the 'wall', thus increasing the potential energy of riders and machines.

Because of the relationship between isobaric slope and potential energy, departures of isobaric surfaces from the horizontal are often discussed in terms of 'dynamic height', i.e. vertical distances are quantified in terms of changes in potential energy (or 'work') rather than simply distance. The units of work that have been adopted for this purpose are known as 'dynamic metres' because they are numerically very close to actual metres. How closely dynamic metres and geometric metres correspond depends on the local value for g: if g is 9.80 m s^{-2}, 1 dynamic metre is equivalent to 1.02 geometric metres.

You have effectively encountered dynamic height already, in Question 3.7(b). When you calculated the value of $h_B - h_A$ as 0.3 m, you could equally well have deduced that the difference in the dynamic height of isobar p_1 between stations A and B was about 0.3 dynamic metres. Such variations in dynamic height are described as **dynamic topography**. Because of the way geostrophic calculations are made (cf. Question 3.7), dynamic heights are always given relative to some depth or pressure at which it is assumed, for the purposes of the calculations, that the isobaric surface is horizontal. In Question 3.7 and Figure 3.16(b), the depth was z_0 and the pressure p_0.

It is possible to determine the dynamic topography of any isobaric surface relative to another. The simplest situation to imagine is that in which the isobaric surface under consideration is the uppermost one – the sea-surface. Figure 3.21 shows the mean dynamic topography of the sea-surface relative to the 1500 dbar isobaric surface, for the world ocean. Clearly, maps like Figure 3.21 may be used to determine geostrophic current flow in the upper ocean.

How does the *direction* of geostrophic current flow relate to the contours of dynamic height?

The current flows *at right angles* to the slope of the isobaric surface, and therefore flows *along the contours* of dynamic height. The direction of flow will be such that the isobaric surface slopes up towards the right in the Northern Hemisphere and up towards the left in the Southern Hemisphere.

Bearing in mind Equation 3.11 (the gradient equation), can you suggest how the contours of dynamic height reflect the geostrophic current velocity?

Figure 3.21   The mean dynamic topography of the sea-surface relative to 1500 dbar. The contours are labelled in dynamic metres and values range from 4.4 to 6.4, corresponding to a relief of about 2 m. (*Note:* Close to certain land masses, the contours suggest flow 'coming out of' the coastline; this is effectively an artefact arising from assumptions made in generating the contours.)
This computed map should be compared with the Frontispiece, which shows the dynamic topography of the sea-surface in December 2000, as determined by satellite-borne radar altimeters.

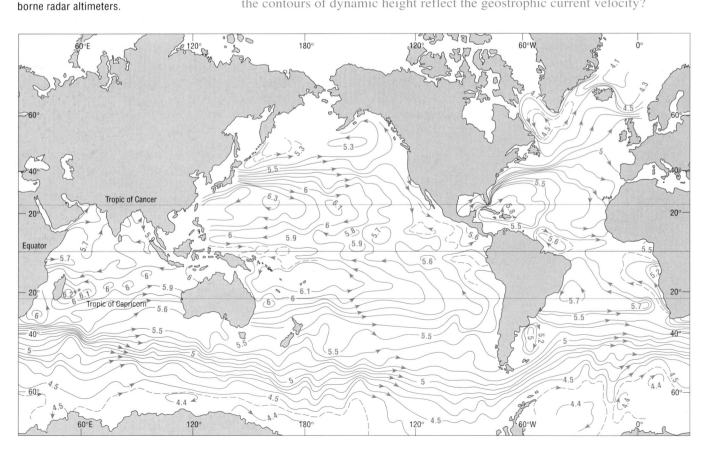

The greater the geostrophic current velocity, the greater the isobaric slope, and – by analogy with ordinary topographic contours – the closer together the contours of dynamic height.

With these points in mind, try Question 3.8.

QUESTION 3.8

(a) What is meant by 'the dynamic topography of the sea-surface relative to 1500 dbar'?

(b) Identify on Figure 3.21 the regions corresponding to flow in (i) the Gulf Stream, and (ii) the Antarctic Circumpolar Current, referring to Figure 3.1 if necessary.

(c) Where is the fastest part of the Antarctic Circumpolar Current, according to Figure 3.21?

(d) At 120° E and about 65° S there is a closed contour corresponding to 4.4 dynamic metres. In the region of this contour, does the sea-surface form a 'hill' or a depression? Given the direction of current flow, is this what you would expect?

Figure 3.22 is a topographic map of the *mean* sea-surface as obtained by satellite altimetry. The mean sea-surface is a very close approximation to the marine geoid because even in regions where there are strong and relatively steady currents (e.g. the Gulf Stream or the Antarctic Circumpolar Current) the contribution to the sea-surface topography from current flow is only about one-hundredth of that resulting from variations in the underlying solid crust (cf. Figure 3.21). In a sense, the topographic surface in Figure 3.22 is the 'horizontal' surface, departures from which are shown by the topographic surface in Figure 3.21.

Figure 3.22   Topographic map of the mean sea-surface (i.e. the marine geoid), as determined using a satellite-borne radar altimeter. The mean sea-surface topography reflects the topography of the sea-floor rather than geostrophic current flow, as the effect of the latter is about two orders of magnitude smaller, even in regions of strong current flow.
Note that uncertainties in the shape of the geoid are the biggest source of possible error in estimates of dynamic topography determined by satellite altimetry (e.g. as in the Frontispiece).

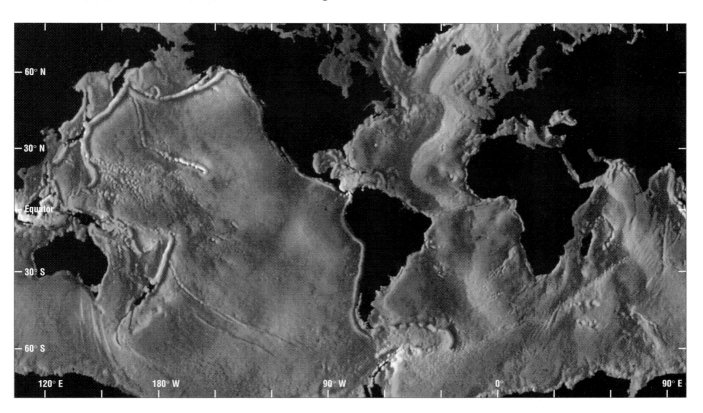

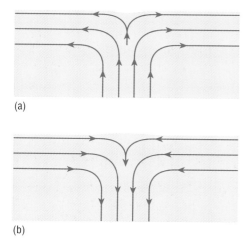

(a)

(b)

Figure 3.23   Schematic diagrams to illustrate how (a) divergence of surface waters leads to upwelling while (b) convergence of surface waters leads to sinking.

3.4   DIVERGENCES AND CONVERGENCES

Wind stress at the sea-surface not only causes horizontal movement of water, it also leads to vertical motion. When the wind stress leads to a **divergence** of surface water, deeper water rises to take its place (Figure 3.23(a)); conversely, when there is a **convergence** of water at the surface, sinking occurs (Figure 3.23(b)). **Upwelling** of subsurface water and sinking of surface water occur throughout the oceans, both at coastal boundaries and away from them. We will consider coastal upwelling in Chapter 4; here we will concentrate on the upwelling that occurs in the open ocean.

Figure 3.24(a) shows the effect on surface waters of a cyclonic wind in the Northern Hemisphere. The Ekman transport – the average movement of the wind-driven layer – is to the right of the wind, causing divergence of surface water and upwelling. Figure 3.24(b) shows how, under these conditions, the sea-surface is lowered and the thermocline raised. This upward movement of water in response to wind stress is sometimes called **Ekman pumping**.

Figure 3.24(c) and (d) show the situation that results from anticyclonic wind stress in the Northern Hemisphere: under these conditions there is convergence and sinking (sometimes referred to as 'downwelling').

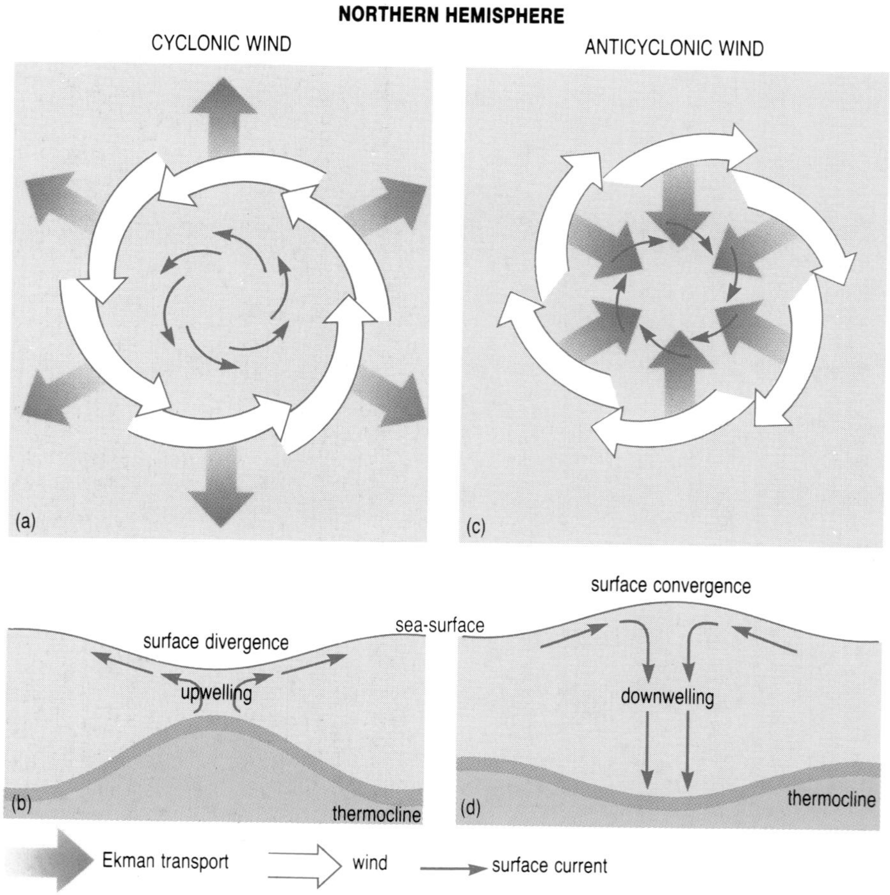

NORTHERN HEMISPHERE

CYCLONIC WIND   ANTICYCLONIC WIND

(a)   (c)

Figure 3.24   The effect of a cyclonic wind in the Northern Hemisphere (a) on surface waters, (b) on the shape of the sea-surface and thermocline. Diagrams (c) and (d) show the effects of an anticyclonic wind in the Northern Hemisphere. (Remember that in the Southern Hemisphere, cyclonic = clockwise and anticyclonic = anticlockwise.)

surface divergence

sea-surface

surface convergence

upwelling

downwelling

(b)

thermocline

(d)

thermocline

Ekman transport   wind   surface current

QUESTION 3.9   Sketch diagrams analogous to those in Figure 3.24 to show the conditions that would lead to convergence and sinking of surface water in the Southern Hemisphere.

The convergence of water as a result of anticyclonic winds thus causes the sea-surface to slope upwards towards the middle of the gyre. As a result, the circulating water will be acted upon by a horizontal pressure gradient force.

In which direction will the horizontal pressure gradient force act?

It will act outwards from the centre, from the region of higher pressure to the region of lower pressure (see Figure 3.25). Under steady conditions the horizontal pressure gradient force will be balanced by the Coriolis force, and a geostrophic ('slope') current will flow in the same direction as the wind. The gyral current systems of the Atlantic and Pacific Oceans between about 10 and 40 degrees of latitude – the **subtropical gyres**, which lie beneath the subtropical highs (cf. Figures 2.2(a) and 2.3) – are large-scale gyres of the type we have been discussing (Figure 3.1).

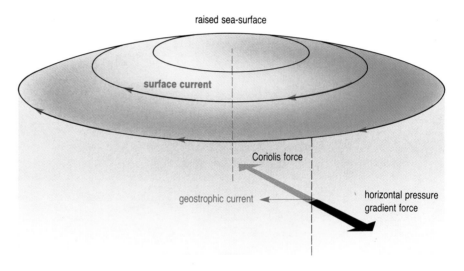

Figure 3.25   The generation of geostrophic current flow in a gyre driven by anticyclonic winds in the Northern Hemisphere. This current is driven by the wind only indirectly and persists below the wind-driven (Ekman) layer.

Having identified the subtropical gyres on Figure 3.1, identify, at higher latitudes, areas characterized by *cyclonic* gyres, within which outward Ekman transport would lead to upwelling as shown in Figure 3.24(a) and (b). According to Figures 2.2(a) and 2.3, what wind systems characteristically affect these regions?

We hope you identified the cyclonic gyres in the northern parts of the North Pacific and the North Atlantic, and in the Norwegian and Greenland Sea. These are the **subpolar gyres**, driven by the subpolar low pressure systems (Figure 2.2(a)), especially in winter (cf. Figure 2.3). You will have noticed that subpolar gyres do not form in the uninterrupted expanse of the Southern Ocean, but further south, off Antarctica (cf. Figure 3.1), are the cyclonic Weddell Sea Gyre and Ross Sea Gyre.

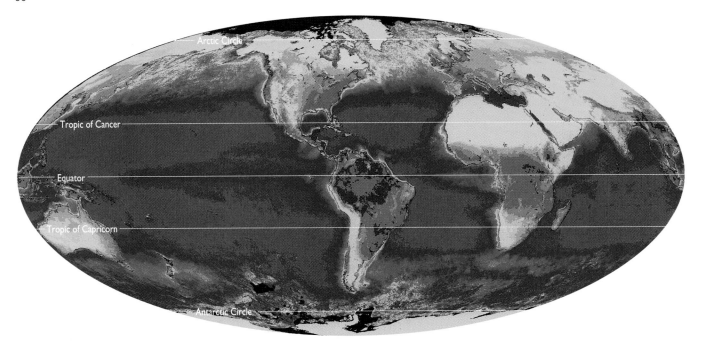

Figure 3.26   Global distribution of the potential for primary (plant) production in surface ocean waters and on land, as indicated by chlorophyll concentration, determined by satellite-borne sensors. In surface waters, regions of highest productivity are bright red, followed by yellow, green and blue, with least productive surface waters shown purplish red. In the Northern Hemisphere, the red areas around coasts correspond to high primary productivity supported by fertilizer run-off and sewage from the land. On land, darkest green = areas with the greatest potential for primary production, yellow = least productive areas.

The vertical motion of water that occurs within wind-driven gyres – whether upwelling or downwelling – has a profound effect on the biological productivity of the areas concerned. In cyclonic gyres, upwelling of nutrient-rich water from below the thermocline can support high primary (phytoplankton) productivity. By contrast, in the subtropical gyres, the sinking of surface water and the depressed thermocline tend to suppress upward mixing of nutrient-rich water. The satellite-derived data in Figure 3.26 show clearly the difference between the productive cyclonic subpolar gyres (green/light blue) and the 'oceanic deserts' (purplish-red) of the subtropical gyres.

The surface waters of the ocean move in complex patterns, and divergences and convergences occur on small scales as well as on the scale of the subtropical and subpolar gyres. Figure 3.27 illustrates schematically several types of flow that would lead to vertical movement of water. Such divergences and convergences may be seen in rivers, estuaries, lakes and shelf seas, as well as in the open ocean.

The position of a divergence may sometimes be inferred from the colour of the surface water: because it is usually richer in nutrients, and therefore able to support larger populations of phytoplankton, upwelled water often becomes greener than the surrounding water. It is also generally colder than surface water and so divergences are sometimes marked out by fog banks. Linear convergences are often known as **fronts**, especially when water properties (e.g. temperature, salinity and productivity) are markedly different on either side of the convergence.

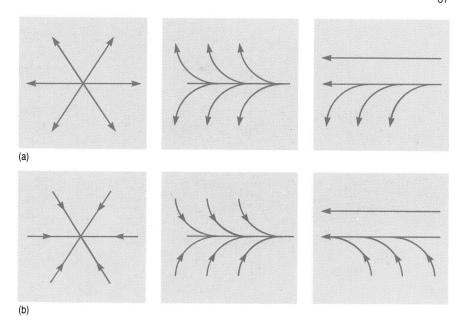

(a)

(b)

Figure 3.27 Schematic representations (as seen from above) of (a) divergent surface flow patterns which would lead to upwelling of subsurface water and (b) convergent flow patterns that would lead to sinking of surface water.

Small-scale convergences are often marked by a collecting together of surface debris, seaweed or foam (Figure 3.28(a)). In certain circumstances, linear convergences form *parallel* to the wind (Figure 3.28(b)). These 'windrows' were first studied by Langmuir, who in 1938 noticed large amounts of seaweed arranged in lines parallel to the wind, in the Sargasso Sea. He proposed that the wind had somehow given rise to a series of helical vortices, with axes parallel to the wind. This circulation system – now called **Langmuir circulation** – is illustrated schematically in Figure 3.28(c). It is thought to result from instability in the well-mixed, and therefore fairly homogeneously dense, surface water. Such 'longitudinal roll vortices' are now known to be common in both the upper ocean and the lower atmosphere, where they give rise to the linear arrangements of clouds, known as 'cloud streets', that are often seen from aircraft (Figure 3.28(d)).

The spatial scales of the circulatory systems of windrows and of basin-wide features like the subtropical gyres are very different. The latter involve horizontal distances of thousands of kilometres; the former extend only for a kilometre or so. It is characteristic of dynamic systems that phenomena with small length-scales also have short time-scales, while phenomena with long length-scales also have long time-scales. This aspect of ocean circulation will be developed further in the next Section.

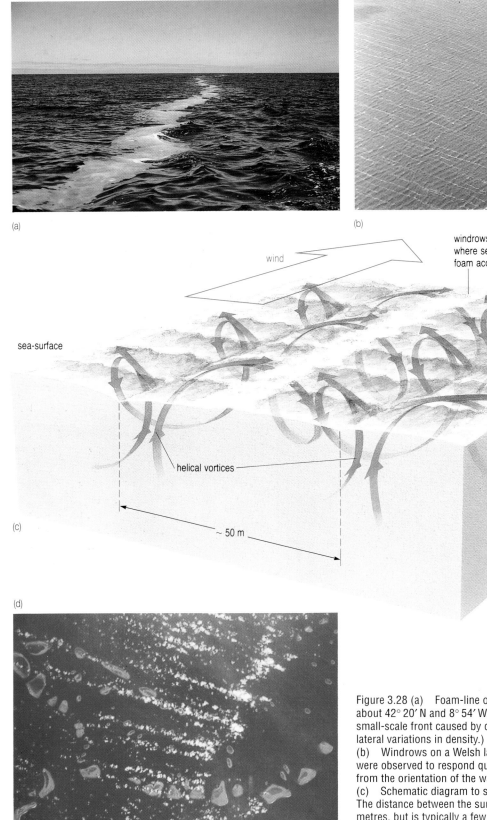

(a)

(b)

(c)

sea-surface

wind

windrows (convergences where seaweed, debris and foam accumulate)

divergence

helical vortices

~ 50 m

(d)

Figure 3.28 (a) Foam-line observed off the north-west coast of Spain at about 42° 20′ N and 8° 54′ W. (As in (b) and (c), this is an example of a small-scale front caused by convergence of surface water rather than by lateral variations in density.)
(b) Windrows on a Welsh lake. The streaks, which were 5–10 m apart, were observed to respond quickly to changes in the wind direction (inferred from the orientation of the wave crests).
(c) Schematic diagram to show Langmuir circulation in the upper ocean. The distance between the surface streaks may be as much as a few hundred metres, but is typically a few tens of metres.
(d) Cloud streets over the reefs of the Maldive Islands in the Indian Ocean.

3.5 THE ENERGY OF THE OCEAN: SCALES OF MOTION

Energy is imparted to the ocean from the Sun, directly through solar radiation (which leads to heating, evaporation, precipitation and changes in density) and indirectly through winds. In the case of tidal currents, the energy results from the continually varying gravitational attractions of the Moon and Sun.

As discussed in Section 3.3.4, the ocean's energy is both kinetic and potential: kinetic energy by virtue of its motion and potential energy as a result of isopycnic and isobaric surfaces being displaced from their position of least energy parallel to the geoid (i.e. horizontal). The potential energy of the world ocean is about *one hundred times* greater than its kinetic energy. It has been calculated that if all the isopycnic surfaces that are presently sloping were allowed to become horizontal, 10^6 joules of potential energy would be released for every square metre of sea-surface. This huge store of potential energy ensures that if the global winds stopped blowing, the ocean circulation would take a decade or so to run down.

3.5.1 KINETIC ENERGY SPECTRA

We have seen that oceanic circulation consists of many types of motion, acting over a range of scales within space and time.

How, then, might it be possible to calculate the total kinetic energy of the ocean?

At any one point in the ocean, at a given time, the observed current may be regarded as the combination of a number of components – perhaps a local northward wind-driven current has been superimposed on an inertia current generated some distance away (rather like the situation illustrated in Figure 3.8), and this is in a region with strong tidal currents. Perhaps all of these motions are occurring in the region of an established current system like the Gulf Stream, which shows fluctuations but is nevertheless a permanent feature of the oceanic circulation. Such a long-term, large-scale flow is referred to as the **mean flow** (or, sometimes, the 'mean motion'). If the varying components of the flow (tides, inertia currents, eddies and other periodic fluctuations in currents) are resolved mathematically into regular oscillations, the relationship between the fluctuating components and the mean flow might look something like Figure 3.29.

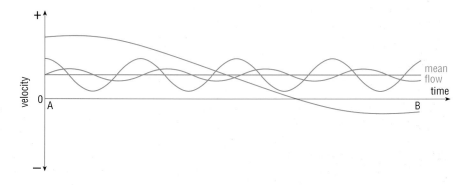

Figure 3.29   A highly schematic representation of how current velocity might vary with time for various fluctuating components. Over a long period (e.g. from A to B), the fluctuating motions average out, leaving only the mean flow represented by the straight line. (Not shown are fluctuations resulting from aperiodic phenomena such as passage of a storm or large-scale eddy through the area in question – see Section 3.5.2.)

The kinetic energy of a mass m moving with speed v is given by $\frac{1}{2}mv^2$. The kinetic energy possessed by a parcel of water at any instant is therefore the sum of the energies of all the components contributing to its motion, *as determined by the squares of their speeds*. For any particular region of the ocean, the fluctuations contributing to the overall current flow may be analysed to produce a **kinetic energy density spectrum**. Perhaps surprisingly, *the fluctuating components account for most of the total kinetic energy of the flow, with only a very small proportion being contained in the mean flow.*

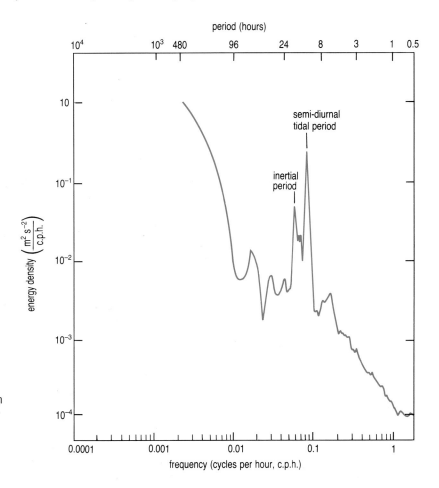

Figure 3.30   The kinetic energy density spectrum for current flow at a depth of 120 m in the north-west Atlantic off Woods Hole, Mass, USA, as determined from current meter measurements made from 24 June to 11 August 1965. (Note that the scales on the axes are logarithmic.)

Figure 3.30 is an energy density spectrum based on current measurements made at 120 m depth at a station in the north-west Atlantic. The components of a flow that have a specific period (specific frequency) typically show up on an energy density spectrum as individual peaks – in this case, a peak corresponding to the semi-diurnal tidal motion is clearly seen at a period of about 12 hours, while the lower-frequency peak to the left has been attributed to inertia currents. In general, however, there is a continuous spectrum containing an infinite number of frequencies because variations in current flow occur over the full range of possible time-scales. This means that it is usually not possible to determine the energy corresponding to a single frequency, as one frequency can only possess an infinitesimal proportion of the total energy. It *is* possible, however, to determine the amount of energy corresponding to a frequency *interval* and the units of energy density are generally given in terms of energy/frequency which is shorthand for 'energy per frequency interval'. The total *area* under an energy density curve corresponds to the total kinetic energy contributed by all the frequencies within the range considered.

Study Figure 3.30 and its caption. What is the longest period (lowest frequency) of current variability that has been extracted from the current-meter data? Why do you think that no attempt was made to analyse the data for longer period (lower frequency) components of the flow?

The graph shows the kinetic energy of fluctuating components of the motion with periods of less than 480 hours (20 days) and frequencies of about 0.002 cycles per hour. According to the caption, the current meter that recorded the raw data was deployed for 48 days (1152 hours), i.e. only about two-and-a-half times the longest period. Although it would have been possible to try to extract information about motions of longer and longer periods, the results would have become increasingly unreliable.

3.5.2 EDDIES

Ocean currents are not continually getting faster and faster, and this is because an equilibrium has been reached whereby the rate at which energy is supplied is being balanced by the rate at which it is being dissipated. Ultimately, the ocean's kinetic energy is converted to heat, through frictional interaction with the sea-bed (this is particularly true of tidal currents in shallow shelf seas) or internal friction at the molecular level, i.e. molecular viscosity. There is a continual transfer of energy from the identifiable currents (i.e. the mean flow) to eventual dissipation as heat, via a succession of eddies of generally decreasing size. The frictional effect of these eddies is the eddy viscosity discussed in Section 3.1.1. Eventually, within small eddies a few centimetres across or less, kinetic energy is converted to heat energy by *molecular* viscosity.

This idea of a 'cascade' of energy flow, from large-scale features down to the molecular level, was neatly summarized by the dynamicist L.F. Richardson (1881–1953), in the following piece of dogerell:

'Big whirls have little whirls
which feed on their velocity.
Little whirls have lesser whirls
and so on to viscosity.'

(A paraphrase of Augustus de Morgan, who had himself paraphrased Jonathan Swift.)

Richardson was in fact referring to atmospheric motions, but could as well have been referring to flow in the ocean.

Eddies form because flowing water has a natural tendency to be turbulent and chaotic. Theoretically, it would be possible for an idealized uniform current to flow smoothly, as long as the current speed was below a certain critical value. In reality, wherever there are spatial variations in flow velocity (i.e. horizontal or vertical current shear), any small disturbances or perturbations in the flow will tend to grow, developing into wave-like patterns and/or eddies. Such effects are described as *non-linear* because they are not predictable simply by adding together flow velocities.

Although the formation of eddies generally results in energy being removed from the mean flow (i.e. from the identifiable current system), eddies may also interact with the mean flow and inject energy into it. They may also interact with one another, sometimes forming jets or plumes, but more often producing a complex pattern of eddies flowing into and around one another, forming more and more intricate swirls (as illustrated by the images in Figure 3.31 and Figure 1.1).

(a)

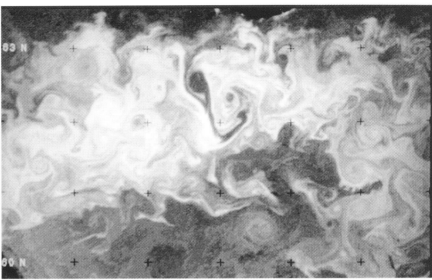

(b)

Figure 3.31   Examples of eddies in the oceans.
(a)   Eddies in the Mediterranean off the coast of Libya. They were photographed from the Space Shuttle and (like those in Figure 1.1) show up because of variations in surface roughness. The picture shows an area about 75 km across; the white line is a ship's wake or bilge dump.

(b)   Eddies in the North Atlantic Current to the south of Iceland. The eddy pattern is made visible by an extensive bloom of cocco-lithophores, phytoplankton with highly reflecting platelets. The distance from top to bottom of the image is about 350 km.

(c)   The complex eddying currents around Tasmania, made visible by means of the Coastal Zone Color Scanner carried aboard the *Nimbus-7* satellite. The colours are false and represent different concentrations of phytoplankton carried in different bodies of water, estimated on the basis of the amount of green chlorophyll pigment in surface water. Australia can be seen at the top of the image, which is about 1000 km across.

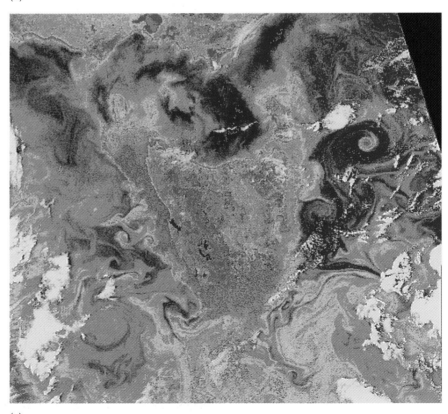

(c)

The pattern of large-scale current systems like the Gulf Stream or the Antarctic Circumpolar Current – i.e. the average current pattern (or mean flow) we attempt to represent geographically in maps like Figure 3.1 – is the oceanic equivalent of climate. Only relatively recently have oceanographers begun to get to grips with the study of the ocean's variability over short time-scales – i.e. with the ocean's 'weather'. In fact, the random fluctuations and eddies so characteristic of the ocean were formerly regarded merely as a nuisance, obscuring the mean flow.

Although eddies occur over a wide range of space- and time-scales, it became clear in the 1970s that variable flows with periods greater than the tidal and inertial periods are dominated by what are now called **mesoscale eddies** (the prefix 'meso-' means 'intermediate'). Mesoscale eddies are the oceanic analogues of weather systems in the atmosphere, but the differing densities of air and water mean that while cyclones (depressions) and anticyclones have length-scales of about 1000 km and periods of about a week, mesoscale eddies generally have length-scales of 50–200 km and periods of one to a few months. Mesoscale eddies travel at a few kilometres per day (compared with about 1000 km per day for atmospheric weather systems), and have rotatory currents with speeds of the order of $0.1 \, \mathrm{m \, s^{-1}}$. In most mesoscale eddies, but not all, flow is in approximate geostrophic equilibrium.

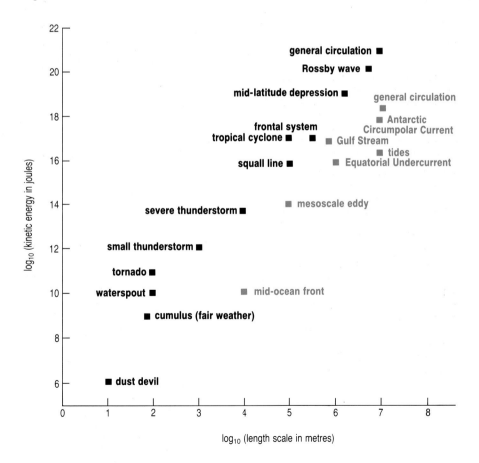

Figure 3.32   The kinetic energy possessed by various oceanic (blue) and atmospheric (black) phenomena. (Atmospheric Rossby waves were discussed in Section 2.2.1.) (Note that for convenience both vertical and horizontal scales are logarithmic.) (For use with Question 3.10, overleaf.)

QUESTION 3.10

(a) Away from intense currents like the Gulf Stream, the mean circulation of the ocean has an average speed of the order of $0.01\ m\ s^{-1}$. If mesoscale eddies typically have currents of about $0.1\ m\ s^{-1}$, explain why the kinetic energy per square kilometre of ocean in the region of an eddy is about 100 times that associated with the mean flow.

(b) Look at Figure 3.32. What atmospheric phenomenon has approximately the same amount of kinetic energy as a mesoscale eddy?

(c) Which ocean current has the greatest kinetic energy, according to Figure 3.32.

The term 'mesoscale eddy' is often used to refer specifically to 'current-rings' that form from meanders in fast currents like the Gulf Stream, the Kuroshio and the Antarctic Circumpolar Current. Such intense currents flow along regions with marked lateral variation in density, i.e. fronts, which are often boundaries between different water masses. Like fronts in the atmosphere (Figure 2.8), oceanic frontal boundaries slope and are intrinsically unstable, with a tendency to develop wave-like patterns and eddies. As a result, frontal currents continually 'spawn' eddies which travel out into the quieter areas of the ocean. (We will come back to this topic in Chapter 4.)

It is clear that mesoscale eddies are an intrinsic part of the ocean circulation. They are particularly numerous in certain regions of the ocean (i.e. close to frontal currents) but no part of the ocean has been shown to be without them. Kinetic energy spectra that extend into periods greater than 30 days nearly always show a bulge in the mesoscale part of the curve. Indeed, it is believed that *most* of the kinetic energy of the ocean – perhaps as much as 99% – is contained in the ocean's 'weather'. The extent to which the energy of mesoscale eddies feeds back into the general large-scale circulation is not clear. Insight into this and other related problems can only be obtained through computer models with the ability to predict basin-wide flow patterns on time-scales and space-scales that are small enough to 'resolve' mesoscale motions. Such computer models (described as 'eddy-resolving') only began to be feasible in the mid-1980s, when computers with sufficient power to run them were developed.

As might be expected, mesoscale eddies play an important role in the transport of heat and salt across frontal boundaries, from one water mass to another. For example, it has been estimated that eddies are responsible for transporting heat polewards across the Antarctic Circumpolar Current at a rate of about $0.4 \times 10^{15}\ W$ (1 watt = $1\ J\ s^{-1}$) (cf. Figure 1.5). Generally, they lead to the dispersal of water properties and marine organisms, and through their 'stirring' motions they contribute to the homogenization of water characteristics within water masses.

Mesoscale eddies are an exciting and relatively new discovery. Understanding them, and how they interact with the mean flow, will immeasurably improve our understanding of the oceanic circulation as a whole. However, mesoscale eddies are only one part of a continuous spectrum of possible motions in the oceans, and by no means all eddies (or even all mesoscale eddies) form from meandering frontal currents. Eddies may be generated by the bottom topography (notably seamounts), and commonly form down-current of islands. They may also be formed as a result of the interaction of a current with the coast, or with other currents or eddies, or as a result of horizontal wind shear. Some of the wide variety of eddy types will appear in later chapters.

3.6 SUMMARY OF CHAPTER 3

1   The global surface current pattern to some extent reflects the surface wind field, but ocean currents are constrained by continental boundaries and current systems are often characterized by gyral circulations.

2   Maps of wind and current flows of necessity represent average conditions only; at any one time the actual flow at a given point might be markedly different from that shown.

3   The frictional force caused by the action of wind on the sea-surface is known as the wind stress. Its magnitude is proportional to the square of the wind speed; it is also affected by the roughness of the sea-surface and conditions in the overlying atmosphere.

4   Wind stress acting on the sea-surface generates motion in the form of waves and currents. The surface current is typically 3% of the wind speed. Motion is transmitted downwards through frictional coupling caused by turbulence. Because flow in the ocean is almost always turbulent, the coefficient of friction that is important for studies of current flow is the coefficient of eddy viscosity. Typical values are 10^{-5} to $10^{-1}\,\mathrm{m^2\,s^{-1}}$ for A_z and 10 to $10^5\,\mathrm{m^2\,s^{-1}}$ for A_h.

5   Moving water tends towards a state of equilibrium. Flows adjust to the forces acting on them so that eventually those forces balance one another. Major forces that need to be considered with respect to moving water are wind stress at the sea-surface, internal friction (i.e. eddy viscosity), the Coriolis force and horizontal pressure gradient forces; in some situations, friction with the sea-bed and/or with coastal boundaries also needs to be taken into account. Although *deflection* by the Coriolis force is greater for slower-moving parcels of water, the *magnitude* of the force *increases* with speed, being equal to *mfu*.

6   Ekman showed theoretically that under idealized conditions the surface current resulting from wind stress will be 45° *cum sole* of the wind, and that the direction of the wind-induced current will rotate *cum sole* with depth, forming the Ekman spiral current pattern. An important consequence of this is that the mean flow of the wind-driven (or Ekman) layer is 90° to the right of the wind in the Northern Hemisphere and 90° to the left of the wind in the Southern Hemisphere.

7   When the forces that have set water in motion cease to act, the water will continue to move until the energy supplied has been dissipated, mainly by internal friction. During this time, the motion of the water is still influenced by the Coriolis force, and the rotational flows that result are known as inertia currents. The period of rotation of an inertia current varies with the Coriolis parameter $f = 2\Omega \sin \phi$, and hence with latitude, ϕ.

8   The currents that result when the horizontal pressure gradient force is balanced by the Coriolis force are known as geostrophic currents. The horizontal pressure gradient force may result only from the slope of the sea-surface, and in these conditions isobaric and isopycnic surfaces are parallel and conditions are described as barotropic. When the water is not homogeneous, but instead there are lateral variations in temperature and salinity, part of the variation in pressure at a given depth level results from the density distribution in the overlying water. In these situations, isopycnic surfaces slope in the opposite direction to isobaric surfaces; thus, isobars and isopycnals are inclined to one another and conditions are described as baroclinic.

9   In geostrophic flow, the angle of slope (θ) of each isobaric surface may be related to u, the speed of the geostrophic current in the vicinity of that isobaric surface, by the gradient equation: $\tan\theta = fu/g$. In barotropic flow, the slope of isobaric surfaces remains constant with depth, as does the velocity of the geostrophic current. In baroclinic conditions, the slope of isobaric surfaces follows the sea-surface less and less with increasing depth, and the velocity of the geostrophic current becomes zero at the depth where the isobaric surface is horizontal. The types of geostrophic current that occur in the two situations are sometimes known as 'slope currents' and 'relative currents', respectively. In the oceans, flow is often a combination of the two types of flow, with a relative current superimposed on a slope current.

10   In baroclinic conditions, the slopes of the isopycnals are very much greater than the slopes of the isobars. As a result, the gradient equation may be used to construct a relationship which gives the average velocity of the geostrophic current flowing between two hydrographic stations in terms of the density distributions at the two stations. This relationship is known as the geostrophic equation or (in its full form) as Helland-Hansen's equation. It is used to determine *relative* current velocities (i.e. velocities relative to a selected depth or isobaric surface, at which it may be assumed that current flow is negligible) at right angles to the section. This method provides information about *average* conditions only, and is subject to certain simplifying assumptions. Nevertheless, much of what is known about oceanic circulation has been discovered through geostrophic calculations.

11   Departures of isobaric surfaces from the horizontal (i.e. from an equipotential surface) may be measured in terms of units of work known as dynamic metres. Variations in the dynamic height of an isobaric surface (including the sea-surface) are known as dynamic topography On a map of dynamic topography, geostrophic flow is parallel to the contours of dynamic height in such a direction that the 'highs' are on the right in the Northern Hemisphere and on the left in the Southern Hemisphere. Dynamic topography represents departures of an isobaric surface from the (marine) geoid, which itself has a relief of the order of 100 times that of dynamic topography.

12   Surface wind stress gives rise to vertical motion of water, as well as horizontal flow. In particular, cyclonic wind systems lead to a lowered sea-surface, raised thermocline and divergence and upwelling, while anticyclonic wind systems give rise to a raised sea-surface, lowered thermocline and convergence and downwelling. Relatively small-scale linear divergences and convergences occur as a result of Langmuir circulation in the upper ocean.

13   Flow in the ocean occurs over a wide range of time-scales and space-scales. The general circulation, as represented by the average position and velocity of well-established currents such as the Gulf Stream, is known as the 'mean flow' or 'mean motion'.

14   Most of the energy of the ocean, both kinetic and potential, derives ultimately from solar energy. The potential energy stored in the ocean is about 100 times its kinetic energy, and results from isobars and isopycnals being displaced from their position of least energy (parallel to the geoid) as a result of wind stress or changes in the density distribution of the ocean. If the ocean were at rest and homogeneous, all isobaric and isopycnic surfaces would be parallel to the geoid. The ocean's kinetic energy is that associated with motion in ocean currents including tidal currents (plus surface waves). The kinetic energy associated with a current is proportional to the square of the current speed, and for any given area of ocean, the total kinetic energy associated with a range of periods/frequencies may be represented by a kinetic energy density spectrum.

15   The ocean is full of eddies. They originate from perturbations in the mean flow, and their formation has the overall effect of transferring energy from the mean flow. There is effectively a 'cascade' of energy through (generally) smaller and smaller eddies, until it is eventually dissipated as heat (through molecular viscosity). Mesoscale eddies, which have length scales of 50–200 km and periods of one to a few months, represent the ocean's 'weather' and contain a significant proportion of the ocean's energy. Current flow around most mesoscale eddies is in approximate geostrophic equilibrium. They are known to form from meanders in intense frontal regions like the Gulf Stream and the Antarctic Circumpolar Current, but may form in other ways too. Eddies of various sizes are generated by interaction of currents with the bottom topography, islands, coasts or other currents or eddies, or as a result of horizontal wind shear.

Now try the following questions to consolidate your understanding of this Chapter.

QUESTION 3.11

(a) The forces acting on water or other fluids may be divided into three categories:

(i) external forces, that arise from outside the fluid;

(ii) internal (or body) forces, that act within the fluid;

(iii) secondary forces, that come into play only because the fluid is in motion relative to the Earth's surface.

Some of the main forces that act on ocean water are:

1   wind stress;

2   viscous forces;

3   the tide-producing forces;

4   horizontal pressure gradient forces; and

5   the Coriolis force.

How would you classify each of forces 1–5 in terms of categories (i)–(iii)?

(b) Motion in the oceans is in equilibrium when the flow has had time to adjust so that the forces acting on the water balance. In the following types of flow, which of the forces 1–5 (above) are balancing one another?

(i) geostrophic flow;

(ii) the mean flow of the whole Ekman layer at right angles to the wind.

QUESTION 3.12   Use information from the end of Section 3.1.2 to show that, theoretically, for a given wind stress, the total volume transport in the wind-driven layer is independent of the value of A_z, the coefficient of eddy viscosity.

QUESTION 3.13   Inertia currents are manifestations of the Coriolis force in action. True or false?

QUESTION 3.14   In the Straits of Dover, water is well mixed by tidal currents and wind. The mean geostrophic current is towards the east.

(a) Are conditions in the Straits of Dover barotropic or baroclinic?

(b) Is the mean sea-level higher on the French or the English side? Sketch a north–south cross-section of water in the Straits of Dover, looking 'down-current'. Show the sloping sea-surface (of necessity, with the slope exaggerated) and isobars. Select the correct symbol for flow direction from Figure 3.17, and add labelled arrows to represent the balance of forces that exists under conditions of geostrophic equilibrium.

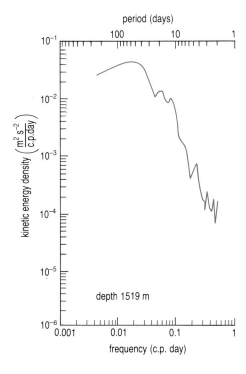

Figure 3.33   Kinetic energy density spectrum for flow in the Drake Passage, between South America and Antarctica.

Figure 3.34   Variability in sea-surface height, as computed from satellite altimetry data. Colours represent different deviations from mean sea-level (for key, see bar along the bottom).

(c) Again assuming geostrophic equilibrium, if the mean geostrophic current through the straits is 0.2 m s^{-1}, what is the slope of the sea-surface, i.e. what is tan θ? What is the difference in sea-level between the French and the English side? (You will need to use $\Omega = 7.29 \times 10^{-5}$ s^{-1}; the Straits of Dover are at 51° N and are about 35 km wide.)

QUESTION 3.15   Until the 1970s, a large proportion of studies of ocean currents had been made using the indirect geostrophic method (Section 3.3.3) combined with a few direct current measurements. Bearing this in mind, explain briefly why it is not surprising that, before then, mesoscale eddies had only rarely been observed.

QUESTION 3.16   In Chapter 2, we mentioned that the surface tracks of tropical cyclones are marked out by cooler water upwelled from a depth of 100 m or so. Using information in Sections 3.1 and 3.4, can you now explain why this happens?

QUESTION 3.17

(a) Why might we expect a kinetic energy density spectrum derived from current measurements made in the Drake Passage, between South America and Antarctica, to show an above-average incidence of mesoscale eddies?

(b) Figure 3.33 is such a kinetic energy density spectrum. Does it in fact show a high incidence of mesoscale eddies?

(c) Explain why *none* of the peaks on Figure 3.33 can be attributed to inertia currents.

QUESTION 3.18   Figure 3.34 shows the *variability* in the level of the sea-surface as computed from altimetry data, collected from the satellites *TOPEX–Poseidon* and *ERS-2*.

(a) Compare Figure 3.34 with Figure 3.1. What do the regions showing the greatest variability in sea-surface have in common? Can you explain this?

(b) Why are the positions of equatorial current systems not strongly visible on Figure 3.34? (*Hint*: Think about *all* the variables in the gradient equation.)

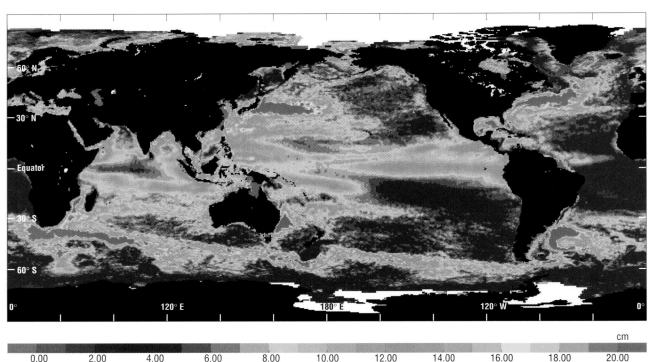

CHAPTER 4 — THE NORTH ATLANTIC GYRE: OBSERVATIONS AND THEORIES

The North Atlantic is the most studied – and most theorized-about – area of ocean in the world. Over the course of many centuries ships have crossed and recrossed it. The effects of currents on sailing times became well known, and no current was more renowned than the Gulf Stream. The Gulf Stream also fascinated natural philosophers, both because of the mystery of its very existence and because of the benign influence it appeared to have on the European climate.

The aim of this Chapter is to use the example of the North Atlantic gyral current system, and of the Gulf Stream in particular, to introduce some actual observations and measurements, and to explore some of the ideas and theories that have been put forward to explain them.

4.1 THE GULF STREAM

Before starting Section 4.1.1, look again at Figure 3.1 to remind yourself of the geographical position of the Gulf Stream and its relationship to the general circulation of the North Atlantic.

4.1.1 EARLY OBSERVATIONS AND THEORIES

European exploration of the eastern seaboard of the New World began in earnest during the sixteenth century. Coastal lands were investigated and much effort was put into the search for a North-West Passage through to the Pacific. The earliest surviving reference to the Gulf Stream was made by the Spaniard Ponce de Leon in 1513. His three ships sailed from Puerto Rico, crossed the Gulf Stream with great difficulty north of what is now Cape Canaveral (in Florida), and then turned south. By 1519 the Gulf Stream was well known to the ships' masters who sailed between Spain and America. On the outgoing voyage, they sailed with the Trade Winds in the North Equatorial Current; on their return, they passed through the Straits of Florida and followed the Gulf Stream up about as far as the latitude of Cape Hatteras (~35° N), and then sailed for Spain with the prevailing westerlies (Figure 2.3).

As early as 1515, there were well-considered theories about the origin of the Gulf Stream. Peter Martyr of Angheira used the necessity for conservation of mass to argue that the Gulf Stream must result from the deflection of the North Equatorial Current by the American mainland. Explorers had not found any passage which would allow the North Equatorial Current to flow through to the Pacific and thence back around to the Atlantic; the only other possibility was that water piled up continuously against the Brazilian coast, and this had not been observed to happen. The North Equatorial Current itself was thought to result in some way from the general westward movement across the heavens of the celestial bodies, which drew the air and waters of the equatorial regions along with them.

During the 1600s, the eastern coast of North America was colonized by Europeans and the Gulf Stream was traversed countless times and at various locations. In the following century the experience gained in the great whaling expeditions added further to the knowledge of currents, winds and bottom topography. This accumulating knowledge was not, however, readily accessible in technical journals, but handed down by word of mouth. Charts indicating currents did exist – the first one to show the 'Gulf Stream' was published in 1665 – but they were of varying quality and showed features that owed more to the imagination than observation.

The first authoritative chart of the Gulf Stream was made by William Gerard De Brahm, an immensely productive scientist and surveyor who, in 1764, was appointed His Majesty's Surveyor-General of the new colony of Florida. De Brahm's chart of the Gulf Stream (Figure 4.1(a)) and his reasoned speculation about its origin were published in *The Atlantic Pilot* in 1772. At about the same time, a chart of the Gulf Stream was engraved and printed by the General Post Office, on the instructions of the Postmaster-General of the Colonies, Benjamin Franklin. This chart was produced for the benefit of the masters of the packet ships which carried mail between London and New England. A Nantucket sea captain, Timothy Folger, had drawn Franklin's attention to the Gulf Stream as one cause of delay of the packet ships, and had plotted the course of the Stream for him (Figure 4.1(b)).

The theories of oceanic circulation that had evolved during the seventeenth century were seldom as well constructed as the charts they sought to explain. During the eighteenth century, understanding of fluid dynamics advanced greatly. Moreover, the intellectual climate of the times encouraged scientific advances based on observations, rather than fanciful theories based in the imagination. Franklin, who had observed the effect of wind on shallow bodies of water, believed that the Trade Winds caused water to pile up against the South American coast; the head of pressure so caused resulted in a strong current flowing 'downhill' through the Caribbean islands, into the Gulf of Mexico, and out through the Straits of Florida.

Franklin was also one of those who hit upon the idea of using the thermometer as an aid to navigation. Seamen had long been aware of the sharp changes in sea-surface temperature that occur in the region of the Gulf Stream. More than a century-and-a-half earlier, Lescarbot had written:

'I have found something remarkable upon which a natural philosopher should meditate. On the 18th June, 1606, in latitude 45° at a distance of six times twenty leagues east of the Newfoundland Banks, we found ourselves in the midst of very warm water despite the fact that the air was cold. But on the 21st of June all of a sudden we were in so cold a fog that it seemed like January and the sea was extremely cold too.'

Franklin made a series of surface temperature measurements across the Atlantic and also attempted to measure subsurface temperatures; he collected water for measurement from a depth of about 100 feet using a bottle, and later a cask, with valves at each end – a piece of equipment not unlike the modern Nansen bottle.

The next intensive study of the Gulf Stream was made by James Rennell, whose authoritative work was published posthumously in 1832. Rennell – who is often referred to as 'the father of oceanography' – studied large amounts of data held by the British Admiralty Office and carefully

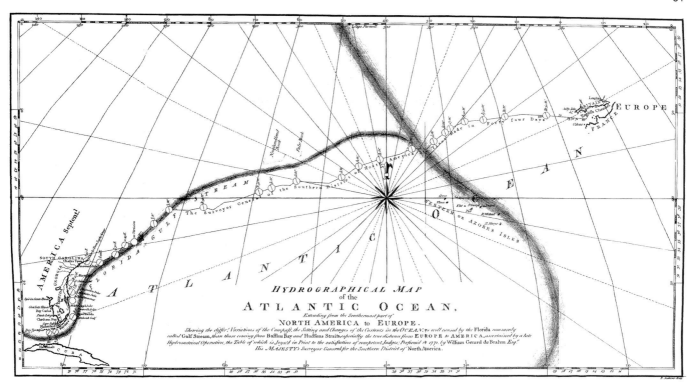

Figure 4.1 (a) De Brahm's chart of Florida and the Gulf Stream. On this projection, lines of latitude (not shown) are parallel to the top and bottom of the map, and the curved lines are lines of longitude. In the small circles along the path of the Gulf Stream are indicated the bearings of the current read against magnetic north.
(b) The chart of the Gulf Stream and North Atlantic gyre (inset), made by Timothy Folger and Benjamin Franklin.

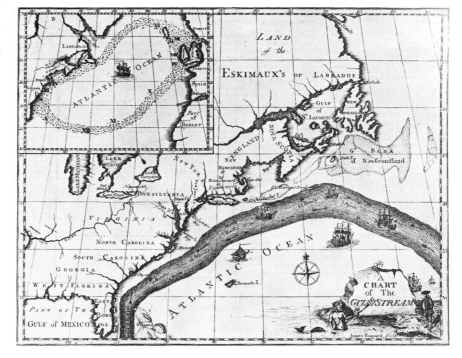

documented the variability found in the Gulf Stream. He distinguished between 'drift currents', produced by direct stress of the wind, and 'stream currents' produced by a horizontal pressure gradient in the direction of the flow (the term 'drift current' is still used occasionally). Rennell agreed with Franklin that the Gulf Stream was a 'stream current'.

The current charts shown in Figure 4.1 were produced mainly from measurements of ships' drift. By the time of Franklin and De Brahm, accurate chronometers had become available and ships' positions could be fixed with respect to longitude as well as latitude. Position fixes were made every 24 hours; between fixes, a continuous record of position was kept by means of *dead-reckoning*, i.e. the ship's track was *ded*uced from the distance travelled from a known (or estimated) position and the course steered using a compass. The accumulated discrepancy after 24 hours between the dead-reckoning position and the accurately fixed position gave an indication of the average 'drift' of the surface water through which the vessel had passed (Figure 4.2). However, sailing vessels were themselves strongly influenced by the wind, and there could be considerable difficulties in maintaining course and speed in heavy seas. These inherent inaccuracies, combined with any inaccuracies in position-fixing, meant that estimates of current speeds so obtained could be very unreliable.

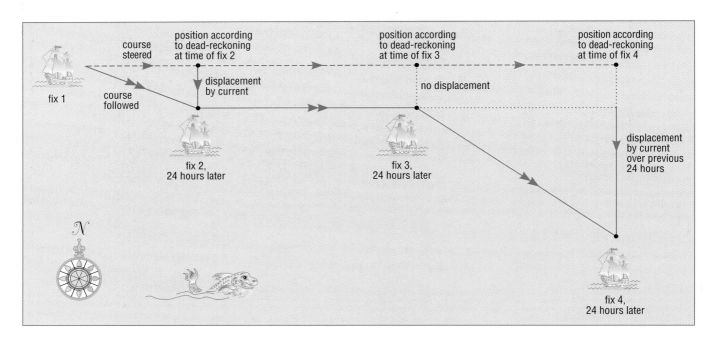

Figure 4.2   Diagram to illustrate how the mean current may be determined from the ship's drift, i.e. the discrepancy between the ship's track as steered using a compass and dead-reckoning, and the course actually followed (determined by celestial navigation, or known coastal features or islands). In this example, the ship has been steered straight according to dead-reckoning, but in reality has been displaced to the south by a current between fixes 1 and 2, and again by a (stronger) current between fixes 3 and 4. In each case, the speed of the mean southerly current was the southward displacement divided by the time between fixes (24 hours). (The displacement caused by the wind ('leeway') was ignored.)

Although single estimates of current speed obtained from ship's drift were unreliable, accumulations of large numbers of measurements, as compiled by Rennell, could be used to construct reasonably accurate charts of current flow, averaged over time and area. Collection of such data from commercial shipping has continued to the present day and this information has been used to construct maps of mean current flow (e.g. Figure 5.12). Current speeds estimated by dead-reckoning are now considerably more accurate, thanks to position-fixing by radar and navigational satellites, and the use of automatic pilots.

Systematic collection of oceanographic data was given a great impetus by the American naval officer and hydrographer, Matthew Fontaine Maury. Maury arranged for the US Hydrographic Office to supply mariners with charts of winds and currents, accompanied by sailing directions; in return, the mariners agreed to observe and record weather and sea conditions and to provide Maury with copies of their ships' logs. In 1853, on Maury's initiative, a conference of delegates from maritime nations was convened in Brussels to devise a standard code of observational practice. It was agreed that observations of atmospheric

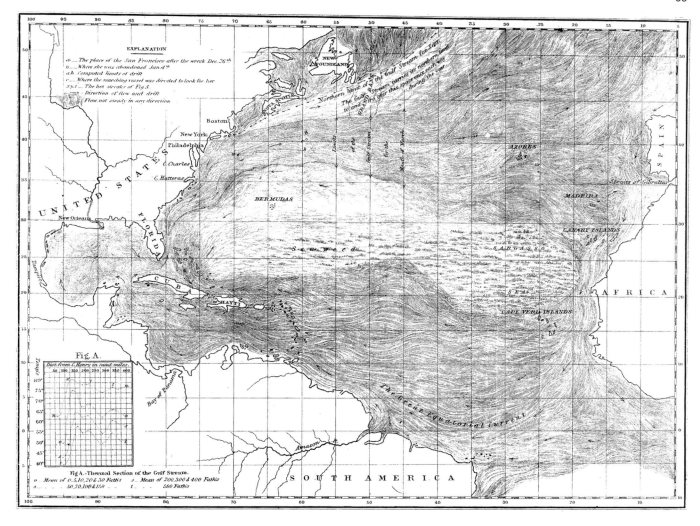

Figure 4.3 Maury's chart of the Gulf Stream and North Atlantic Drift (now usually referred to as the North Atlantic Current). Note also the temperature traverses of the Gulf Stream at different latitudes, shown at bottom left.

and oceanic conditions (including sea-surface temperature, specific gravity and temperature at depth) should be made at intervals of two hours. This was soon increased to four hours, but apart from that the system has remained largely unchanged up to the present time.

Maury's contribution to practical navigation can be immediately appreciated by looking at his chart of the Gulf Stream and North Atlantic Drift (Figure 4.3). He was not, however, a great theoretician. He rejected the theory that the Gulf Stream owes its origin to the wind on the basis of the following argument. The Gulf Stream is wider at Cape Hatteras than in the Straits of Florida; so, given that its total volume transport stays the same and that the flow extends to the bottom, the sea-bed must be shallower at Cape Hatteras. The Gulf Stream therefore has to flow uphill and must be maintained by some mechanism other than wind stress.

QUESTION 4.1

(a) By reference to the idea about Gulf Stream generation shared by Franklin and Rennell, explain briefly why Maury's argument is seriously flawed. About what basic principle was he confused?

(b) In fact, Maury was not justified in assuming that the volume of water transported in the Gulf Stream is the same at Cape Hatteras as it is in the Straits of Florida. By reference to Figure 3.1, can you explain why?

Maury suggested that the Gulf Stream was driven by a kind of peristalsis, and believed that the general ocean circulation resulted from the difference in seawater density between the Equator and the Poles. This latter idea had been put forward in 1836 by François Arago, as an alternative to the theory that the Gulf Stream owed its existence ultimately to the wind. Arago pointed out that the drop in the level of the Gulf Stream from the Gulf of Mexico side of Florida to the Atlantic side, as determined by a levelling survey across Florida, was only 19 mm at most, and this was surely much too small to drive the Gulf Stream. In fact, more recent calculations (assuming greater frictional forces in the Gulf Stream than were thought of then) have shown that 19 mm is enough of a head to drive the Gulf Stream. The Stream is still sometimes thought of as a 'jet' being squirted out through the Straits of Florida.

Modern hydrographic surveying of the Gulf Stream can be said to have begun in 1844 with the work of the United States Coast and Geodetic Survey, under the direction of Franklin's great-grandson, Alexander Dallas Bache. Since that time there have been a number of surveys of both current velocity and temperature distribution. Perhaps the most impressive work was done by John Elliott Pillsbury who made observations along the Gulf Stream at a number of locations, including several in the Straits of Florida. He measured the temperature at a number of depths and also recorded the direction and speed of the current using a current meter of his own design. These observations made in the 1890s were not repeated until relatively recently and have been invaluable to oceanographers in the twentieth century (as you will see in Section 4.3.2).

Like most seagoing men of his time, Pillsbury was convinced that ocean currents were generated by the wind. There were, however, severe difficulties in explaining the mechanism whereby the wind could drive the ocean. Advances in understanding were held up for two main reasons. The first of these was that the significance of the turbulent nature of flow in the oceans was not fully appreciated.

QUESTION 4.2   One objection to the idea of wind-driven currents (put forward by a contemporary mathematical physicist) was that the wind would need to blow over the sea-surface for hundreds of thousands of years before a current like the Gulf Stream could be generated. From your reading of Chapter 3, can you suggest why this objection would have seemed reasonable at the time, and why we now know otherwise?

The other great difficulty hindering progress was that the effect on ocean currents of the rotation of the Earth was not properly understood, with the result that the role of the Coriolis force in balancing horizontal pressure gradients in the oceans was overlooked. One of the first scientists to understand the effect of the Coriolis force on ocean currents was William Ferrel. Ferrel derived the relationship between atmospheric pressure gradient and wind speed, i.e. the equation for geostrophic flow in the atmosphere. Unfortunately, oceanographers were not aware of Ferrel's work and so it was not immediately applied to the oceans. The formula for computing ocean current speeds from the slopes of isobaric surfaces – in other words, the gradient equation (3.11) – was derived by Henrik Mohn in 1885, a few years later. It was not until the first decade of the twentieth century that Sandström and Helland-Hansen seriously investigated the possibility of using the density distribution in the oceans to deduce current velocities (cf. Section 3.3.3).

It is now clear that the Gulf Stream *is* wind-driven and is not so much a warm current as a ribbon of high-velocity water forming the boundary between the warm waters of the Sargasso Sea and the cooler waters over the continental margin. Armed with an understanding of geostrophic currents, we can see that the sharp gradient in temperature (and hence density) across the Stream is an expression of a balance between the horizontal pressure gradient force and the Coriolis force, an equilibrium that has been brought about indirectly by the wind. (We will return to geostrophic equilibrium in the Gulf Stream in Section 4.3.2.)

4.2 THE SUBTROPICAL GYRES

As discussed in Section 3.4, current flow in the subtropical gyres is related to the overlying anticyclonic wind systems, which blow around the subtropical high pressure regions. The centres of the atmospheric and oceanic gyres are not, however, coincident: the centres of the atmospheric gyres tend to be displaced towards the eastern side of the oceans, while the centres of the oceanic gyres tend to be displaced towards the western side. As a result, the currents that flow along the western sides of oceans – the **western boundary currents** – are characteristically fast, intense, deep and narrow, while those that flow along the eastern sides – the **eastern boundary currents** – are characteristically slow, wide, shallow and diffuse.

The western boundary current in the North Atlantic is the Gulf Stream; its counterpart in the North Pacific is the Kuroshio. Both of these currents are typically only some 100 km wide and in places have surface velocities in excess of $2 \, \text{m s}^{-1}$. By contrast, the Canary Current and the California Current are over 1000 km wide and generally have surface velocities less than $0.25 \, \text{m s}^{-1}$. In the South Atlantic and the South Pacific, the difference between the western boundary currents (the Brazil Current and the East Australian Current, respectively) and the eastern boundary currents (the Benguela Current and the Humboldt, or Peru, Current) is not so marked. This may be because the South Atlantic and South Pacific are open to the Southern Ocean so that their gyres are strongly influenced by the Antarctic Circumpolar Current; furthermore, in the case of the South Pacific there is no continuous barrier along its western side. By contrast, the southern Indian Ocean has the powerful Agulhas Current flowing south along its western boundary; the northern Indian Ocean has a seasonal western boundary current in the form of the Somali Current (both of these will be discussed in Chapter 5).

4.2.1 VORTICITY

You will already be familiar with the idea – essential to an understanding of dynamic systems – that energy and mass must be conserved. Another property that must be conserved is momentum – both linear momentum associated with motion in straight lines and angular momentum associated with rotatory movement. In oceanography, it is more convenient to view the conservation of angular momentum as the conservation of a tendency to rotate, or the conservation of **vorticity**, the tendency to form vortices.

Ocean waters have rotatory motions on all scales, from the basin-wide subtropical and subpolar gyres down to the smallest swirls and eddies (e.g. Figures 4.4 and 3.31). Fluid motion does not have to be in closed loops to be rotatory: whenever there is **current shear** (a change in velocity at right angles to the direction of flow), there is a tendency to rotate and the water has vorticity. For mathematical convenience, *a tendency to rotate anticlockwise is referred to as positive, and a tendency to rotate clockwise is referred to as negative*. These aspects of vorticity are illustrated schematically in Figure 4.5.

Figure 4.4 Small-scale eddies forming in the vicinity of a rocky coastline.

NEGATIVE (CLOCKWISE) VORTICITY

POSITIVE (ANTICLOCKWISE) VORTICITY

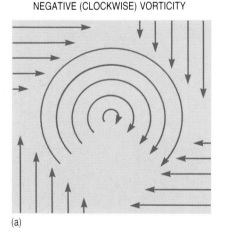 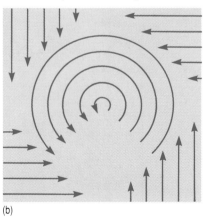

(a) (b)

Figure 4.5 Diagrams to show examples of flow with (a) negative and (b) positive vorticity. The speed and direction of the flow are indicated by the lengths and directions of the arrows.

Note that we say 'a *tendency* to rotate' rather than simply 'rotatory motion'. This is because water may be acquiring positive vorticity by one mechanism at the same time as it is acquiring negative vorticity by another. For example, water could be acquiring positive vorticity as a result of current shear caused by friction with adjacent bodies of water, or a coastal feature such as a headland or spit (Figure 4.6), while at the same time acquiring negative vorticity from a clockwise wind. The actual rotatory motion that results will depend on the relative sizes of the two effects. In theory, positive and negative vorticity tendencies could be exactly equal so that *no* rotatory motion would result.

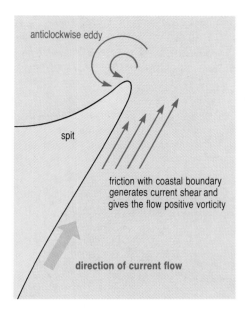

Figure 4.6 Diagram to illustrate how an eddy can be generated off a spit; the lengths and directions of the arrows indicate the speed and direction of the current. (The eddy on the left-hand side of Figure 4.4 is forming in the same way, though on a smaller scale.)

Water that has a rotatory motion in relation to the surface of the Earth, caused by wind stress and/or frictional forces, is said to possess **relative vorticity**. However, the Earth is itself rotating. The vorticity possessed by a parcel of fluid by reason of its being on the rotating Earth is known as its **planetary vorticity**.

Planetary vorticity and the Coriolis force

In Chapter 1, you saw how the rotation of the Earth about its axis results in the deflection of currents and winds by the Coriolis force. These deflections were explained in terms of the poleward decrease in the eastward velocity of the surface of the Earth. Figure 1.2 showed how a missile fired northwards from the Equator is deflected eastwards in relation to the surface of the Earth because, with increasing latitude, the surface of the Earth travels eastwards at a progressively decreasing rate. This explanation is valid but it is only part of the story. In addition to a linear eastward velocity, the surface of the Earth also has an *angular velocity*, so that in the Northern Hemisphere it turns anticlockwise about a local vertical axis, and in the Southern Hemisphere it turns clockwise (Figure 4.7(a)). In other words, a cross marked on the Earth's surface, and viewed from a satellite in space positioned directly above it, would be seen to rotate.* Because the angular velocity of the Earth's surface is latitude-dependent, there would be relative motion between a hypothetical missile and the surface of the Earth, *regardless of the direction in which the missile was fired* (as long as it was not fired along the Equator), and this relative motion would increase with increasing latitude. Similarly, winds and currents moving eastwards and westwards experience deflection in relation to the Earth, as well as those moving northwards and southwards.

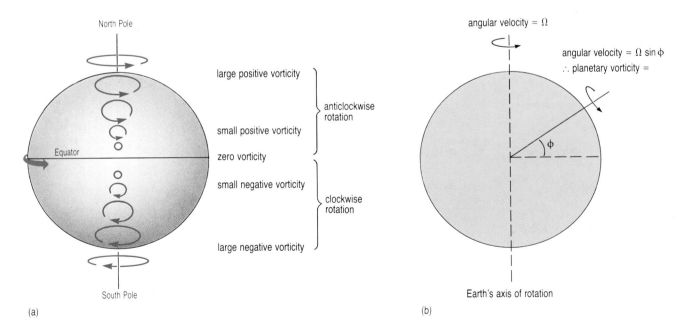

(a) (b)

Figure 4.7 (a) Schematic diagram to show the variation of planetary vorticity with latitude. The blue arrows represent the component of the Earth's rotation about a local vertical axis (viewed from above) at different latitudes.
(b) Diagram to show the derivation of the expression for the planetary vorticity of a fluid parcel on the surface of the Earth at latitude ϕ. (For use with Question 4.3(a).)

*This is easiest to imagine for a location close to one of the poles – you can try it yourself by marking a cross on a globe, and turning it slowly while keeping your viewpoint directly above the cross.

While the linear eastward velocity of the Earth's surface decreases with latitude, its angular velocity about a local vertical axis *increases* with latitude (Figure 4.7). At the poles, the angular velocity of the surface of the Earth is simply Ω (= $7.29 \times 10^{-5}\,\text{s}^{-1}$ (cf. Section 3.2)). At increasingly lower latitudes, it is a smaller and smaller proportion of Ω: the bigger the angle between the Earth's axis of rotation and the local vertical axis, the smaller the angular velocity of the surface of the Earth about this local vertical axis. At the Equator, where a vertical axis is at right angles to the axis of rotation of the Earth, the angular velocity of the surface is zero. In general, the angular velocity of the surface of the Earth about a vertical axis at latitude ϕ is given by $\Omega \sin \phi$ (Figure 4.7(b)).

If the surface of the Earth at latitude ϕ has an angular velocity of $\Omega \sin \phi$, so too does any object or parcel of air or water on the Earth at that latitude. In other words, any parcel of fluid on the surface of the Earth shares the component of the Earth's angular rotation appropriate to that latitude.

The vorticity of a parcel of fluid is defined mathematically to be equal to twice its angular velocity.

QUESTION 4.3

(a) What, then, is the planetary vorticity possessed by a parcel of fluid on the surface of the Earth at latitude ϕ? Insert your answer where indicated on Figure 4.7(b). Do you recognize the expression you have written?

(b) What is the planetary vorticity (i) at the North Pole, (ii) at the South Pole, and (iii) at the Equator?

In answering Question 4.3(a), you will have seen that planetary vorticity is given by the same expression as the Coriolis parameter. This is appropriate because they are both aspects of the same phenomenon, i.e. the rotation of the Earth. Like the Coriolis parameter, planetary vorticity is given the symbol f, while relative vorticity is given the Greek symbol ζ (zeta).

At the beginning of this Section it was stated that vorticity – effectively, angular momentum – must be conserved. But it is vorticity *relative to fixed space* (i.e. with respect to an observer outside the Earth) that must be conserved. In other words, it is the **absolute vorticity** ($f + \zeta$) of a parcel of fluid – its vorticity as a result of being on the Earth *plus* any vorticity relative to the Earth – that must remain constant, in the absence of external forces such as wind stress and friction. Imagine, for example, a parcel of water that is being carried northwards from the Equator in a current. As it moves northwards it will be moving into regions of increasingly large positive (anticlockwise) planetary vorticity (see Figure 4.7(a)). However, as it is only weakly bound to the Earth by friction, it tends to be 'left behind' by the Earth rotating beneath it* and, relative to the Earth, it acquires an increasingly negative (clockwise) rotational tendency. In the extreme case where the only influence on the flow (apart from friction) is the Earth's rotation, we end up with a simple inertia current, with clockwise flow in the Northern Hemisphere (Figure 3.7(a)). More generally, if there are no external influences imparting positive or negative relative vorticity to the parcel of water (e.g. it does not come under the influence of cyclonic or anticyclonic winds), the increase in positive planetary vorticity will be exactly equal to the increase in negative relative vorticity, and the absolute vorticity ($f + \zeta$) will remain constant.

---

*If you find this difficult to imagine, fill a glass with water and twirl it. You will find that the water remains stationary while the glass rotates outside it. Of course, *in relation to the glass*, the water *is* moving.

QUESTION 4.4

(a) A body of water is carried *southwards* from the Equator in a current.
(i) As the water moves south, how is does its planetary vorticity change?
(ii) How does this affect its relative vorticity?

(b) What happens to the relative vorticity of a body of water if it is acted upon by:
(i) winds blowing in a clockwise direction?
(ii) cyclonic winds in the Southern Hemisphere?

However, the situation is not quite as straightforward as this because the vorticity of a body of water is the sum of the vorticity of all the constituent particles of water. Consider, for convenience, a column of water moving in a current. For the purposes of this discussion, we can imagine the column of water spinning about its own axis, although its vorticity could of course be due to any type of rotatory motion (Figure 4.5). For simplicity, we can also assume that the column is rotating anticlockwise in the Northern Hemisphere, i.e. it has positive relative vorticity.

What will happen if the rotating column of water becomes longer and thinner as a result of stretching, because it moves into a region of deeper sea-floor, for example? (Consider what happens when spinning ice-skaters bring their arms in close to their sides.)

The simple answer is that, like the ice-skater, the column of water will spin faster, i.e. its relative vorticity will increase (become more positive). Looked at from the point of view of angular momentum, the angular momentum of each particle of water is $mr^2\omega$, where r is the distance from the particle (of mass m) to the axis of rotation, and ω is the angular velocity (Figure 4.8(a)). When the column stretches, the average radius r decreases and so, for angular momentum to be conserved, ω the speed of rotation must increase (Figure 4.8(b)). Because of the effect of changes in the length (D) of the water column, the property that is actually conserved is therefore not $f + \zeta$, the absolute vorticity, but $(f + \zeta)/D$, the **potential vorticity**.

Figure 4.8 (a)   The angular momentum of a particle of mass m moving with angular velocity ω in a circle of radius r is given by $mr^2\omega$.
(b)   For the total angular momentum (vorticity) of a stretched column of water to remain constant, the angular velocity ω of the particles in the column must increase.

The example given above is unrealistic because we have omitted consideration of variations in f, the planetary vorticity. In fact, in the real oceans, away from coastal boundaries and other regions of large current shear, f is very much greater than ζ. This means that $(f + \zeta)/D$ is effectively equal to f/D and it is f that must change in response to changes in D; and as f is simply a function of latitude, it can only be changed by water changing its latitude.

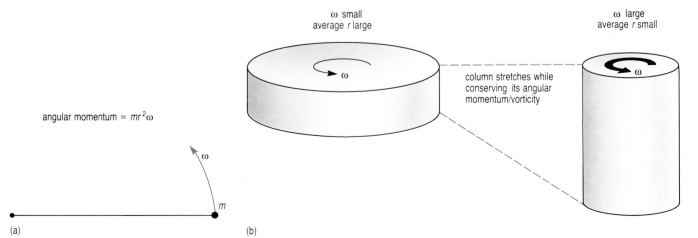

angular momentum = $mr^2\omega$

(a)

(b)

ω small
average r large

ω large
average r small

column stretches while conserving its angular momentum/vorticity

QUESTION 4.5

(a) Imagine a current flowing from east to west in the Northern Hemisphere. What will happen if the current flows over a shallow bank, so that the depth D is reduced? (You may assume that this occurs in the open ocean and that $f \gg \zeta$.)

(b) What will happen to a zonal* current flowing over a shallow bank in the Southern Hemisphere?

The fact that conservation of f/D causes currents to swing equatorwards and polewards over topographic highs and lows, respectively, is sometimes referred to as 'topographic steering'.

In studies of potential vorticity in the real oceans, D, the 'depth of the water column', need not be the total depth to the sea-floor. It is the thickness of the body of water under consideration and so it could, for example, be the depth from the sea-surface to the bottom of the permanent thermocline or, for deep-water movements, the depth from the bottom of the permanent thermocline to the sea-floor.

Since the mid-1980s, potential vorticity has become an important tool in the study of the movement of oceanic water masses. This application of vorticity will be discussed in Chapter 6.

4.2.2 WHY IS THERE A GULF STREAM?

We will now summarize briefly, with as little mathematics as possible, some of the developments in ideas that have led to an increased understanding of the dynamics of ocean circulation in general, and of the subtropical gyres in particular. In the course of this discussion, we will go a substantial way towards answering the question posed above: Why *is* there a Gulf Stream?

Ekman's initial work on wind-driven currents, intended to explain current flow at an angle to the wind direction, was published in 1905. In 1947, the Scandinavian oceanographer Harald Sverdrup used mathematics to demonstrate another surprising relationship between wind stress and ocean circulation. In constructing his theory, Ekman had assumed a hypothetical ocean which was not only infinitely wide and infinitely deep, but also had no horizontal pressure gradients because the sea-surface and all other isobaric surfaces were assumed to be horizontal. Sverdrup's aim was to determine current flow in response to wind stress *and* horizontal pressure gradients. Unlike Ekman, Sverdrup was not interested in determining how horizontal flow varied with depth; instead, he derived an equation for the *total* or *net* flow resulting from wind stress.

Consider a simple situation in which the winds blowing over a hypothetical Northern Hemisphere ocean are purely zonal and vary in strength with latitude sinusoidally, as shown in Figure 4.9(a).

QUESTION 4.6

(a) How closely does this hypothetical wind field resemble that over the real oceans between 15° and 45° N (cf. Figure 2.3)?

(b) On Figure 4.9(b), draw two arrows (one for the northern half of the ocean, one for the southern half) to indicate the Ekman transport that would result from the wind field shown in (a).

*Zonal means *either* from east to west *or* from west to east; the equivalent word for north–south flow is 'meridional'.

(c) What effect will this have on the shape of the sea-surface and on the horizontal pressure gradients at depth? On Figure 4.9(b) draw two more pairs of arrows, (1) to indicate the horizontal pressure gradient forces in the northern and southern halves of the ocean, and (2) to show the direction(s) of the geostrophic currents that would be associated with these horizontal pressure gradient forces.

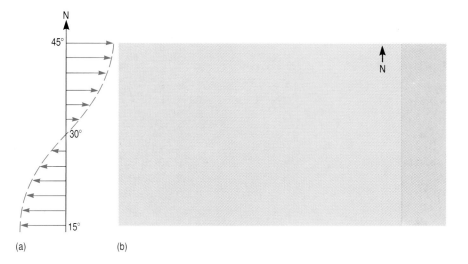

Figure 4.9 (a)   A hypothetical wind field in which the wind is purely zonal (i.e. easterly or westerly) and varies in strength with latitude sinusoidally.
(b)   Plan view of a rectangular ocean in the Northern Hemisphere, showing the direction of the Ekman transports, horizontal pressure gradients and geostrophic currents that result from the wind field shown in (a). (To be completed for Question 4.6(b) and (c).)

In answering Question 4.6(b) and (c), you considered two types of current flow resulting from wind stress:

(i) Ekman transport at right angles to the direction of the wind, and

(ii) geostrophic flow, in response to horizontal pressure gradients caused by the 'piling up' of water through Ekman transport.

Sverdrup combined these components mathematically and obtained the flow pattern shown in Figure 4.10(b). The 'gyre' is not complete because in deducing this flow pattern, Sverdrup considered the effect of an eastern boundary but could not also include the effect of a western boundary. As a result, there could be no northward flow in a 'Gulf Stream'.

The most interesting aspect of Sverdrup's results, however, was that the net amount of water transported by a given pattern of wind stress depends not on the absolute value of the wind stress but on its *torque*, i.e. its tendency to cause rotation – or, if you like, its ability to supply relative vorticity to the ocean. In particular, Sverdrup showed that the net amount of water transported meridionally (i.e. north–south or south–north) is directly proportional to the torque of the wind stress. You can see this for yourself, in a qualitative way, by studying Figure 4.10. The first thing to note is that the pattern of wind stress shown in part (a) of Figures 4.9 and 4.10 will result in a clockwise torque on the ocean (cf. Question 4.6(a)). This torque is greatest at latitude 30°, and it is here that the southward motion, as shown by the flow lines on Figure 4.10(b), is greatest. The torque is least (effectively zero) at latitudes 45° and 15°, and here the flow is almost entirely zonal (eastwards and westwards, respectively).

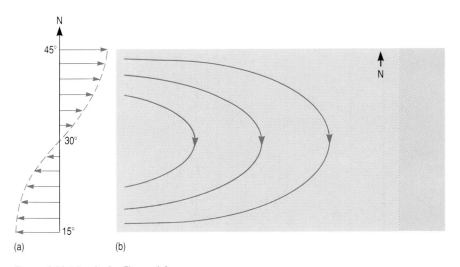

Figure 4.10 (a) As for Figure 4.9.
(b) The 'depth-integrated' (i.e. net, when summed over the depth) flow pattern that results from combining Ekman transports at right angles to the wind with geostrophic flow in response to horizontal pressure gradient forces, according to Sverdrup.

In summary then, Sverdrup showed that at any location in the ocean, the total amount of water transported meridionally in the wind-influenced layer is proportional to the torque of the wind stress. If the meridional transport is given the symbol M, this can be written as:

$$M = b \times \text{torque of } \tau \qquad (4.1a)$$

where τ is the wind stress, b is a latitude-dependent constant and the units of M are kg s^{-1} per m, or kg m^{-1} s^{-1}. The word used in mathematics for 'torque' is 'curl', so Equation 4.1a can be written as:

$$M = b \times \text{curl } \tau \qquad (4.1b)$$

Sverdrup found that the constant in Equation 4.1 is the reciprocal of the rate of change of the Coriolis parameter with latitude. The rate of change of f with latitude is commonly given the symbol β, so Equation 4.1 is usually written as:

$$M = \frac{1}{\beta} \text{ curl } \tau \qquad (4.1c)$$

You are not expected to manipulate this equation, but you should appreciate that it is a convenient way of summarizing Sverdrup's theoretically determined relationship, namely that at any location in the ocean the *total* meridional flow – i.e. the *net* north–south flow, taking into account the speed and direction of flow at all depths, from the surface down to the depth at which even indirect effects of the wind cannot be felt – is determined by the rate of change of f at the latitude concerned and the torque, or curl, of the wind stress.

Sverdrup's theory was an advance on Ekman's in that the ocean was not assumed to be unlimited. However, as mentioned above, it could not take account of a western boundary, and certainly could not explain the existence of intense western boundary currents like the Gulf Stream (Figure 4.11). This problem was solved by the American oceanographer Henry Stommel, in a paper published in 1948. Stommel considered the effect of a symmetrical anticyclonic wind field on a rectangular ocean in three different situations:

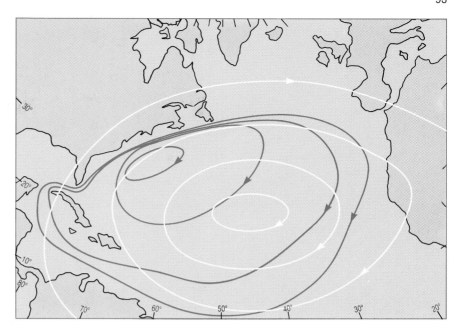

Figure 4.11 Schematic illustration of the asymmetrical North Atlantic gyre (blue) and the more or less symmetrical wind field which overlies it.

1   The ocean is assumed to be on a non-rotating Earth.

2   The ocean is rotating but the Coriolis parameter f is constant.

3   The ocean is rotating and the Coriolis parameter varies with latitude (for the sake of simplicity, this variation is assumed to be linear).

Unlike Sverdrup, Stommel included friction in his calculations and he worked out the flow that would result when the wind stress and frictional forces balanced (i.e. when there was a steady state, with no acceleration or deceleration) for each of the three situations described above.

Figure 4.12 (overleaf) summarizes the results of Stommel's calculations. The left-hand diagrams represent the patterns of flow that result from a symmetrical anticyclonic wind field in situations (1)–(3), and in each case the right-hand diagram shows the sea-surface topography that accompanies this flow.

Look first on the diagrams on the right-hand side, which show the sea-surface topography. What topographic feature may be seen clearly in situation (2), but is not present in situation (1)?

In (2), the sea-surface has a large (150 cm) topographic high in the middle of the gyre, whereas there is no such feature in (1). The reason for the difference is that each diagram represents a situation of equilibrium, where forces balance. In (2), unlike (1), the ocean is on a rotating Earth and a Coriolis force has come into existence. For there to be equilibrium, this must be balanced by horizontal pressure gradient forces, which are brought about by the sea-surface slopes.

94

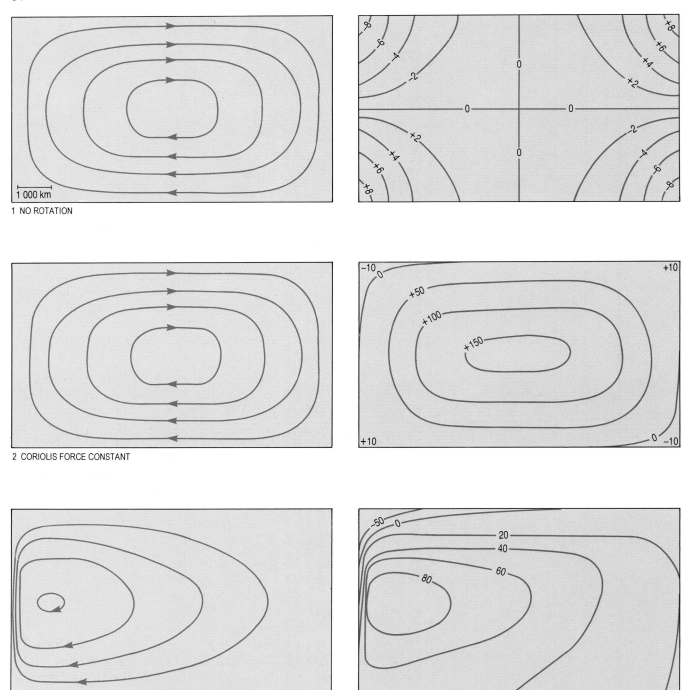

1 NO ROTATION

2 CORIOLIS FORCE CONSTANT

3 CORIOLIS FORCE INCREASES LINEARLY WITH LATITUTE

Figure 4.12   Summary of the results of Stommel's calculations. The diagrams on the left-hand side show the streamlines parallel to which water flows in the wind-driven layer (assumed to be 200 m deep). The volume of water transported around the gyre per second between one streamline and the next is 20% of the total flow. The diagrams on the right-hand side show contours of sea-surface height in cm. In (1), the ocean is assumed to be on a non-rotating Earth; in (2), the ocean is on a rotating Earth but the Coriolis parameter is assumed to be constant with latitude; in (3), the Coriolis parameter is assumed to vary linearly with latitude.

QUESTION 4.7 Now concentrate on the diagrams on the left-hand side of Figure 4.12. What do they tell us about the cause of intensification of western boundary currents? Is it simply the result of the *existence* of the Coriolis force?

This is how Stommel showed that the intensification of western boundary currents is, in some way, the result of the fact that the Coriolis parameter *varies with latitude*, increasing from the Equator to the poles.[*] Stommel also explained intensification of western boundary currents in terms of vorticity balance. (In some ways, it is more convenient to work with vorticity than with linear current flow, because horizontal pressure gradient forces, which lead to complications, do not need to be considered.)

Imagine a subtropical gyre in the Northern Hemisphere, acted upon by a symmetrical wind field like that in Figure 4.11. If the wind has been blowing for a long time, the ocean will have reached a state of equilibrium, tending to rotate neither faster nor slower with time; at every point in the ocean, the relative vorticity has a fixed value. This means that, over the ocean as a whole, those factors which act to change the relative vorticity of the moving water must cancel each other out. Assuming for convenience that the depth of the flow is constant, we will now consider in turn each of the factors that affect relative vorticity.

The most obvious factor affecting the relative vorticity of the water in a gyre is the *wind*. The wind field is symmetrical and acts to supply negative (clockwise) vorticity over the whole region.

The next factor to consider is *change in latitude*. Water moving northwards on the western side of the gyre is moving into regions of larger positive planetary vorticity and hence acquires negative relative vorticity (cf. Figure 4.7(a)); similarly, water moving southwards on the eastern side of the gyre is moving into regions of smaller positive planetary vorticity and so loses negative relative vorticity (or gains positive relative vorticity). However, because as much water moves northwards as moves southwards, the net change in relative vorticity of water in the gyre as a result of change in latitude is zero.

What, then, is acting to counteract the negative (clockwise) tendency supplied by the wind stress, so that the gyre does not speed up indefinitely?

The answer is *friction*. We may assume that the wind-driven circulation is not frictionally bound to the sea-floor, but there will be significant friction with the coastal boundaries as a result of turbulence in the form of (horizontal) eddies. Using appropriate values of A_h, the coefficient of eddy viscosity for horizontal motion (Section 3.1.1), we may make vorticity-balance calculations for a symmetrical ocean circulation. Such calculations show that for the frictional forces to be large enough to provide sufficient positive relative vorticity to balance the negative relative vorticity provided by wind stress, the gyral circulation would have to be many times faster than that observed in the real oceans.

---

[*] The variation of the Coriolis force with latitude is often referred to as the *β-effect*.

Figure 4.13(a) is a pictorial representation of the vorticity balance of a symmetrical Northern Hemisphere gyre (cf. Figure 4.12(2)) that carries water around at the rate observed in the real oceans. In this gyre, there is an approximate vorticity balance in the eastern part of the ocean. The negative relative vorticity supplied to the water by the wind is nearly cancelled out by the positive relative vorticity that results from friction with the eastern boundary combined with that gained by the water as a result of moving into lower latitudes. In the western part of the ocean, however, the combined effect of the negative relative vorticity supplied by the wind and that resulting from the movement of water into higher latitudes far outweighs the positive relative vorticity provided by friction. There is a continual gain of negative relative vorticity in the western part of the ocean and the gyre will accelerate indefinitely.

Figure 4.13(b) shows the corresponding diagram for a strongly asymmetrical current system in which water flows north in a narrow boundary current in the western part of the ocean and flows south over most of the rest of the ocean (cf. Figure 4.12(3)). In this asymmetrical gyre, the vorticity balance may be maintained in both the eastern *and* the western part of the ocean. In the eastern part of the ocean, where friction with the boundary is small, the vorticity balance is hardly affected at all, but on the western side, flow must now be much faster and the effects of both friction and change in latitude increase significantly. The negative relative vorticity resulting from change in latitude is acquired at a greater rate because water is now changing latitude much faster, and the gain of positive relative vorticity through friction is increased because both velocity and velocity shear have increased.* Both the planetary and frictional relative vorticity tendencies (which oppose one another) increase by an order of magnitude at least, and a vorticity balance is attained in the western ocean.

The result demonstrated in Question 4.7 and the vorticity-balance explanation given above are not mutually exclusive. Both are concerned with the effect on ocean circulation of the latitudinal variation in the angular velocity of the surface of the Earth. In the first case, this variation in angular velocity is represented by the variation of the Coriolis force with latitude, and in the second it is represented by the variation in planetary vorticity with latitude (cf. Figure 4.7(a)).

The theoretically derived pictures – or models – of ocean circulation derived by Sverdrup and Stommel (Figures 4.10 and 4.12(3)) bear a strong resemblance to the circulatory systems of the subtropical gyres. These circulation models were extended and made even more realistic by Walter Munk (1950). Like Stommel, Munk used a rotating rectangular ocean and assumed that the Coriolis force varied linearly with latitude, but he extended the ocean to include latitudes up to 60° and the equatorial zone. He also balanced the relative vorticity supplied by wind, change of latitude and friction, but improved on the representation of both friction and the wind. As far as friction was concerned, he considered friction with coastal boundaries and friction associated with both lateral *and* vertical current shear, thus including the effect of eddy viscosity in both the horizontal and the vertical dimensions (i.e. both A_h and A_z; cf. Section 3.1.1).

---

*The frictional force between moving water and a boundary is approximately proportional to the *square* of the current speed. This is analogous to the relationship between the frictional force of the wind on the sea-surface (the wind stress, τ) and the wind speed, W (Equation 3.1, $\tau = cW^2$) although, as mentioned later, in constructing the diagrams in Figure 4.12, Stommel used a simpler relationship, with friction *directly proportional* to current velocity.

SYMMETRICAL GYRE

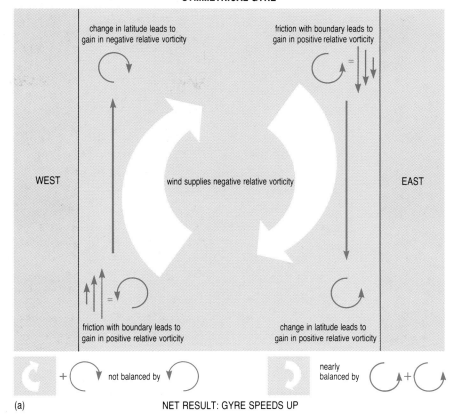

ASYMMETRICAL GYRE

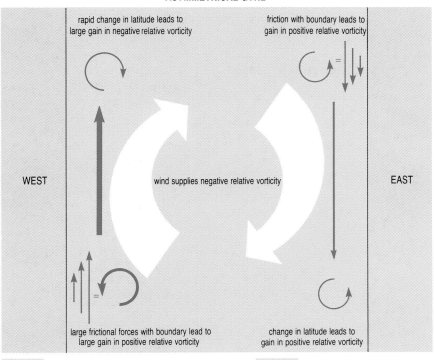

Figure 4.13 Pictorial representation of the various contributions to the vorticity of (a) a symmetrical subtropical gyre (cf. Figure 4.12(2)), and (b) a strongly asymmetrical subtropical gyre with an intensified western boundary current (cf. Figure 4.12(3)). The flow pattern in (b), in contrast to that in (a), enables a vorticity balance to be attained, so that the gyre does not rotate faster and faster indefinitely.

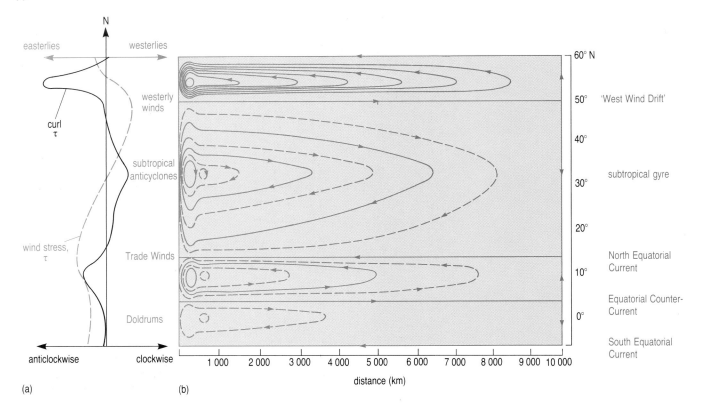

Figure 4.14 (a)   The blue curve is the averaged annual zonal wind stress τ for the Pacific and Atlantic Oceans; the black curve is the curl or torque of this wind stress, curl τ. By definition, curl τ is at a maximum at those latitudes where the wind stress curve shows the greatest change with latitude, e.g. at about 55° N where the wind stress changes from easterly to westerly, and at about 30° N where it changes from westerly to easterly (cf. Figure 4.10(a)).
(b)   The circulation pattern that Munk calculated using the values of curl τ. shown in (a). The volume transport between adjacent solid lines is 10^7 m^3 s^{-1}. The greatest meridional flow occurs where curl τ is at a maximum, i.e. at about 55°, 30° and 10° N; at these latitudes, flow is either southwards or northwards, rather than eastwards or westwards. (The 'West Wind Drift' is the old name for the Antarctic Circumpolar Current, which is of course, in the Southern Hemisphere – hence the quotation marks.)

Figure 4.14(b) shows the pattern of ocean circulation that Munk calculated. If you compare this pattern with Figure 3.1 you will see that it bears a strong resemblance to the actual circulation observed in the North Pacific and the North Atlantic, so much so that it is possible to identify major circulatory features, as indicated on the right-hand side of the diagram.

Thus, in many ways, Munk's model of ocean circulation reproduces the circulation patterns seen in the real ocean. What this means is that the oceanic process that he and others considered important – namely flow of water into different latitudes (i.e. regions in which the Coriolis force/planetary vorticity is different), and horizontal and vertical frictional forces – really are important in determining the large-scale current patterns that result from the wind in the real oceans.

Models of ocean circulation are being improved and refined all the time. Nevertheless, the basic 'tools' which oceanographers use to construct these models remain the same: they are the 'equations of motion'. The aim of Section 4.2.3 is to explain what these equations are and to convey something of how they are used. We have simplified the mathematical notation, but the general principles outlined form the basis for solving even very complicated dynamic problems.

4.2.3   THE EQUATIONS OF MOTION

It may surprise you to know that many of the calculations that led to significant advances in the understanding of ocean circulation were made without the help of computers. The diagrams in Figures 4.10, 4.12 and 4.14 were drawn by solving the appropriate mathematical equations and plotting the results by hand.

The **equations of motion** – the equations that physical oceanographers need to solve in order to be able to describe the dynamics of the ocean – are simply Newton's Second Law of Motion:

$$\text{force} = \text{mass} \times \text{acceleration} \tag{4.2a}$$

applied to a fluid moving over the surface of the Earth.

Equation 4.2a can be rearranged to give:

$$\text{acceleration} = \frac{\text{force}}{\text{mass}} \tag{4.2b}$$

In dealing with moving fluids, it is convenient to consider the forces acting 'per unit volume'. The mass of unit volume of fluid is given numerically by its density, ρ . We can therefore write Equation 4.2b as:

$$\text{acceleration} = \frac{\text{force per unit volume}}{\text{density}}$$

$$= \frac{1}{\rho} \times \text{force per unit volume} \tag{4.2c}$$

In order to determine the characteristics of flow in three dimensions, we must apply Equation 4.2c in three directions at right angles to one another. The x, y, z coordinate system used by convention in oceanography is shown in Figure 4.15. The velocities in the x-, y- and z-directions are u, v and w, respectively, and so the corresponding accelerations (rates of change of velocity with time) are $\mathrm{d}u/\mathrm{d}t$, $\mathrm{d}v/\mathrm{d}t$ and $\mathrm{d}w/\mathrm{d}t$.

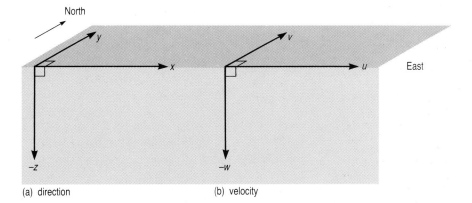

Figure 4.15 (a) The coordinate system commonly used in oceanography. The x-axis is positive eastwards and negative westwards; the y-axis is positive northwards and negative southwards; and the z-axis is positive upwards and negative downwards.
(b) The components of the current velocity in the x-, y- and $-z$-directions are, by convention, u, v and $-w$, respectively.

(a) direction   (b) velocity

Before moving on, consider for a moment what forces might lead to acceleration in the horizontal x- and/or y-directions, and therefore need to be included in the equations of motion for horizontal flow.

We hope you came up with the Coriolis force and the horizontal pressure gradient force, as well as perhaps wind stress and other frictional forces. The equation of motion appropriate to flow in the x-direction (i.e. easterly or westerly flow) may therefore be written:

$$\frac{\mathrm{d}u}{\mathrm{d}t} = \frac{1}{\rho}\left(\begin{pmatrix} \text{horizontal pressure} \\ \text{gradient force in} \\ \text{the } x\text{ - direction} \end{pmatrix} + \begin{pmatrix} \text{Coriolis force} \\ \text{resulting in motion} \\ \text{in the } x\text{ - direction} \end{pmatrix} + \begin{pmatrix} \text{other forces} \\ \text{related to motion} \\ \text{in the } x\text{ - direction} \end{pmatrix} \right)$$

For flow in the *y*-direction (i.e. northwards or southwards flow), the left-hand side of the equation is d*v*/d*t* and the right-hand side is identical to that above, except that 'in the *x*-direction' is replaced in each case by 'in the *y*-direction'.

In mathematical terms, the equations of motion for flow in the *x*- and *y*-directions may therefore be written:

$$\underset{\text{acceleration}}{\frac{\mathrm{d}u}{\mathrm{d}t}} = \frac{1}{\rho}\left(\underset{\substack{\text{pressure}\\ \text{gradient force}\\ \downarrow}}{-\frac{\mathrm{d}p}{\mathrm{d}x}} + \underset{\substack{\text{Coriolis}\\ \text{force}\\ \downarrow}}{\rho f v} + \underset{\substack{\text{contributions}\\ \text{from other forces}\\ \downarrow}}{F_x} \right) \tag{4.3a}$$

$$\frac{\mathrm{d}v}{\mathrm{d}t} = \frac{1}{\rho}\left(-\frac{\mathrm{d}p}{\mathrm{d}y} - \rho f u + F_y \right) \tag{4.3b}$$

F_x and F_y may include wind stress, friction or tidal forcing, depending upon the problem being investigated and the simplifications that can be made.

Note that the mathematical expressions used here have all come up already. The expressions for the horizontal pressure gradient forces are the same as those used in Chapter 3, and they have minus signs because the flow resulting from a horizontal pressure gradient is in the direction of *decreasing* pressure.

The expression for the Coriolis force has also been used already, although here *mfu* and *mfv* have been replaced by $\rho f u$ and $\rho f v$. because we are considering the forces per unit volume.

Why does the equation for flow in the *x*-direction have $\rho f v$ as the Coriolis term rather than $\rho f u$, and *vice versa* for flow in the *y*-direction?

The reason is, of course, that the Coriolis force acts at right angles to the current, so the component of the Coriolis force acting in the *x*-direction is proportional to the velocity in the *y*-direction and *vice versa*. (The minus sign before $\rho f u$ is a consequence of the coordinate system in Figure 4.15; for example, in the Northern Hemisphere, the Coriolis force acting on water flowing in the positive *x*-direction (i.e. towards the east) is in the *negative* *y*-direction (i.e. towards the south).)

As stressed already, the easiest situations to consider are those in which the ocean has reached an equilibrium or steady state, in which all the forces acting on the flowing water are in balance. In such situations, there is no acceleration and d*u*/d*t* and d*v*/d*t* in Equation 4.3 are zero. This means that the equations become very much easier to solve, especially if they can be simplified further.

The easiest way to simplify the equations is to assume that one or more of the forces concerned may be ignored altogether. For example, in his work on wind-driven currents Ekman assumed that the ocean was homogeneous and that the sea-surface was horizontal. This meant that there were no horizontal pressure gradients to worry about, and d*p*/d*x* and d*p*/d*y* were zero.

Another way of keeping the equations simple is to express the contributions to F_x and F_y simply. For example, we may decide to assume that friction is directly proportional to current speed, in which case it can be written as *Au* (for the *x*-direction) and *Bv* (for the *y*-direction), where *A* and *B* are constants. This is the approach that Stommel adopted in calculating the flow patterns in Figure 4.12, which clearly showed that the intensificaton of western boundary currents can be explained by the variation of the Coriolis parameter with latitude.

The examples of the work done by Ekman and Stommel have been quoted to illustrate a particular point. It is this: breakthroughs in understanding of dynamic situations often come about as a result of someone realizing what the most important variable(s) might be and then simplifying the situation under consideration (and hence the equations of motion) so that the effect of the particular aspect being investigated can be seen more clearly.

So far, we have not considered the vertical or z-direction. The main forces that must be considered here are the pressure gradient force in the vertical direction and, of course, the force resulting from gravity, written ρg rather than mg because it is the weight per unit volume that we are interested in. The equation of motion for the z-direction may therefore be written as:

$$\frac{dw}{dt} = \frac{1}{\rho}\left(-\frac{dp}{dz} - \rho g + F_z\right) \tag{4.3c}$$

Except beneath surface waves, vertical accelerations in the ocean are generally very small, so for many purposes, dw/dt may be assumed to be zero.

QUESTION 4.8   Rewrite Equation 4.3c assuming that $dw/dt = 0$, and that there are no forces acting vertically other than the weight of the water and the vertical pressure gradient force. Do you recognize the equation you have written?

Thus, when there are no vertical accelerations (other than g), the equation of motion appropriate to the z-direction becomes simply the hydrostatic equation (Equation 3.8). Indeed, for large-scale motion, the hydrostatic equation is a very good approximation to the 'vertical' equation of motion.

At this point we should mention the **principle of continuity**, another important tool in dealing with moving water. The principle of continuity expresses the fact that mass must be conserved, i.e. the mass of water flowing into a given space must equal the mass of water flowing out of that space. As seawater is virtually incompressible, continuity of mass is effectively continuity of volume. It follows that if the dimensions of a parcel of water change in a particular direction (e.g. the y-direction in Figure 4.16), they must also change in one or both of the other directions. Thus, a wide shallow parcel of water will, on passing through narrow straits, become elongated and, if possible, deeper; alternatively, if a flow of water is obstructed, it may 'pile up', causing a local rise in sea-level.

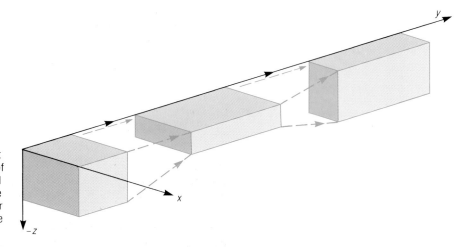

Figure 4.16   Continuity of volume during flow (here in the y-direction). Dimensions change but volume remains the same. Note that the parcel of water may change its shape as a result of spatial constraints (e.g. bottom topography) or because of change in current speed; e.g. a parcel of water entering a region of increased speed will become elongated in the direction of flow.

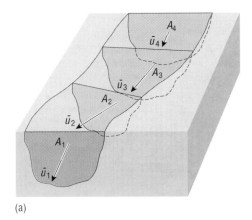

(a)

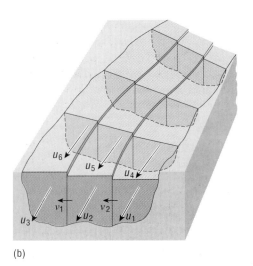

(b)

Figure 4.17   Schematic diagrams to illustrate one- and two-dimensional models for relatively simple, small-scale situations. Note that in these circumstances, the coordinate system shown in Figure 4.15 is applied differently, in that x and u are used for flow *along* the channel, and y and v are used for flow *across* the channel.
(a)   A one-dimensional model for investigating flow through a channel. Such a model would require information about cross-sectional area (A_1, A_2, ...) and the average current speed ($\bar{u}_1$, $\bar{u}_2$, $\bar{u}_3$, ...) through each cross-section.
(b)   A two-dimensional model of the channel can take account of cross-channel flow, v_1, v_2, v_3, ..., between adjacent grid boxes, as well as along-channel flow, u_1, u_2, u_3, ... , which can vary across the channel. The model takes account of flow into and out of each side of each grid box (with the exception of those corresponding to fixed boundaries), but flow arrows are shown only for the nearest row of boxes. For clarity, flows v_1, v_2, v_3 are showing at the front, but in reality they would be through the centres of the sides of the boxes. For a situation like an estuary, the grid boxes might have sides of ~10–100 m.

As the amount of water flowing into a space must equal the amount of water flowing out of it per unit time, the rate of flow, i.e. velocity, is also important in continuity considerations. For example, a broad, shallow current entering narrow straits will become faster as well as perhaps becoming deeper.

QUESTION 4.9   Explain briefly how the flow pattern of the subtropical gyres exemplifies the principle of continuity.

The mathematical equation used to express the principle of continuity is:

$$\frac{du}{dx} + \frac{dv}{dy} + \frac{dw}{dz} = 0$$

(4.3d)

which simply means that any change in the rate of flow in (say) the x-direction must be compensated for by a change in the rate of flow in the y- and/or z-direction(s). This continuity equation is used in conjunction with the equations of motion and provides extra constraints, enabling the equations to be solved for whatever dynamic situation is being investigated.

4.2.4   INVESTIGATING THE OCEAN THROUGH COMPUTER MODELLING

As mentioned earlier, the simple square oceans driven by simple wind fields, envisaged by Sverdrup, Stommel and Munk (Figures 4.10, 4.13 and 4.14) are representations or *models* of the real ocean. They are not attempts to replicate the real ocean in all its complexity, but experimental mathematical constructs intended to reveal the fundamental factors that determine the pattern of the global ocean circulation. Furthermore, the imaginary oceans they represent are essentially two dimensional, and have no bottom topography.

With the advent of powerful computers in the 1970s, it became possible to model more complex imaginary oceans (or parts of oceans) by dividing the volume occupied by water into 'boxes' by a grid. The rate of change of flow through the sides of each box may be calculated using the equations of motion (Equation 4.3), then calculated again on the basis of the results of the first calculation, and so on, the repeated recalculations following the evolution of the flow over time. For a very simple situation, such as flow through a channel where the water is assumed to be well mixed and therefore homogeneous, a one-dimensional model may be sufficient (Figure 4.17(a)); for more complex situations, two-dimensional and three-dimensional models may be needed (Figures 4.17(b) and 4.18). A three-dimensional model can take account of variations of flow with depth (i.e. not just depth-averaged flow) and can include the effect of the sea-bed (albeit relatively crudely). The choice of the model grid to be used (i.e. how the 'boxes' are defined) will depend on the problem being addressed and the computing power available. Figure 4.18 shows three types of grids that might be used, each of which has its own advantages and disadvantages. Whatever the type of grid used, the number of 'boxes' needed will increase with the complexity of the situation being investigated and the resolution (i.e. degree of detail) required; and the greater the number of boxes, the more powerful the computer needed to run the model.

Coastlines and the sea-bed are fixed boundaries of any model of an ocean, or part of an ocean.

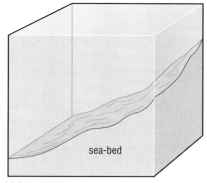

(a) REAL OCEAN

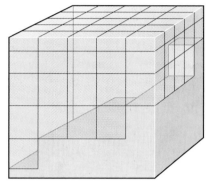

(b) 'LEVEL' MODEL

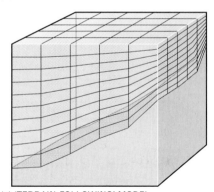

(c) 'TERRAIN-FOLLOWING' MODEL

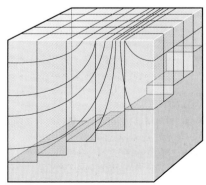

(d) ISOPYCNIC MODEL

Would the sea-surface be a fixed boundary?

Not really. Models may be kept more simple by assuming that the sea-surface behaves as if a solid barrier is held against it (a so-called 'rigid lid'), but as you know, winds cause horizontal motion of surface water, so a 'rigid lid' is generally considered an unacceptable simplification. However, as most models are concerned with bulk movements resulting from ocean currents, the sea-surface is often assumed to be flat, because its ups and downs would average out over the duration of a model 'time-step'.

At a horizontal sea-bed, what can you say about the vertical velocity, w?

At a horizontal sea-bed, w must be zero. More generally (as the sea-bed might be sloping), the velocity *perpendicular* to the sea-bed must be zero.

Models are driven or 'forced' by mathematically applying the effect of whatever is/are considered to be the principal driving force(s). The results are constrained by including in the model known, or assumed, values for one or more parameters – for example, in the simple case of flow through straits, it could be the total volume transport through the straits. If the aim of the exercise is to simulate the average situation – the 'mean flow' – the model is run until the flow pattern generated stops changing, i.e. has reached an equilibrium situation.

The early models of Sverdrup, Stommel and Munk, discussed earlier, were driven simply by the frictional influence of wind stress on the sea-surface. Most modern three-dimensional models also include information about the transport of heat and salt, both into and out of the ocean, and within it. The importance of fluxes of heat and salt for the ocean circulation will be discussed in Chapter 6, but for now we should simply note that heat and salt are **conservative properties**, in that once a body of water has moved away from the sea-surface, its temperature, T, and salinity, S, can only be changed by mixing with another body of water with different T and S characteristics. This means that if we can assign initial values of T and S to each grid box, and feed in information about addition and removal of heat and salt and/or freshwater at the sea-surface, we can predict fluxes of heat and salt through the faces of the subsurface grid boxes by using equations for the conservation of heat and salt, similar in type to the continuity equation (4.3d) for the conservation of mass. In this way, we can produce a pattern for the transport of heat and salt around our model ocean, enabling us not only to model a more realistic density distribution, but also to tackle problems of environmental and climatic relevance.

Figure 4.18   Schematic diagram to illustrate three types of grids used in three-dimensional models. (a) Section showing the bathymetry of the region being modelled. In (b) to (d) the horizontal coordinate system is represented by fine vertical lines, while the vertical coordinate system is indicated by fine horizontal, sloping or curved lines. The most conventional system is (b), known as a 'level' model, because the vertical levels are fixed; in (c), the 'terrain-following' model, the heights of the boxes are a fixed proportion of the ocean depth; and in (d), known as an isopycnic model, the vertical coordinates are isopycnals. In each case, the equations relating to flow velocity are solved for points at the centres of the faces of the grid boxes (cf. Figure 4.17), while those relating to transport of heat and salt are solved for points in the centres of the grid boxes themselves.

How close a modelled situation comes to reality can only be judged by how closely the flow patterns, flow speeds, and T and S distributions generated by the model resemble those actually observed (although, of course a realistic flow pattern *could* be generated by a modelled situation in which the effects of two or more incorrect assumptions have cancelled out). Paradoxically, insights into the processes really at work are sometimes provided by those parts of a modelled flow pattern that diverge most strongly from what is observed in reality. These may indicate that factors that were ignored (i.e. terms left out of the equations) because they were thought to be unimportant were, in fact, significant after all. Stommel's modelling of the North Atlantic gyre (Figure 4.12) provides a clear example of this: the fact that the model (2), which included *no* variation of the Coriolis parameter with latitude, also showed no western intensification of the flow, strongly suggested that variation with latitude was the significant factor (which was, of course, confirmed by model (3)).

Models like those of Stommel, Sverdrup and Munk, which omit unnecessary detail in order to reveal the fundamental processes at work, are known as *process models*. Another type of model includes as many factors as practicable, so as to be as realistic as possible. Such models are known as *predictive models*.

An example of a predictive model
Models are sometimes used to predict current patterns over a particular time period, under certain circumstances. For example, in 1988, oceanographic modellers at the Proudman Oceanographic Laboratory were asked by the British Olympic Sailing Team if they could predict racing conditions in the waters off Pusan in the Korea Strait (Figure 4.19(a)). Although current flow in the area is quite complex, the dominant features seem to be (1) strong tidal currents and (2) a mean flow to the north-east in the Tsushima Current, an offshoot of the Kuroshio (the equivalent of the Gulf Stream in the North Pacific).

Because time was short, it was decided to construct a simple two-dimensional model, in which current flow (the Tsushima Current plus tidal currents) would, of necessity, be depth-averaged. This was considered to be an acceptable simplification, but resulted in local wind-driven currents being under-represented, because the flow was spread over the whole depth instead of confined to the surface Ekman layer. It was decided to compensate for this effect at the time of racing by adding a surface current of 3% of the wind speed (from local weather forecasts), at a small angle to the right of the wind (cf. Section 3.1).

The aim was to predict the main features that were likely to occur in the race area off Pusan, particularly any effects of changes in the strength of the Tsushima Current. The model consisted of two parts. The first was a fine-scale model of the actual race area, with a resolution of 450 m (i.e. with grid boxes of 450 m × 450 m), sufficiently small to cope with the eddies generated off headlands and the sharp divergences of surface currents observed in the area. This was 'nested' within a coarser-scale model of the Strait of Korea, with a grid size of about 10 km, which was used to provide tidal conditions and mean flow consistent with available observations. The Tsushima Current was represented by 'inputting' a north-easterly geostrophic current (with appropriate sea-level slope) entering through the southern and western sides of the model, and then ensuring that the same amount of water was removed through the northern and eastern sides of the model. Interpolations of results from this coarse-scale model were used to force the fine-scale model along its open boundaries, and flow patterns within this smaller area were produced for different states of the tide and different strengths of the Tsushima Current.

The models were generally successful. They reproduced many of the features actually observed, and provided useful extra information. In particular, they predicted that while the flow direction affecting race courses A, B and C changed with the tides, despite the influence of the Tsushima Current, the flow affecting course D was always towards the north-east (Figure 4.19(c)). The British Olympic Sailing team were pleased with their overall performance, which included winning a Gold Medal. They were not the only team to make use of computer modelling at the Seoul Olympics, and since then it has become routine.

You do not need to remember the details of the previous subsection, but hopefully it has conveyed something of how oceanographers go about tackling predictive problems using computer modelling. We will return to the theme of modelling later – for now, we turn to modern observations of the North Atlantic gyre.

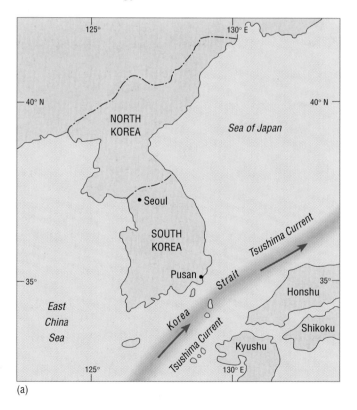

(a)

Figure 4.19 (a)   The oceanographic setting of the 1988 Olympic sailing races in the Korea Strait.

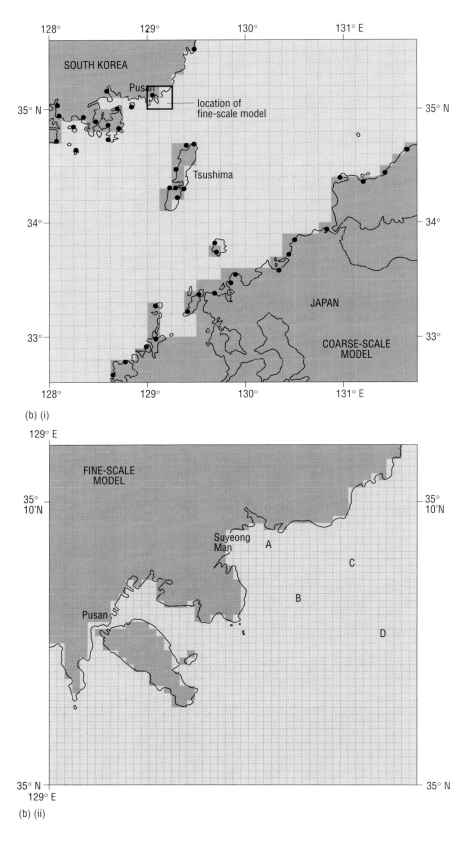

(b) (i)

(b) (ii)

Figure 4.19 (b) (i) The coarse-scale model, results from which were used to force the fine-scale model. (ii) The fine-scale model of the area off Pusan. The black dots are locations of tide gauges which provided sea-level data against which the model could be checked or verified, i.e. run assuming conditions different from those specific ones used to calibrate the model. A, B, C and D were the centres of the different sail race courses.

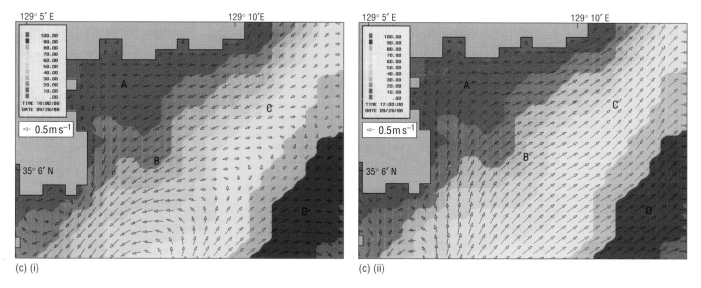

Figure 4.19 (c)   Current patterns off Pusan from the northern part of the fine-scale model, (i) 3 hours before and (ii) 3 hours after high water, assuming a Tsushima Current of 1 knot (~0.5 m s⁻¹). The colours indicate the water depth; red is shallowest, blue deepest (≥100 m).

4.3   MODERN OBSERVATIONS AND STUDIES OF THE NORTH ATLANTIC GYRE

In the South Atlantic Ocean, the shape of the Brazilian coastline diverts a large proportion of the water carried by the South Equatorial Current northwards so that it crosses the Equator as the Guyana Current and joins water from the North Equatorial Current (Figures 3.1 and 4.20(a) (overleaf)). Much of this water enters the Caribbean Sea through the passages between the islands of the Lesser Antilles, and continues through the Yucatan Channel into the Gulf of Mexico; the remainder continues to flow north-westwards in the Antilles Current outside the chain of islands. Most of the flow entering the Gulf of Mexico takes a direct path from the Yucatan Channel to the Straits of Florida, although some takes part in the circulation within the Gulf itself (Figure 4.20(a)).

4.3.1   THE GULF STREAM SYSTEM

The Gulf Stream proper may be considered to extend from the Straits of Florida to the Grand Banks off Newfoundland. It does, however, have two fairly distinct sections, upstream and downstream of Cape Hatteras (at ~ 35° N).

Between the Straits of Florida and Cape Hatteras, the current flows along the Blake Plateau, following the continental slope (Figure 4.20(b)), so that its depth is limited to about 800 m. In this region, the flow remains narrow and well defined. Its temperature and salinity characteristics show that it is supplemented by water from: (1) the Antilles Current (which includes deep water from the South Atlantic, kept out of the Caribbean Sea by the shallowness of the passages between the Lesser Antilles); and (2) water that has recirculated in the Sargasso Sea.

108

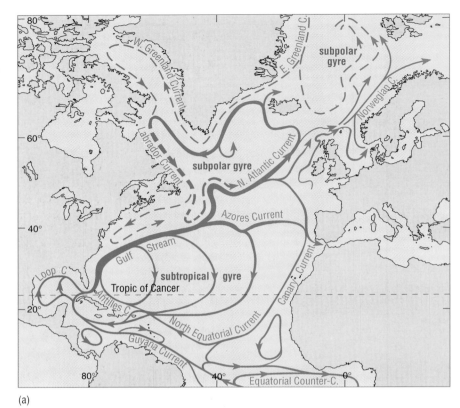

Figure 4.20 (a)   The Gulf Stream in relation to the surface circulation of the Atlantic. The Stream consists of water from the equatorial current system (much of which comes via the Caribbean/Gulf of Mexico) and water that has recirculated in the North Atlantic subtropical gyre. The broken lines represent cold currents.

(b)   Map to show the sea-floor topography off the east coast of the United States, and geographical locations of places mentioned in the text.

(a)

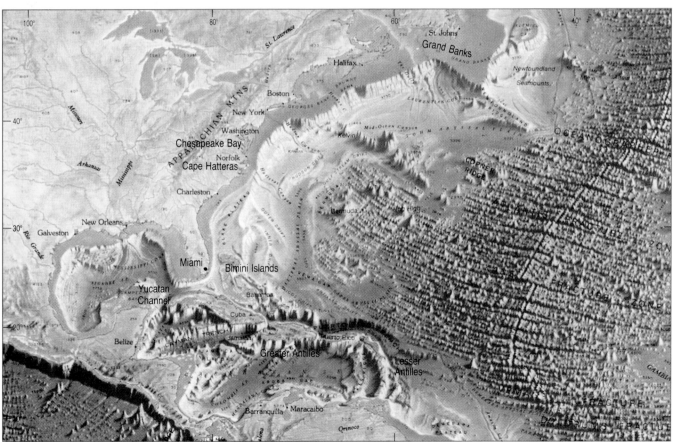

(b)

As the Gulf Stream continues beyond Cape Hatteras, it leaves the continental slope and moves into considerably deeper water (4000–5000 m). While the current was following the continental slope, any fluctuations in its course had been limited and meanders had not exceeded about 55 km in amplitude. Beyond Cape Hatteras there are no topographic constraints, the flow becomes more complex, and meanders with amplitudes in excess of 350 km are common. These meanders often give rise to the Gulf Stream 'rings' or eddies, mentioned in Section 3.5 (and discussed again shortly).

By the time it has reached the Grand Banks off Newfoundland (Figure 4.20(b)), the Gulf Stream has broadened considerably and become more diffuse. Beyond this area it is more correctly referred to as the **North Atlantic Current** (or, in older literature, the North Atlantic Drift). Much of the water in the North Atlantic Current turns south-eastwards to contribute to the Canary Current and circulate again in the subtropical gyre (Figures 3.1 and 4.20(a)); other flows become part of the subpolar gyre, or continue north-eastwards between Britain and Iceland.

Continuity and recent ideas about how the Gulf Stream is driven

Because of the contributions from the recirculatory flow and the Antilles Current, the volume transport of the Gulf Stream increases as it flows northwards (cf. Question 4.1(b)). The average transport in the Florida Straits is about $30 \times 10^6 \, m^3 \, s^{-1}$; by the time the Gulf Stream leaves the shelf off Cape Hatteras this has been increased to $(70–100) \times 10^6 \, m^3 \, s^{-1}$. The maximum transport of about $150 \times 10^6 \, m^3 \, s^{-1}$ is reached at about 65° W, after which the transport begins to decrease again because of loss of water to the Azores Current and other branches of the recirculatory flow (Figure 4.20(a)). We should note here that volume transport values are often quoted in 'sverdrups' (after the distinguished oceanographer), where 1 sverdrup (Sv) $= 10^6 \, m^3 \, s^{-1}$.

Perhaps not surprisingly, these volume transports are much greater than the values that are obtained using Sverdrup's relationship to wind stress curl (Section 4.2.2), which cannot take account of any recirculatory flow, either at the surface or at depth. Another reason for the discrepancy is that, in winter, Gulf Stream/North Atlantic Current water sinks at subpolar latitudes, forming dense deep water which then flows equatorwards. Not only does the dense deep water contribute to the deep recirculatory flow but, for reasons of continuity (Section 4.2.3), the sinking of surface water 'draws' more Gulf Stream water polewards to take its place. In other words, the Gulf Stream is driven not only by the wind, but also by the deep thermohaline circulation. (Do not worry if you do not fully understand this – we will be considering the thermohaline circulation and formation of deep water masses in Chapter 6.)

4.3.2 GEOSTROPHIC FLOW IN THE GULF STREAM

Figure 4.21(a) and (b) show the distributions of temperature and salinity for a section across the narrowest part of the Straits of Florida, based on measurements made in 1878 and 1914. Figure 4.21(c) shows the velocity of the geostrophic current, as calculated from these temperature and salinity data. Part (d) shows the distribution of current velocity based on *direct* measurements made by Pillsbury in the 1890s (Section 4.1.1); the correspondence between (c) and (d) is remarkable. This single example of agreement between observed current speeds and the values calculated using the geostrophic method (Section 3.3.3) did much to increase confidence in the use of geostrophic calculations. (It is worth mentioning that modern hydrographic sections across the Straits of Florida look very similar to those in Figure 4.21.)

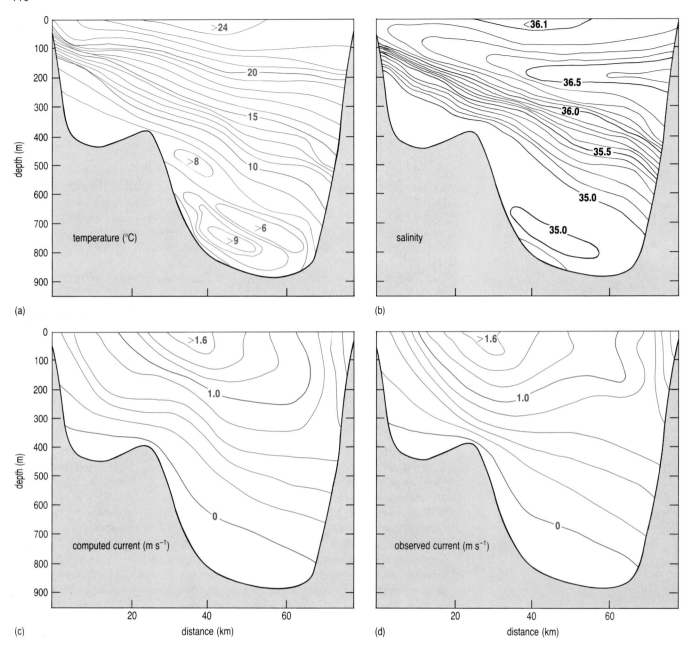

Figure 4.21   The distribution of (a) temperature (°C) and (b) salinity, plotted from measurements made in 1878 and 1914, for an east–west section in the narrowest part of the Straits of Florida, between Fowey Rocks, a short distance south of Miami, and Gun Cay, south of Bimini Islands (see Figure 4.20(b)). *Note:* Salinity values are effectively parts per thousand by weight, but for reasons explained later are usually given without units. (c) The velocity distribution as calculated from the temperature and salinity data, on the assumption that the current has decreased to zero by the depth of the 0-contour; and (d), the velocity distribution on the basis of direct measurements by Pillsbury.

Study Figure 4.21, then answer Question 4.10, bearing in mind that although the warmer water is also more saline, the ranges of temperature and salinity in the Straits of Florida are such that the density distribution is determined mainly by *T* rather than *S*. (If it were determined by salinity alone, the water column in the Straits of Florida would be unstable.)

QUESTION 4.10

(a) Do the sections in Figure 4.21 indicate that flow through the Florida Straits is barotropic or baroclinic?

(b) In which direction does the sea-surface slope, therefore? Does this mean that flow is 'into the page' or 'out of the page' in Figure 4.21(c) and (d)? In other words, on which side of the section is Florida and which Bimini?

(c) Does the sea-surface simply slope up from one side to the other, or does it have a more complex shape?

Calculations of the sea-surface slope indicate that the sea-level is about 45 cm lower on the Florida side of the Straits than on the Bimini side.

Figure 4.22 shows the geostrophic velocity as calculated from *T* and *S* measurements made at hydrographic stations along an east–west line out from Cape Hatteras (cf. Figure 4.20(b)). The section illustrates how the current is no longer a relatively coherent flow, but instead typically consists of narrow vertical filaments separated by counter-currents – i.e. contrary flow (shaded blue).

In the regions shaded blue, does the horizontal pressure gradient force act towards the east (strictly south-east) or towards the west (north-west)?

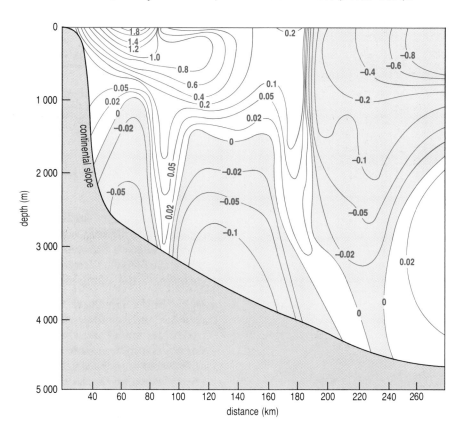

Figure 4.22 Geostrophic current velocity (in m s⁻¹) in the Gulf Stream off Cape Hatteras; the section is east–west so the blue shaded areas (with negative velocity values) represent flow with a southerly component (probably south-westerly). The horizontal scale is smaller than that in Figure 4.21.

Towards the east. In the Gulf Stream, the flow is to the north (or strictly, north-east), and the Coriolis force acting to the right of the flow, i.e. towards the east, balances the horizontal pressure gradient acting to the left (west). The density distribution is such that at the depth corresponding to the zero-velocity contour, the horizontal pressure gradient has become zero because the effects of the sea-surface slope and the density distribution have balanced out with depth. Below the zero-velocity contour, the horizontal pressure gradient reverses. Flow is to the south (or strictly south-west), and the horizontal pressure gradient force to the left of the flow (i.e. to the east) is balanced by the Coriolis force to the right of the flow (i.e. to the west).

The velocity section shown in Figure 4.22 is one of many that show a deep counter-current flowing south-westwards beneath the Gulf Stream. However, until the 1960s many oceanographers found the idea of a significant current close to the deep sea-bed, at depths of 3000–5000 m, hard to believe. The determination of distributions of temperature, salinity and velocity is fraught with difficulty, especially if the *T*, *S* and direct current measurements are widely spaced, so the sceptics could reasonably argue that other interpretations of the data, not involving counter-currents, were equally valid.

In 1965, Stommel developed a theory of the global thermohaline circulation that supported the idea of such equatorward deep currents. However, it was freely drifting floats that in the 1950s and 1960s first provided direct evidence of Gulf Stream counter-currents. We will consider these direct current measurements in Section 4.3.3.

First, however, look at Figure 4.23 which shows *T* and *S* sections across the Gulf Stream between Chesapeake Bay and Bermuda, i.e. downstream of Cape Hatteras. The horizontal scale is much smaller than those of Figures 4.21 and 4.22, and the oceanographic stations were too far apart to allow any filaments or counter-currents to be resolved.

Where is the Gulf Stream on Figure 4.23?

The Gulf Stream is the region where isotherms and isohalines are close together and slope steeply down to the east, about 300 km offshore; in Figure 4.23(a) its 'warm core' (20–22 °C) may be clearly seen extending to depths of 200–300 m.

Why does Figure 4.23 show the Gulf Stream to be flowing along a frontal boundary?

The zone of steeply sloping isotherms and isohalines indicates a boundary between two different water masses – cooler coastal water to landward and warmer Sargasso Sea water on the seaward side. As discussed in Section 3.5.2, fronts like this, with large lateral variations in density, have intense geostrophic currents flowing along them.

Note that the Sargasso Sea water is not only warmer but also more saline than the water on the coastal side of the Stream, which is influenced by cool freshwater input from land. As in the Straits of Florida (Figure 4.21), the salinity distribution *on its own* would result in an unstable situation. However, as is common in the oceans, the temperature distribution here has by far the greater effect on the density distribution and the slopes of the isopycnals.

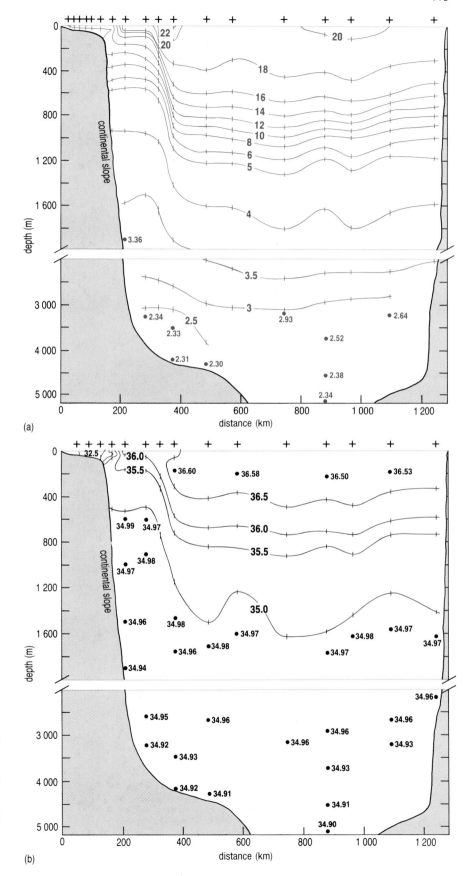

Figure 4.23 (a) Temperature (°C) and (b) salinity sections across the Gulf Stream between Chesapeake Bay and Bermuda, based on measurements made between 17 and 23 April, 1932. These cross-sections, like those in Figures 4.21 and 4.22, were plotted using T and S measurements of water collected at widely spaced hydrographic stations (shown as crosses along the top), and at specific depths: the contours are interpolations based on the spot measurements. Note that as it was expected that there would be more variability in the western part of the section, the hydrographic stations were positioned closer together there.

4.3.3 INSIGHTS FROM MODE

Up until the 1970s, it was generally believed that the deep ocean was a steady, unchanging environment with large bodies of water moving slowly and coherently. There was very little evidence to the contrary, because available oceanographic techniques meant that current, temperature and salinity data came from discrete measurements (or a relatively short series of measurements), widely separated in space and (generally) time. As a result, it was impossible to get any meaningful information about how these properties might vary over relatively short distances or over relatively short time periods (cf. Section 3.5). Then during 1971–73, an international expedition was mounted to study intensively an area of ocean several hundred kilometres across, in the western North Atlantic to the east of the Gulf Stream and to the south of Bermuda, at ~28° N, 70° W (cf. Figure 4.20(b)). The aim of the expedition was to reveal relatively small-scale variability within the ocean by combining results obtained from fixed current meters, freely drifting floats, and dynamic topography calculated from temperature and salinity data. This project, which in its design was a turning point in oceanographic experimentation, was named the **Mid-Ocean Dynamics Experiment (MODE).***

The freely drifting floats used during MODE were of the same type as those which provided the first direct evidence for counter-currents within the Gulf Stream (Section 4.3.2). The floats consisted of aluminium tubes, about 6 m long (the prototypes were constructed out of lengths of scaffolding!), each containing a battery, an acoustic beacon, and a precisely determined amount of ballast. At the surface, such a float is negatively buoyant and so initially it sinks; however, as it is less compressible than seawater its density increases more slowly with increasing pressure than does the density of the seawater. Eventually, at a density level that can be predetermined by adjusting the ballast, the float has the *same* density as the surrounding seawater. It is then neutrally buoyant and stops sinking; thereafter, it drifts with the water that surrounds it. Neutrally buoyant floats are also known as Swallow floats, after John Swallow, the British oceanographer who invented them.

The original Swallow floats were designed to sink to about 1000 m, but in the 1970s they began to be deployed in the **sound channel**, the level of minimum sound velocity where sound waves become trapped by refraction, which varies between about 1000 and 2000 m depth. Large numbers of these floats could be continuously tracked over ranges of 1000 km or more, from acoustic receiving stations on the shore or the sea-bed. **Sofar** (*SO*und *F*ixing *A*nd *R*anging) floats like these were used in MODE, along with the conventional neutrally buoyant floats.

* Around that time, Soviet oceanographers were also investigating intermediate-scale variability in the ocean, and in 1970 they carried out Polygon-70 in the Atlantic North Equatorial Current. Then, during 1974–75 (at the height of the Cold War), the Americans and Soviets collaborated in PolyMODE which investigated mesoscale activity over much of the western part of the North Atlantic subtropical gyre. These experiments had results consistent with those of MODE, but unfortunately we do not have room to discuss them here.

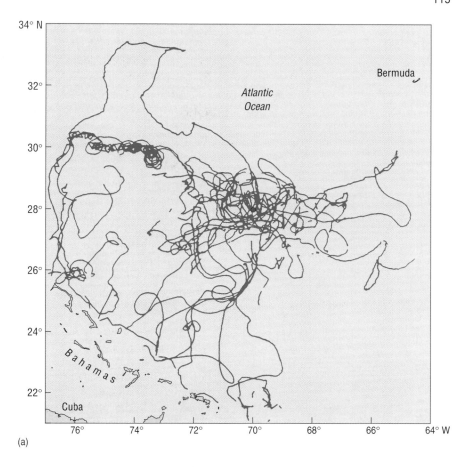

(a)

Figure 4.24 (a)   A compilation of the tracks of all the Sofar floats launched in the MODE area (in the vicinity of the red dot), as recorded from August 1972 to June 1976. Note how the floats became dispersed from the original area, so that after a number of years they had become scattered over the western part of the North Atlantic subtropical gyre.

The most exciting result to come out of analysis of the MODE records was the ubiquity of mesoscale eddies, which were found to dominate flow with periods greater than the tidal and inertial periods (Section 3.5.2). This was despite that fact that – as has since become clear – the MODE site is relatively 'quiet' eddy-wise, when compared with areas closer to intense frontal currents (Figure 3.34).

Figure 4.24(a) is a compilation of the tracks recorded during MODE. Most of the floats seem to have been carried in eddy currents at some time or another, and at about 30° N, a number of floats were trapped in an eddy (or eddies) for a considerable time. The track of one of these floats is shown in Figure 4.24(b) (overleaf), which shows that before being caught in a fast southward flowing current, the float spent most of the monitoring period circulating within a cyclonic eddy which was moving due west.

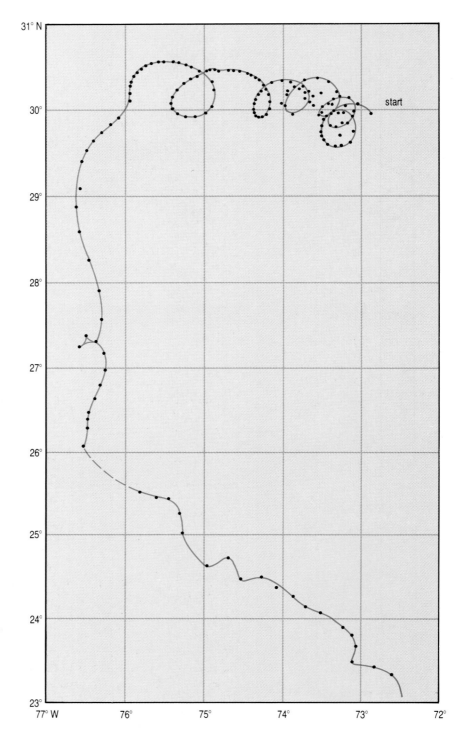

Figure 4.24 (b)   The path of one of the Sofar floats tracked during the MODE experiment. The float was drifting in the sound channel at 2000 m. Dots show daily positions, so the track shows the path taken by the float over 5½ months. The float first circulates at speeds of about 0.5 m s^{-1} within a westward-moving eddy. It is then caught in a fast southward-flowing current.

Figure 4.25 illustrates geostrophic flow patterns at a depth of 150 m in the MODE area, on the basis of density determined from T and S data.

day 105

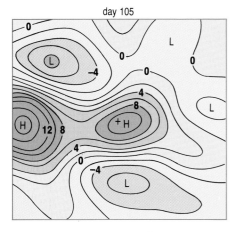

day 135

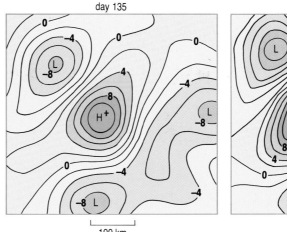

day 165

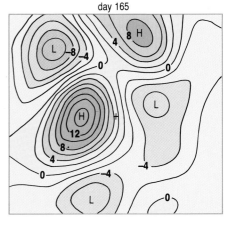

100 km

Figure 4.25 The mesoscale structure in the region of 69° 40′ W, 28° N (marked by the cross in the centre of each map), as revealed by the dynamic topography at 150 m depth, calculated from *T* and *S* data collected during MODE. The numbers on the contours are a measure of the total flow between the contour in question and the zero contour. The 'highs' (H) correspond to warmer water and the 'lows' (L) to cooler water. The three pictures show the situation at intervals of 30 days.

QUESTION 4.11 Given that the contours on Figure 4.25 represent dynamic topography (at a depth of 150 m, rather than at the surface as in Figure 3.21), use your knowledge of geostrophic currents to determine the directions of flow around the 'highs' and 'lows', respectively.

Maps similar to Figure 4.25 were drawn for different depths in the water column and it was found that although the fastest speeds occurred near the surface (being about 0.15 m s^{-1} at 150 m depth) the eddies persisted down to at least 1500 m. It was also found that the 'axes of rotation' of the eddies were not all vertical.

4.3.4 MEASURING CURRENTS DIRECTLY

At this point, it would be useful to make another short diversion away from our Gulf Stream theme, to briefly consider current measurement techniques in general.

Surface buoys that could be tracked visually or by radar (Figure 4.26(a), left) were in use before neutrally buoyant floats were developed, but drifting buoys in general only came into their own with the advent of electronic acoustic and radio tracking systems and, later, satellite navigation (Figure 4.26(a), middle and right). Surface buoys are inevitably affected by the wind. In order to reduce its effect, they generally have most of their volume below water and in addition are equipped with some type of drogue, at depth.

Sofar floats and surface buoys may, theoretically, be tracked until their batteries run down, which may be as long as several years. Other simpler types of drifters are not tracked at all. The current flow is simply deduced from the time and place at which they are found, either at sea or on the shore. In the nineteenth and early twentieth centuries, the 'drift bottle' was commonly used. Today, a variety of cheap 'drifters' (e.g. plastic cards) are used for small-scale local studies in estuaries, coastal waters, or semi-enclosed seas such as the North Sea and the Irish Sea. Each drifter carries a message asking the finder to inform a central agency of the position and time at which it was found, so that its trajectory may be deduced. Of the large number of drifters released in any one experiment, only a small proportion will be returned, and of those that are returned some may have been washed ashore many weeks earlier.

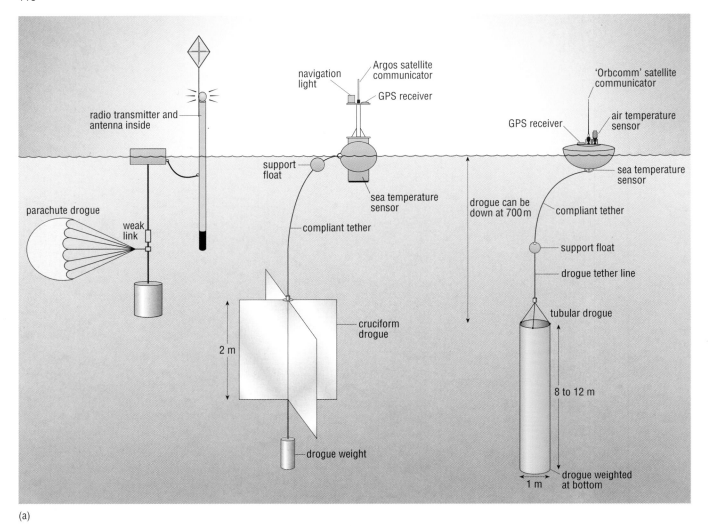

(a)

Figure 4.26 (a)   Examples of freely drifting surface buoys. *Left:* simple spar buoy, with radar reflector and parachute drogue. *Middle and right:* more sophisticated buoys with satellite-tracking devices, for use in shallow seas (with cruciform drogue) and the deep ocean (with tubular drogue at ~ 700 m depth).

Apart from these obvious disadvantages, it is difficult to interpret results obtained using simple drifters. Drifters moving near the surface may be directly affected by the wind; others, designed to be just negatively buoyant at the sea-bed so that they trail along the bottom, are affected by friction with the sea-bed (as is the layer of water in which they move) and so their movement cannot be taken to represent that of the main body of water. Also, it is hard to assess the effects on the floats' trajectories of tidal flows and other fluctuating currents.

In tracking or deducing the paths of drifting objects, we are effectively following the path taken by a parcel of water as it moves relative to the Earth; the velocity of the parcel of water may be calculated from its change of position with time. Such methods of current measurement are described as **Lagrangian***. Much of our knowledge about oceanic circulation patterns and current velocities has been obtained by Lagrangian methods, including use of ship's drift (e.g. Figure 5.12). An enormous amount of information has been obtained from the tracks of Sofar floats and, later, from tracks of more sophisticated neutrally buoyant floats, including ALACE (*A*utonomous

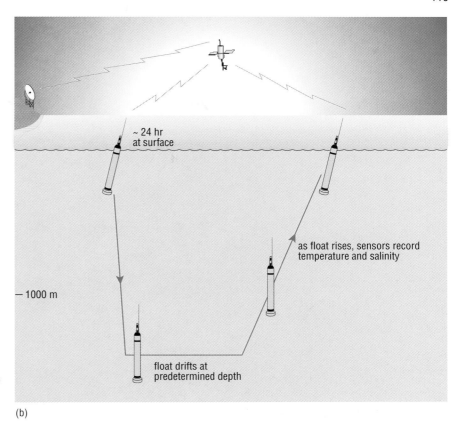

Figure 4.26 (b)   Typical operational cycle of a PALACE float. Some more sophisticated versions actively 'seek' the required depth, rather than simply drifting at the depth for which they have been ballasted.

(b)

*La*grangian *C*irculation *E*xplorer) floats. These were developed during the 1990s to fulfil the needs of the World Ocean Circulation Experiment (WOCE, Section 6.6.1), which required about a thousand subsurface floats to be deployed worldwide, for which acoustic tracking was not feasible. Like Sofar floats, ALACE floats are designed to become neutrally buoyant at a specified depth, remain submerged for a fixed period (typically a few weeks) and then return to the surface where they transmit their position (accurate to 1–3 km) via the *Argos* satellite. They remain at the surface, transmitting, for 24 hours and then return to their operating depth. Those deployed in the 1990s could undertake 70 or more such cycles over a five-year period. (For examples of tracks from ALACE floats, see Figures 5.34 and 6.45.) There are now versions of ALACE and other autonomous floats that carry sensors for temperature, conductivity (for salinity) and pressure (for depth). These collect data on water properties as they rise or sink and are known as profiling floats. Figure 4.26(b) shows the operational cycle of an upward-profiling PALACE (*P*rofiling *A*utonomous *La*grangian *C*irculation *E*xplorer) float.

Alternatively, currents may be measured by **Eulerian*** methods, in which the measuring instrument is held in a fixed position, and current flow past that point is measured. The greatest challenge with Eulerian methods of current measurement is keeping the measuring instrument fixed. Nowadays, this is commonly done by anchoring a subsurface float to the sea-bed and securing

---

*Both these terms are named after mathematicians: Joseph Louis Lagrange (1736–1813) and Leonard Euler (1707–83), respectively.

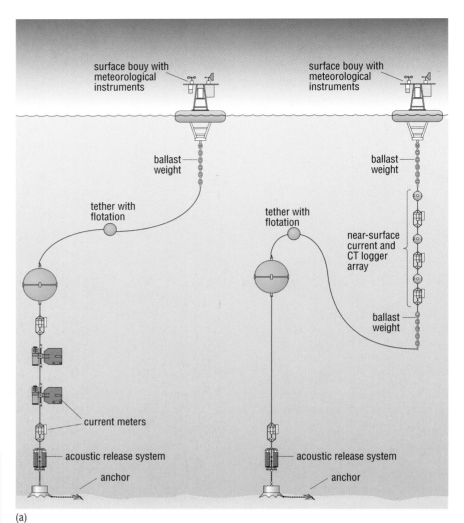

Figure 4.27 Some examples of current-meter moorings (vertical scale greatly compressed). (a) Surface buoys carrying meteorological instruments (anemometer etc.) may either support current-meter arrays directly, or act as marker buoys for a bottom mooring. (b) A deep-ocean bottom mooring incorporating an ADCP (see Section 4.3.7). The 'logger' internally records data on current velocity, salinity (as conductivity, C) and temperature, T.

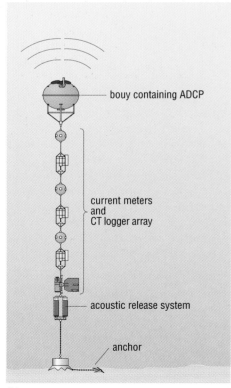

one or more current meters to the taut mooring cable (Figure 4.27(a), left). Current-meter arrays can also be suspended from a securely anchored surface buoy (Figure 4.27(a), right), but such moorings are affected by surface wave motions which are transmitted to the meters. They are also particularly vulnerable to damage by shipping or fishing trawls. Even bottom-mounted current-meter arrays (Figure 4.27(b)) may be damaged by trawling activity, which now extends to depths of well over 1000 m. Occasionally, current meters are deployed directly from anchored ships.

In the most common type of current meter, current flow causes a propeller to rotate at a rate that is proportional to the current speed. Some current meters are designed to align themselves in the direction of flow so that the propeller rotates about an axis parallel to the current direction (Figure 4.28(a)). By contrast, in the Savonius rotor, two hollow S-shaped rotors are mounted about a *vertical* axis (Figure 4.28(b)). This type of meter is very sensitive and responds even to very weak currents, but it only rotates in one direction, so that rapidly reversing flows caused, for example, by surface waves, all contribute to the total number of rotations recorded. In the other type of meter, the propeller rotates in both directions so that the effects of rapidly reversing flows cancel out.

(a)   (b)

Figure 4.28 (a) A typical propeller-type current meter. (b) A Savonius rotor. The meters are oriented with respect to the current by their large vanes.

In most modern current meters, current speed and direction (measured by means of a magnetic compass) are internally recorded electronically. The record is retrieved at the end of the experiment, when the moorings are released from the anchor (generally by means of an acoustic signal) and the current meter bobs up to the surface. Sometimes, current data are transmitted to shore stations via satellite, so that they can be analysed immediately and used for forecasting.

Figure 4.29 (a) The paths taken by a number of Sofar floats at a depth of 700 m in the western North Atlantic. The arrows are 100 days apart; note the high velocity of the floats caught in the Gulf Stream. (b) The same region as in (a), showing the mean current flow as it would have been measured by fixed current meters.

When moored current meters were first used, their recovery rate was often as low as 50%. Technological advances have permitted the development of robust mooring systems and extremely strong cables that can withstand the effects of corrosion, extreme tension, and even fish, which have been known to bite through nylon mooring lines. Modern moorings can be deployed in almost any part of the ocean for a considerable period (over a year), although eventually they may become affected by 'biofouling' (colonization by marine organisms), which alters their drag characteristics.

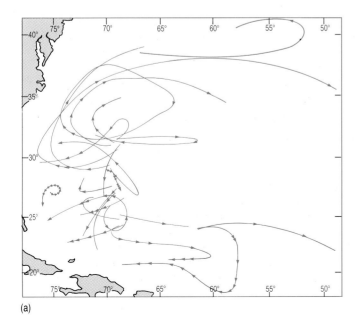

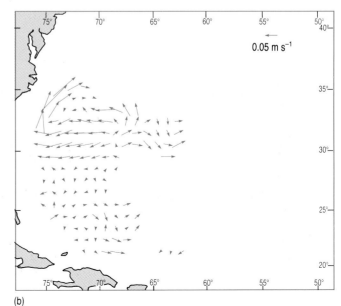

(a)   (b)

Figure 4.29 shows the paths taken by a number of Sofar floats in the region of the Gulf Stream and, for comparison, the mean current flow as it *would* have been measured by means of moored current meters (although, in fact, such measurements were not made).

QUESTION 4.12 By comparing Figure 4.29(a) and (b), can you suggest some advantages and disadvantages of the Eulerian and Lagrangian approaches to current measurement?

For completeness, we should mention the use of coloured dye – in a sense, the ultimate Lagrangian method. This approach is generally used to study relatively small-scale turbulence, and to investigate rates of dispersion, i.e. the extent to which a patch of initially coherent water becomes widely spread, or dispersed, as a result of mixing and turbulence (cf. Figure 4.24(a)).

All the methods mentioned above provide information about horizontal velocity but nothing at all about the vertical component of current velocity. Vertical flow velocities are generally very much smaller than horizontal velocities but in certain unusual circumstances have been measured using floats with fins so arranged that the float rotates at an angular velocity proportional to the vertical current velocity.

An instrument which measures both vertical and horizontal velocities is the Acoustic Doppler Current Profiler (ADCP). ADCPs are now widely used, and can be deployed in various ways (e.g. on moorings, cf. Figure 4.27(b)). The way in which they exploit sound waves is explained in Section 4.3.7.

4.3.5 MAPPING THE GULF STREAM USING WATER CHARACTERISTICS

As discussed in Section 4.1.1, seafarers have long been aware of the high temperatures associated with the Gulf Stream. They have also noted that the edge of the Gulf Stream is often marked by accumulations of *Sargassum* (a floating seaweed, endemic to the Sargasso Sea), and that the waters of the Stream are a clear blue, contrasting strongly with the relatively murky waters between the Stream and the coast.

Since Franklin's time, there have been various surveys of the temperature distribution in the region of the Gulf Stream. These have greatly added to knowledge of the flow: for example, it was through measurement of surface-water temperatures that the north-easterly extension of the Gulf Stream towards Britain and Scandinavia was discovered by Captain Strickland in 1802.

Nevertheless, measurements made directly from ships are time-consuming and expensive, and so are relatively few in number and widely spaced. That this can lead to difficulties in interpreting the results is graphically illustrated by Figure 4.30(a)–(c). These maps show plots made from temperature data collected in 1953 in the vicinity of the Gulf Stream, and the important point to note is that *all three have been drawn using the same data,* collected while the ship moved along the tracks indicated by the red lines. Of course, obtaining measurements at sea has become easier and quicker since the 1950s, but the inherent problem of large gaps in the data (in both space and time) still remains.

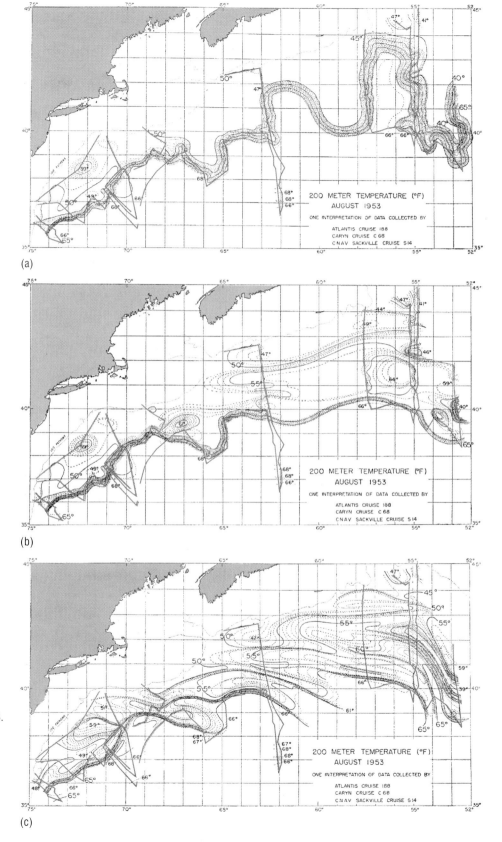

Figure 4.30 Three interpretations of temperature data collected in August 1953. The tracks along which measurements were made are shown as red lines. Interpretation (a) shows a single, simple stream, while (b) shows a double stream (one part stronger than the other) with some branching. The third interpretation (c) shows the Gulf Stream as a series of disconnected fragments.

To add to such difficulties of interpretation, scientists using observations made from ships have to cope with the problem that measurements are made over periods of perhaps weeks and it is therefore impossible to obtain a 'snapshot' of the flow at any one time. This is not the case with satellite measurements, which enable us to see complex spatial variations of surface waters over a wide area, effectively instantaneously (i.e. they provide *synoptic* information). Nevertheless, satellite orbits are such that a given area of ocean is 'viewed' relatively infrequently (e.g. every ten days for *TOPEX–Poseidon*).

Figure 4.31(a) and (b) show the distributions of sea-surface temperature and phytoplankton off the eastern coast of North America, as measured by the satellite-borne Coastal Zone Color Scanner (CZCS).

QUESTION 4.13   In Figure 4.31(a), the blue end of the colour range represents the coldest water and the orange–red the warmest water; the green region corresponds to cool water over the continental shelf and slope. Bearing this in mind, and referring to Figure 3.1 and/or Figure 4.20(a) if necessary, can you identify: (i) the water in the Labrador Current; and (ii) the Sargasso Sea water and water flowing in the Gulf Stream?

In interpreting images like Figure 4.31, it is important to remember that the boundaries between areas of different colours are, in a sense, arbitrary, because colours have been assigned to particular ranges of temperature or chlorophyll content, often in order to bring out certain features. Changing either the ranges or the colours themselves could significantly alter the appearance of an image. More fundamentally, the patterns seen in images like Figure 4.31 are a *result* of the flow pattern, not the actual flow pattern itself, and other factors are always at work. In particular, chlorophyll distributions (Figure 4.31(b)) are significantly affected by nutrient supply, rates of growth of the phytoplankton species present, etc.

4.3.6   GULF STREAM 'RINGS'

Perhaps the most dramatic aspect of Figure 4.31 is the complexity of the (implied) flow pattern, which no conventional method of current measurement could ever have revealed. Various types of eddies can be seen, but the most striking is the circular feature between the Gulf Stream and Cape Cod to the north. This is an example of a 'Gulf Stream ring' – an eddy that formed from a meander which broke off to form an independent circulatory system.

Figure 4.32(a) (overleaf) shows the evolution of Gulf Stream rings on both the landward and Sargasso Sea side of the Stream, over about a month. Gulf Stream eddies are often described as 'cold-core' or 'warm-core'.

Given the way in which Gulf Stream rings form (Figure 4.32(a)), would you expect the continental-margin side of the Gulf Stream to be characterized by warm-core or cold-core eddies? Is this borne out by Figure 4.31(a)?

As eddies form rather in the manner of river meanders cutting off ox-bow lakes, enclosing and pinching off volumes of water so that they end up on the opposite side of the Stream, eddies on the continental-margin side of the Gulf Stream must be warm-core eddies. This is borne out by Figure 4.31(a), in which the central region of the eddy is yellow, corresponding to warm Gulf Stream and Sargasso Sea water. (Note that the rotatory flow in the eddy extends over an area considerably larger than that shown in yellow.)

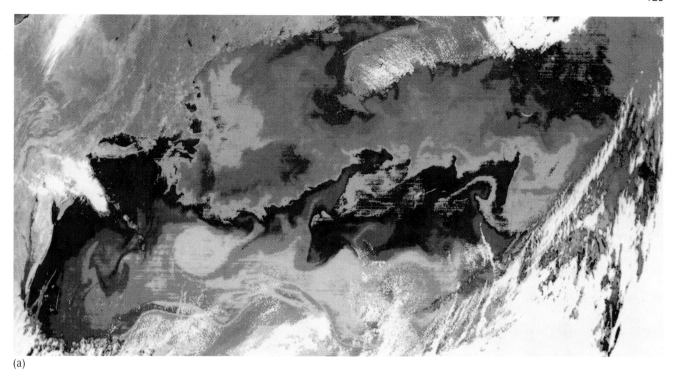

(a)

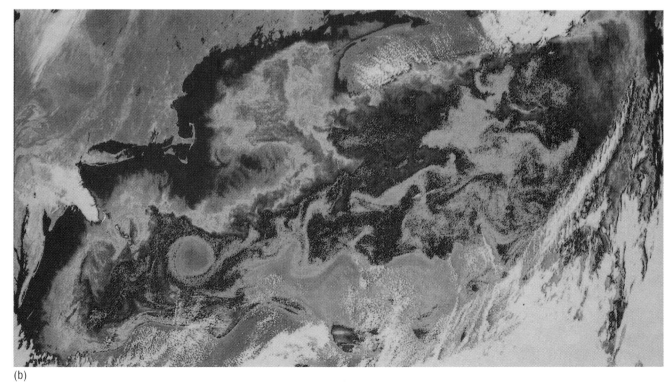

(b)

Figure 4.31   Distributions of (a) sea-surface temperature and (b) phytoplankton pigments, off the eastern coast of the United States and Canada (Cape Cod and Long Island may be seen two-thirds of the way up the image). The data were collected on 14 June 1979, by the Coastal Zone Color Scanner on the *Nimbus-7* satellite. The image in (a) is based on measurement of infrared radiation: the warmest water (shown red) is about 25 °C and the coldest (shown dark blue) is about 6 °C. The brown colour is the land and the white streaks are cloud (which often limits the usefulness of such images). The image in (b) is the same as that shown on the front cover. The highest concentrations of phytoplankton pigment are shown in brown; intermediate concentrations in red, yellow and green; and lowest levels in blue (concentrations have been deduced from the relative absorption and reflection of red and green light by organic pigments).

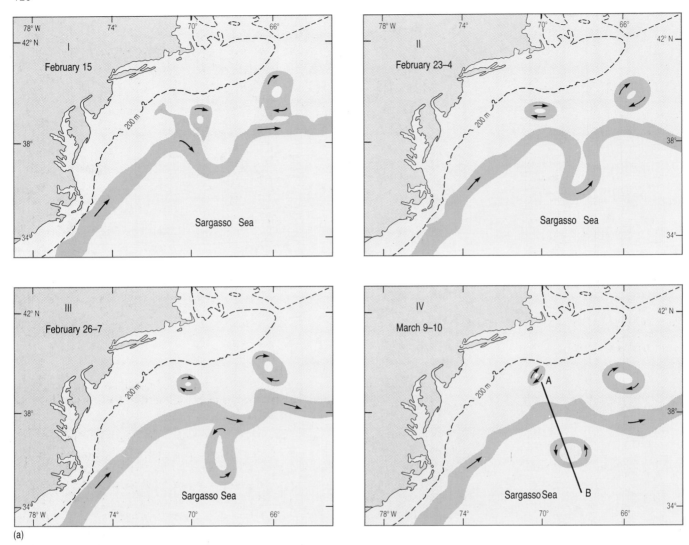

Figure 4.32 (a)   The evolution of Gulf Stream eddies or 'rings', as deduced from infrared satellite images made in February–March 1977. The warm Sargasso Sea water is shown as light pink, the cool continental shelf water as blue and the Gulf Stream as darker pink.

Incidentally, the Gulf Stream meanders that tend to develop into rings are sometimes referred to as 'baroclinic instabilities', because they are perturbations of a flow with strong density gradients (Figure 4.32(b)) and hence velocity gradients, vertical as well as horizontal. Their kinetic energy is believed to be derived from the potential energy of the mean flow, i.e. from the 'relaxation' of sloping isopycnals in the Gulf Stream (cf. Section 3.5). Other types of eddies develop as a result of large lateral variations in velocity, or lateral current shear (cf. Figure 4.6), in which case the original disturbances are referred to as barotropic instabilities.

Figure 4.31(a) and (b) show vividly the role that eddies play in transferring water properties across frontal boundaries. Together, the two images show how the formation of a warm-core eddy results in warm, relatively unproductive Sargasso Sea water being transferred across the Gulf Stream into the cool, productive (because nutrient-rich) waters over the continental margin. Similarly, cold-core eddies will carry cool productive coastal water into the Sargasso Sea. Eddy generation may also be important in transferring water characteristics *between oceans*. Eddies similar to Gulf Stream rings

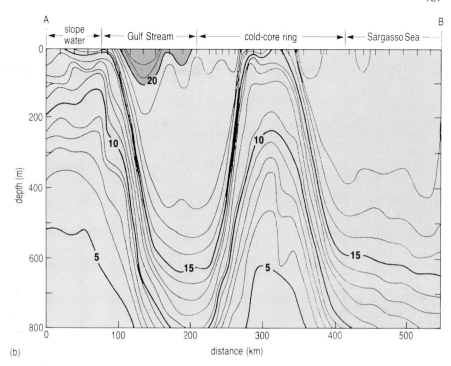

Figure 4.32 (b)   Temperature section along the black line in (a)(iv), showing that the eddies extend to significant depths.

form from the Agulhas Current 'loop' off the tip of South Africa (see Figure 3.1), and are believed to be an important agent in the transfer of water between the Indian and Atlantic Oceans.

The temperature section in Figure 4.32(b) shows that, like the western boundary current from which they form, Gulf Stream rings extend to considerable depths. Cold-core eddies may extend to the sea-floor at a depth of 4000–5000 m, while warm-core eddies impinge on the continental slope and rise when, after forming, they drift erratically towards the south-west. Gulf Stream rings tend to move westwards and/or equatorwards (as do similar eddies elsewhere in the ocean) rather than polewards and/or eastwards. Their survival times seem to depend to a large extent on the path they take: warm-core eddies often last until they are entrained back into the large-scale north-easterly flow of the Gulf Stream, and may have lifetimes of anything from a few months to a year; cold-core eddies, which can more easily escape being caught up in the Gulf Stream again, generally survive somewhat longer.

Gulf Stream eddies are not only deep, they also extend over large areas. A newly formed cold-core eddy typically has a diameter of 150–300 km; a warm-core eddy has a diameter of about 100–200 km. Furthermore, at any one time as much as 15% of the area of the Sargasso Sea may be occupied by cold-core eddies, and as much as 40% of the continental shelf water by warm-core eddies. They have a significant influence on the North Atlantic as a whole, continually exchanging energy, heat, water, nutrients and organisms with their surroundings. Locally, they also greatly affect exchanges of heat and water between the ocean and the overlying atmosphere.

Returning for a moment to the Gulf Stream rings in Figure 4.32, what can you say about the directions of rotation of cold-core and warm-core eddies?

Cold-core eddies are always cyclonic (anticlockwise in the Northern Hemisphere) and warm-core eddies are always anticyclonic – this is true of *all* mesoscale eddies, not just Gulf Stream rings. You have already encountered this idea in Question 4.11: the 'highs' on Figure 4.25 correspond to eddies with warm central regions and the lows to eddies with cold central regions.

All Gulf Stream eddies, whether warm-core or cold-core, contain a ring of Gulf Stream water. Rotational velocities are highest in this ring – as much as 1.5–2.0 m s^{-1} – and decrease both towards the centre of the eddy and towards the outer 'rim'. Such information was initially obtained largely through direct current measurements. Satellite images like those in Figure 4.31 are extremely effective in portraying horizontal variations in water properties, and they suggest flow patterns whose complexity could not have been fully appreciated through traditional oceanographic techniques, but they cannot provide information about current velocity.

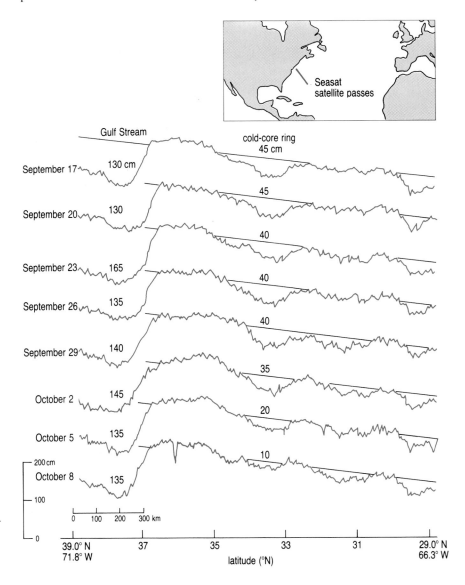

Figure 4.33   Variation in the height of the sea-surface along a satellite track (see inset map) in the western North Atlantic. The measurements were made by the *Seasat* radar altimeter from 17 September to 8 October 1978. The black line represents the local height of the marine geoid, and the distances in centimetres are departures from this level.

A remote-sensing technique that *can* provide information about current velocity is *satellite altimetry*, which you have already encountered in Section 3.3.4. The Frontispiece shows the dynamic topography of the sea-surface – i.e. the sea-surface height *minus* the geoid (Figure 3.22) – for one pass of the *TOPEX–Poseidon* satellite. Such instantaneous pictures of the sea-surface may be used to deduce the velocities of surface currents at that time, if geostrophic equilibrium is assumed. Figure 4.33 shows how the shape of the sea-surface along a south-east–north-west satellite track in the region of the Gulf Stream changed over the course of 21 days. The Gulf Stream itself shows up clearly, as does a cold-core ring, which was moving away to the side of the satellite track during the period in question.

Satellite altimetry is very exciting to physical oceanographers as it enables them to *see* the dynamic topography of the sea-surface (Section 3.3.4). Also, comparison of directly measured current velocities with values calculated from the observed sea-surface slopes enables the depths at which geostrophic current velocities become zero (i.e. the depths at which isobaric surfaces become horizontal) to be accurately determined (Section 3.3.3).

This almost concludes our survey of recent measurements and observations of the Gulf Stream – an example of an intense western boundary current. It is likely that as more becomes known about the other western boundary currents, they will be found to share many of the characteristics of the Gulf Stream. Before moving on to look at the equatorward-flowing eastern limbs of the subtropical gyres – the eastern boundary currents – we will briefly mention some methods of current determination that have not been discussed so far, and look at some of the results of a project to model the circulation in the North Atlantic.

4.3.7 OTHER METHODS OF CURRENT MEASUREMENT

The methods of current measurement discussed in Section 4.3.4 all depend on motion caused by the current itself – either the motion of floats drifting with the current, or the rotatory motion of propellers within fixed current meters. An alternative approach is to use the various ways in which moving water affects the passage of electromagnetic and acoustic waves.

In the 1980s, observational oceanography was revolutionized by the introduction of the Acoustic Doppler Current Profiler (ADCP). This instrument uses sound to measure current velocity by exploiting the Doppler effect whereby the measured frequency of vibration is affected by relative movement between the source of the vibration and the point of measurement. The ocean is full of small floating particles, including plankton, which scatter sound. ADCPs emit sound at a fixed frequency and 'listen' to echoes returning from the sound-scatterers. Most of the sound travels forwards, but that which bounces off particles is scattered in all directions (Figure 4.34) and has its frequency Doppler-shifted. This Doppler shift in the frequency (i.e. pitch) of the sound received by the ADCP is proportional to the current speed, which can therefore be determined. The sound frequencies commonly used are ~70 kHz and ~150 kHz; the higher-frequency sound does not penetrate so far, but provides greater accuracy.

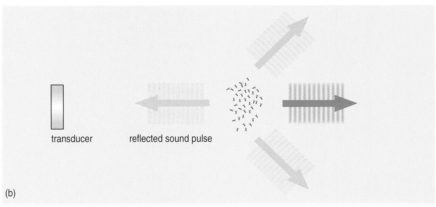

Figure 4.34   Schematic diagrams to show how the Acoustic Doppler Current Profiler works.
(a)   A sound pulse is emitted by a transducer (usually one of three, at right angles to one another), usually in the hull of a ship or on the sea-bed.
(b)   The sound is scattered in all directions by small particles and plankton in the water and the echoes are Doppler-shifted according to the direction of movement of the scatterers. If, for example, the patch is moving away from the ADCP, the reflected sound pulse will be shifted towards longer wavelengths and lower frequencies. (If the ADCP is ship-mounted, the velocities recorded by the ADCP must be corrected to remove the velocity of the ship relative to the water.)

It has become routine practice for hull-mounted ADCPs to be used for continuous measurement of current velocities to depths of several hundred metres, while the ship is underway (see Figure 5.5(c) and (d)). The great advantage of ADCPs is that current profiles and sections can be obtained very quickly, so the amount of data that can be acquired is orders of magnitude greater than would be possible by conventional methods. However, the returned sound pulses each carry information about the average velocity of a finite volume of water (~1 m thick in shallow water, ~10 m thick in deep water) rather than about the water at a particular point. This means ADCPs are not that useful for resolving small-scale features. In addition, data from hull-mounted ADCPs are affected by errors associated with the moving ship, so their absolute accuracy might be low. This problem is avoided with ADCPs deployed on the sea-bed, and largely avoided with ADCPs moored on the bottom (Figure 4.27(b)) or lowered from the surface (see Figure 5.32). Nevertheless, because of the limitations in accuracy, ADCPs are generally used in concert with other methods – conventional current meters and/or geostrophic calculations. Because data can be collected so quickly, ADCPs (and a similar technique which employs lasers – i.e. light – instead of sound waves) are particularly useful for measuring the short-period fluctuations in velocity associated with turbulence.

ADCPs depend on there being small organisms in the water; if, for some reason, there are none present, the instrument becomes ineffective. In physical oceanography research, it is generally assumed that the planktonic scatterers are floating passively in the water, and on average move at the same velocity as the water. For large zooplankton, this is not always the case, and ADCPs are therefore becoming increasingly useful in biological as well as physical investigations.

The Doppler shift may be used in a completely different context to measure surface currents. The technique known as Ocean Surface Current Radar, or OSCR, exploits the fact that radio waves reflected from surface waves that have half the radio wavelength, and are travelling either directly towards or away from the radio transmission station, interfere constructively to produce a large signal, analogous to the Bragg reflections observed in crystallography. At any one time and location, waves at the sea-surface have a range of wavelengths which includes very small waves of the wavelengths that interact with radio waves. The frequency spectrum of the reflected radio signals therefore includes two 'Bragg lines', which in the absence of a current are symmetrical about zero Doppler shift. When the surface water on which the waves are propagating is itself moving, carrying the small waves with it, the frequency of the reflected radio waves is further shifted and the two Bragg peaks are no longer symmetrically disposed about zero Doppler shift. The magnitude of the current is derived by measuring the displacement of the Bragg lines from their 'zero current' positions.

We have described only some of the ways in which currents can be measured, but the selection given here should give you some idea of the wide variety of techniques available.

4.3.8 MODELLING THE CIRCULATION OF THE NORTH ATLANTIC

To simulate the circulation of the North Atlantic well enough to have a chance of re-creating some of the features seen in the real ocean (e.g. Figure 4.20(a)), we would need a three-dimensional model, with sea-floor topography (Figure 4.18) and variations with depth in current flow, temperature and salinity, and hence density (Section 4.2.4). The details of how such a model would be 'constructed' are outside the scope of this book, but modelling is an intrinsic part of modern oceanography and our review of how ideas about the North Atlantic gyre have developed would be incomplete without a brief look at some recent models of the North Atlantic.

One way that modellers can learn about the processes governing ocean circulation at the same time as improving modelling techniques is to compare different types of models with each other and with what is observed in reality. In 1999, as part of a European Community project known as DYNAMO (*Dy*namics of *N*orth *A*tlantic *Mo*dels), three different models of the North Atlantic were compared in detail. One was a 'level' model, one a 'terrain-following' model, and one an isopycnic model – the coordinate systems for these types of model were shown in Figure 4.18.

All three were forced in the same way, using realistic datasets derived from information about wind patterns, and exchanges of heat and freshwater at the ocean surface, provided by the European Centre for Medium-range Weather Forecasting. The starting conditions were (1) a state of rest, and (2) T, S and density values for each box consistent with the observed mean hydrographic state for the ocean, as detailed in the *Climatological Atlas of the World Ocean*.* The models were run forward for the equivalent of 20 years, to allow a near-equilibrium state to be reached. The winter sea-surface temperature distributions that were produced by the three models are shown in Figure 4.35.

*This atlas was first produced by NOAA in 1982 and was revised in 1994; it is usually referred to just as 'Levitus' (or 'Levitus climatology'), after its original compiler, S. Levitus (the 1994 edition is by S. Levitus and T.P. Boyer).

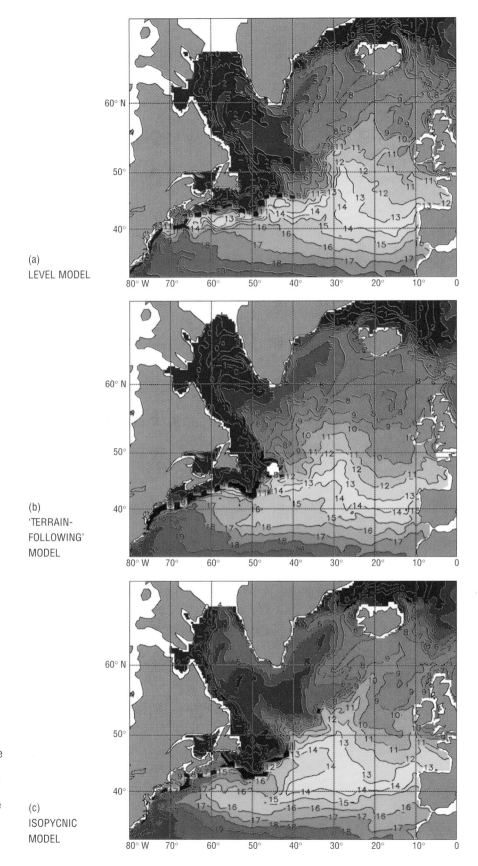

(a)
LEVEL MODEL

(b)
'TERRAIN-
FOLLOWING'
MODEL

(c)
ISOPYCNIC
MODEL

Figure 4.35   Winter sea-surface temperatures
(°C) predicted by the three models tested in the
DYNAMO project: (a) the level model, (b) the
'terrain-following' model, and (c) the isopycnic
model. All three models produce a Gulf Stream
carrying warm water up the western side of the
ocean, but the way the Gulf Stream/North
Atlantic Current crosses the ocean is slightly
different in each case.

All the models simulated the circulation of the North Atlantic with some considerable success. Perhaps the most obvious feature matching the real world is a Gulf Stream carrying warm water up the western side of the ocean (although in all three models it leaves the coast at about 40 °N, rather than at Cape Hatteras at ~35° N). In all three models, there is also a more diffuse North Atlantic Current crossing the ocean and splitting, so some warm water is carried into the Nordic Seas and the rest is carried around into the subpolar gyre (whose shape in each case bears a resemblance to what is observed in reality; cf. Figure 4.20(a)).

Not only did the models simulate many features of the surface circulation, but they also reproduced important aspects of the deep circulation as well. However, there are some interesting differences between the models. For example, only the isopycnic model generates a feature like the Azores Current (cf. Figure 4.20(a)), branching off from the Gulf Stream at about 40° N. Also, the path of the North Atlantic Current is different in the three models. According to observations, it should follow round the Grand Banks (cf. Figure 4.20(b)), pass through 40° W and 50° N, and flow eastwards to reach the Mid-Atlantic Ridge near 30° W before breaking up into a series of branches. Again, only the isopycnic model is reasonably realistic. Why this should be so is not (at the time of writing) clear, but the point is that by working out how and why models such as these differ from one another and from reality, modellers gain insights into the processes that control the ocean circulation.

4.4 COASTAL UPWELLING IN EASTERN BOUNDARY CURRENTS

Eastern boundary currents are less spectacular than western boundary currents and have received less attention. Nevertheless, the upwelling which is often associated with them has biological implications of considerable ecological and economic importance.

In Chapter 3, you saw how the average motion of the wind-driven layer is 90° *cum sole* of the wind direction (Figure 3.6(b)), and hence how cyclonic winds lead to a surface divergence and upwelling in mid-ocean (Figure 3.24(a) and (b)). Coastal upwelling is the result of a divergence of surface water away from the coastal boundary.

In which direction must the wind blow in order to cause most upwelling?

Because Ekman transport is 90° *cum sole* of the wind direction, the most offshore transport of surface water, and the most upwelling, occurs in response to *longshore equatorward winds,* not offshore winds.

The situation is complicated by the fact that the offshore movement of surface water leads to a lowering of sea-level towards the coast. This results in a landward horizontal pressure gradient, which in turn generates geostrophic flow towards the Equator. The equatorward geostrophic current, combined with the offshore wind-driven current, results in a surface current directed offshore and towards the Equator. How this arises is explained in detail in Figure 4.36 and its caption, which you should study carefully.

134

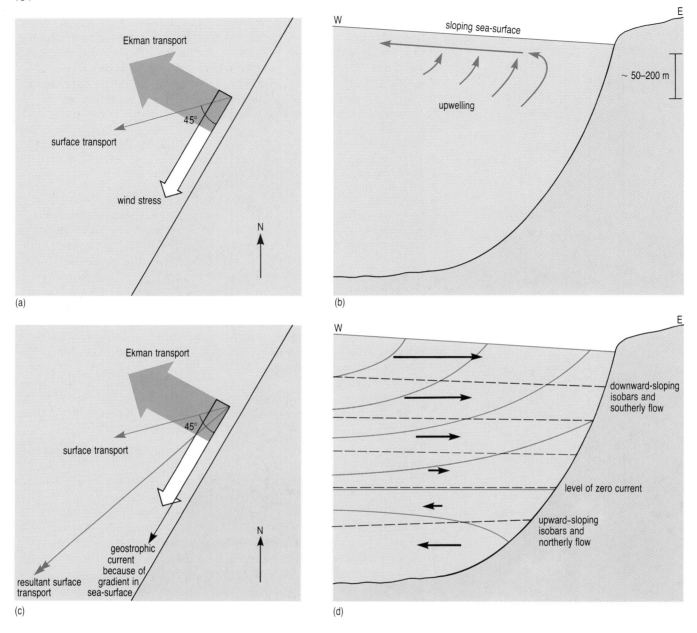

Figure 4.36 Diagrams (*not to scale*) to illustrate the essentials of coastal upwelling (here shown for the Northern Hemisphere).
(a) Initial stage: wind stress along the shore causes surface transport 45° to the right of the wind, and Ekman transport (average motion in the wind-driven layer) 90° to the right of the wind (cf. Figure 3.6(b)). (*Note:* this shows the idealized situation which is never observed in reality.)
(b) Cross-section to illustrate the effect of conditions in (a): the divergence of surface waters away from the land leads to their replacement by upwelled subsurface water, and to a lowering of sea-level towards the coast.
(c) As a result of the sloping sea-surface, there is a horizontal pressure gradient directed towards the land (black arrows in (d)) and a geostrophic current develops 90° to the right of this pressure gradient. This 'slope' current flows along the coast and towards the Equator. The resultant surface transport, i.e. the transport caused by the combination of the surface transport at 45° to the wind stress and the slope current, still has an offshore component so upwelling continues.
(d) Cross-section to illustrate the variation with depth of density (the blue lines are isopycnals) and pressure (the dashed black lines are isobars and the horizontal arrows represent the direction and relative strength of the horizontal pressure gradient force). Isopycnals slope up towards the shore as cooler, denser water wells up to replace warmer, less dense surface waters. The shoreward slope of the isobars decreases progressively with depth until they become horizontal; at this depth the horizontal pressure gradient force is zero, and so the velocity of the geostrophic current is also zero. At greater depths, isobars slope up towards the coast indicating the existence of a *northerly* flow; a deep counter-current is a common feature of upwelling systems.

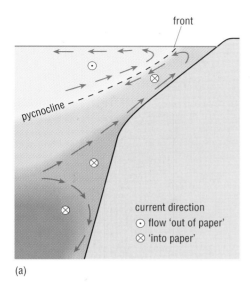

(a)

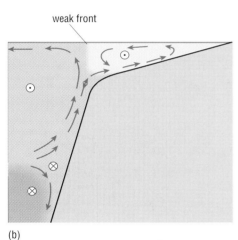

(b)

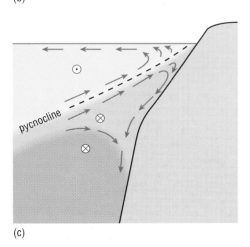

(c)

Note that the current arrows on Figure 4.36(a) and (c) are idealized representations of a steady-state situation, assuming a fully developed Ekman spiral, resulting in Ekman transport at right-angles to the wind (Section 3.1.2). Even if the wind was steady for some time, a fully developed Ekman spiral would not be possible in shallow coastal waters, and in reality, upwelling occurs in response to particular wind events, which might be quite short-lived. Thus, the actual pattern of isopycnals and along-shore current flow varies from time to time, depending on the direction and strength of the wind, and is also affected by local factors like the topography of the sea-bed and the shape of the coastline. Three examples of upwelling regimes are shown in Figure 4.37 – note that they all include a poleward-flowing counter-current, which is found in most upwelling regions in eastern boundary currents.

The ecological importance of coastal upwelling lies in the fact that – like upwelling in cyclonic gyres (Figure 3.24) – it usually replenishes surface waters with nutrient-rich sub-thermocline water, stimulating greater productivity of phytoplankton (and hence supporting higher trophic levels). Figure 4.38 illustrates the marked effect that wind has on upwelling, and hence primary productivity, in the surface waters of the Canary Current off north-west Africa. Upwelling off the coast of north-west Africa occurs in response to the North-East Trade Winds. To the north of about 20° N, there is a region where the Trade Winds blow along the coast all year, but to the north and south of this, the wind direction changes seasonally (cf. Figure 2.3). As a result, although upwelling occurs all year round to the north of Cape Blanc, further to the north and to the south of Cape Blanc it varies seasonally. Similar seasonal variations are observed in all eastern boundary currents, and there are also marked differences between one year and another – compare Figure 4.38(b) and (c) for the month of November in 1982 and 1983.

Coastal upwelling is most marked in the Trade Wind zones, but it can occur wherever and whenever winds cause offshore movement of water. Nevertheless, it is a difficult process to investigate directly because it occurs episodically, and because the *average* speeds of upward motion are very low – generally of the order of 1–2 metres per day though sometimes approaching 10 metres per day. Indirect methods must be used and, as with indirect methods of measuring currents, these may be based on either the *causes* or the *effects* of upwelling.

The cause of coastal upwelling is the offshore movement of water in response to wind stress. Since the upwelled water rises to replace that moved offshore by the wind, the rate at which water upwells is the *same* as that with which it moves offshore. Hence, the rate of upwelling may be calculated using Equation 3.4, which tells us that it must be directly proportional to the wind stress and inversely proportional to the sine of the latitude. This method of calculating the rate of upwelling gives reasonable results only if the assumption that a steady state has been attained is valid, *and* there is adequate information about local winds. However, as mentioned earlier, the average *speed* of the wind is not the best indicator of the amount of upwelling it will induce, because the wind *stress* is proportional to something like the square of the wind speed (Equation 3.1). Thus, occasional strong winds have a disproportionately large influence on water movement, and upwelling rates fluctuate greatly in response to fairly small changes in wind speed.

Figure 4.37 Three examples of upwelling regimes over different continental margins. In each case, the wind is equatorward and out of the page. The darker the blue–green tone, the denser the water. The diagrams show the uppermost 200 m or so of the water column; the vertical scale is greatly exaggerated. Note the fronts; these tend to develop wave-like instabilities, eddies and filaments (cf. Figure 4.38).

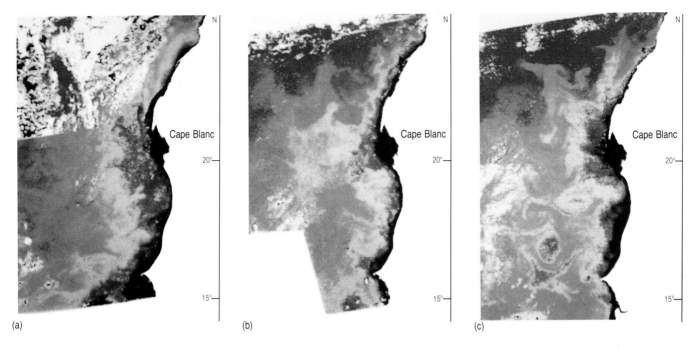

Figure 4.38 CZCS images showing the concentration of chlorophyll-*a* pigment in surface waters off the coast of north-west Africa, in each case averaged over about a month: (a) March–April 1983; (b) November 1982; (c) November 1983. In these images, clouds and land are shown as white and turbid coastal waters are black. The colours represent pigment concentration according to a logarithmic scale: dark blue is smallest concentration; dark green largest. Note the wave-like undulations, eddies and filaments, made visible by their higher chlorophyll content.

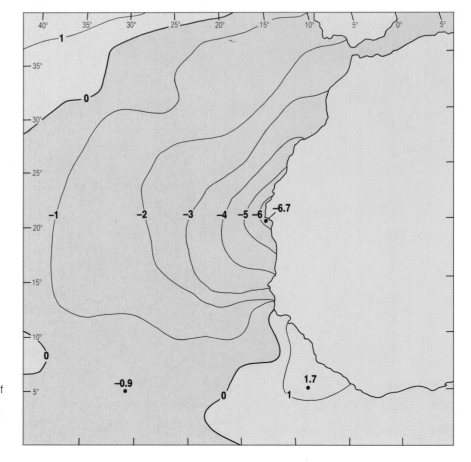

Figure 4.39 Mean anomaly in the sea-surface temperature off north-west Africa for April. The large area with a negative anomaly (i.e. region of significant difference between actual surface temperatures and average surface temperatures for these latitudes) is mainly attributable to upwelling (cf. Figure 4.38).

Although there are chemical and biological indicators of upwelling (Figure 4.38), in practice it is the physical characteristic of temperature that is most often used to identify and investigate regions of upwelling. Water that upwells to the surface comes from only ~100–200 m depth, but it is nevertheless significantly colder than surface water (Figure 4.39). Relatively cold surface water does not always imply upwelling, however; it may simply result from advection of water from colder regions, by currents or in mesoscale eddies. When *subsurface* measurements are available, the clearest indication of coastal upwelling is the upward slope of isotherms (and associated isopycnals) towards the coast (Figure 4.36(d)).

Before leaving the topic of upwelling, we should emphasize that upwelling does not only occur in response to longshore or cyclonic winds. It may also occur, on a local scale, as a result of subsurface currents being deflected by bottom topography. The most extensive areas of upwelling occur in mid-ocean, in response to wind-driven divergence of surface waters. These areas of upwelling, which are at high southern latitudes and along the Equator, will be discussed in Chapters 5 and 6.

4.5 THE NORTH ATLANTIC OSCILLATION

Since 1966, the annual mean position of the Gulf Stream off Cape Hatteras has been determined from observations of surface temperature from satellites, aircraft and ships. These observations have been used to provide statistics on the position (latitude) of the so-called 'north wall' of the Gulf Stream (the boundary between the Stream and the colder coastal water to the west or north-west) at particular longitudes eastward from Cape Hatteras. Given what you have read about the variability of current systems, it will probably not surprise you to know that some years the path of the Gulf Stream is further south, some years further north. What *would* probably surprise you is that the position of the Gulf Stream may be correlated with various climate-related ecological responses on the *eastern* side of the Atlantic – including the size of certain zooplankton populations in parts of the North Sea, the size of the spring zooplankton population in Lake Windermere in Cumbria and – believe it or not – rates of growth of certain terrestrial plant species in controlled plots in Gloucestershire! Such long-distance climatic links – known as **teleconnections** – are becoming easier to study, with an increasingly multidisciplinary approach to climate-related research, and continual improvements in archiving and electronic access to databases world-wide.

These correlations between the path of the Gulf Stream in the western Atlantic and climatic conditions in north-western Europe almost certainly arise because, rather than one being caused by the other, both are manifestations of the same climatic phenomenon – the **North Atlantic Oscillation**, or **NAO**. The NAO is a continual oscillation in the difference in atmospheric pressure between the Iceland Low and the Azores High (cf. Figure 2.3), and is the most important factor affecting climatic conditions over the northern Atlantic Ocean and the Nordic Seas, particularly in winter.

The state of the NAO at any one time is described in terms of the NAO Index. This is usually expressed simply as the wintertime pressure difference between certain meteorological stations in Iceland and the Azores (or, sometimes, Lisbon in Portugal) (Figure 4.40). The Index may also be expressed in terms of the departure of the observed Iceland–Azores pressure difference from the average. In this case, when the pressure difference is higher than usual, the NAO Index is described as positive, and when the pressure difference is lower than usual the Index is described as negative. Generally, when the pressure difference is large the NAO is described as strong, when it is small the NAO is described as weak.

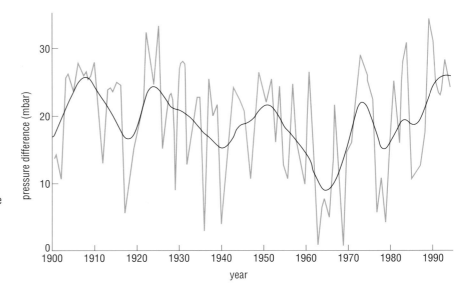

Figure 4.40   Record of the wintertime difference in atmospheric pressure (reduced to sea-level) between the Azores High and the Iceland Low over the course of the twentieth century. The black line is a 5-year running mean, i.e. a line joining average values of the Index through successive 5-year periods.

The state of the NAO affects not just wind speed and direction – westerlies are stronger when the NAO is strong – but also the paths followed by winter storms, rainfall patterns and transport of heat and moisture between the North Atlantic and the surrounding land masses, and even the strength of the North-East Trades. The importance of the state of the NAO is illustrated by Figure 4.41 which compares conditions during the winter of 1994–95, when the NAO was strong (positive Index), with conditions during the winter of 1995–96, when it was weak (negative Index).

Returning to the apparently mysterious links between the position of the Gulf Stream in the western Atlantic and climate-related ecological responses in the British Isles and North Sea, it seems likely that the linking factor is heat transfer between the Gulf Stream/North Atlantic Current and the overlying atmosphere far downstream of Cape Hatteras. When the NAO is strong (positive Index) and surface westerlies are strong, the Gulf Stream and North Atlantic Current flow more strongly than usual, and on average their paths are further north; as a result, more heat and moisture is transferred to north-west Europe, which becomes warmer and wetter. The reverse is true when the NAO is weak. However, as Figure 4.41 suggests, the NAO is more complex, and its effects more wide-ranging, than this simple explanation suggests.

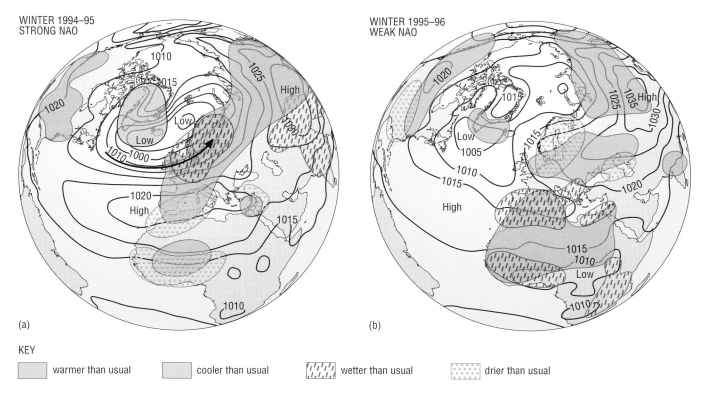

(a)

(b)

KEY

▨ warmer than usual   ▨ cooler than usual   ▨ wetter than usual   ▨ drier than usual

Figure 4.41   Schematic maps to show the pattern of sea-level atmospheric pressure over the North Atlantic region, and associated extreme conditions, in (a) the winter of 1994–95, when the NAO was positive, and (b) the winter of 1995–96, when the NAO was negative. In 1994–95, the Icelandic Low and Azores High were both more intense than usual. As a result, strong westerly winds (black arrow) steered winter Atlantic storms eastward and brought unusually wet conditions to northern Europe and warmer conditions to a band extending from North Africa to Siberia. In 1995–96, the Icelandic Low and Azores High were both less intense than usual. These weak pressure systems led to a reduction in the influence of the Atlantic over northern Europe, with less rainfall and much lower temperatures.

As far as the atmosphere is concerned, the NAO does not seem to have an obvious periodicity (Figure 4.40) or, rather, it is a *mix* of periodicities. Over the second half of the 20th century, it remained in one or other state for five or more years. During most of the 1950s and 1960s, the NAO was weak, and during most of the 1980s and (especially) 1990s the NAO was strong; sometimes, however, the change from a strong NAO to a weak NAO took place from one winter to the next (cf. Figure 4.41). The ocean (which tends to retain the effects of winter atmospheric conditions) has a slower response time than the atmosphere, and a longer 'memory', so it responds to relatively short time-scale fluctuations in the atmosphere on a roughly decadal (i.e. ten-year) time-scale. It is generally considered (as will be discussed in Chapter 6) that – except in specific circumstances such as El Niño events – the large-scale pattern of sea-surface temperature is affected by the overlying atmosphere, rather than *vice versa*. Researchers interested in the NAO have the task of determining whether the ocean might play an *active* role in atmosphere–ocean coupling at higher latitudes, too.

4.6 SUMMARY OF CHAPTER 4

1   The subtropical gyres are characterized by intense western boundary currents and diffuse eastern boundary currents. In the North Atlantic, the western boundary current is the Gulf Stream, and the eastern boundary current is the Canary Current.

2   Exploration of the east coast of America, followed by colonization, trading and whaling, led to the western North Atlantic in general, and the Gulf Stream in particular, being charted earlier than most other areas of ocean. Some of the most notable charts were made by De Brahm and by Franklin and Folger (late eighteenth century) and Maury (mid-nineteenth century); Maury was also the first to encourage systematic collection and recording of oceanographic and meteorological data.

3   The Gulf Stream consists of water that has come from equatorial regions (largely via the Gulf of Mexico) and water that has recirculated within the subtropical gyre. The low-latitude origin of much of the water means that the Gulf Stream has warm surface waters, although the warm core becomes progressively eroded by mixing with adjacent waters as the Stream flows north-east.

4   The prevailing Trade Winds cause sea-levels to be higher in the western part of the Atlantic basin than in the eastern part, and the resulting 'head' of water in the Gulf of Mexico provides a horizontal pressure gradient acting downstream. The flow leaving the Straits of Florida therefore has some of the characteristics of a jet.

5   The Gulf Stream follows the continental slope as far as Cape Hatteras where it moves into deeper water and has an increasing tendency to form eddies and meanders; the flow also becomes more filamentous, with cold counter-currents. Beyond the Grand Banks, the current becomes even more diffuse and is generally known as the North Atlantic Current.

6   Flow in the Gulf Stream is in approximate geostrophic equilibrium, and the strong lateral gradients in temperature and salinity mean that the flow is baroclinic. Confidence in practical application of the geostrophic method was greatly increased when it was successfully used to calculate geostrophic current velocities in the Straits of Florida.

7   The fast, deep currents in the Gulf Stream are associated with the steep downward slope of the isotherms and isopycnals towards the Sargasso Sea. The Gulf Stream may be regarded as a ribbon of high velocity water forming a front between the warm Sargasso Sea water and the cool waters over the continental margin. This is a frontal region with large lateral variations in density, and so is subject to wave-like perturbations known as 'baroclinic instabilities'. These lead to the formation of mesoscale eddies, especially downstream of Cape Hatteras. Eddies formed from meanders with an anticyclonic tendency are known as warm-core eddies, and those formed from meanders with a cyclonic tendency are called cold-core eddies. The importance of mesoscale eddies only began to be appreciated in the 1970s, as a result of MODE and other similar projects.

8   One of the tools for studying fluid flow is the concept of vorticity, or the tendency to rotate. All objects on the surface of the Earth of necessity share the component of the Earth's rotation appropriate to the latitude: this is known as planetary vorticity and, because vorticity is defined as $2 \times$ angular velocity,

is equal to $2\Omega \sin \phi$ and given the same symbol as the Coriolis parameter, f. Rotatory motion *relative* to the Earth is known as relative vorticity, ζ. The vorticity of a fluid parcel relative to fixed space – its absolute vorticity – is given by $f + \zeta$. In the absence of external influences, potential vorticity $(f + \zeta)/D$ (where D is the depth of the water parcel) remains constant. Away from coastal waters and other regions of strong velocity shear, f is much greater than ζ. By convention, an anticlockwise rotatory tendency is described as positive, and a clockwise one as negative (regardless of hemisphere).

9   Early theories about oceanic circulation were restricted because the effects of the Earth's rotation – the Coriolis force and hence geostrophic currents – were not appreciated. During the 20th century, ideas about large-scale ocean circulation developed dramatically. Stommel demonstrated that the intensification of the western boundary currents of subtropical gyres is a consequence of the increase in the Coriolis parameter with latitude, and that western intensification can be explained in terms of vorticity balance.

10   Sverdrup showed that when horizontal pressure gradient forces, caused by sea-surface slopes, are taken into account, the total wind-driven, meridional (north–south) flow is proportional to the torque, or curl, of the wind stress. Sverdrup's ideas were extended by Munk who used real wind data (rather than a sinusoidally varying wind distribution, as used by Stommel and Sverdrup) and allowed for frictional forces resulting from turbulent mixing in both vertical and horizontal directions. The circulation pattern he derived bears a close resemblance to that of the real oceans.

11   The equations of motion – i.e. the mathematical equations used to investigate water movements in the oceans – are simply Newton's Second Law, force = mass × acceleration, applied in each of the three coordinate directions. They are most easily solved by considering equilibrium flow, in which there is no acceleration. When this is done for the equation of motion in a vertical direction, it becomes the hydrostatic equation.

12   Another important principle governing flow in the oceans is the principle of continuity, which expresses the fact that the mass (and, because water is virtually incompressible, volume) of water moving into a region per unit time must equal that leaving it per unit time.

13   Today, computer modelling is a valuable oceanographic tool. Process models (e.g. the simulations of the North Atlantic circulation by Sverdrup, Munk and Stommel) are not attempts to replicate the real situation in all its complexity, but experimental mathematical constructs intended to reveal the fundamental factors that determine the ocean circulation. Models may also be used in a predictive mode.

14   Direct methods of measuring currents divide into two categories. Lagrangian methods provide information about circulation *patterns* and involve tracking the motion of an object on the assumption that its motion represents that of the water surrounding it. Eulerian methods are those in which the measuring instrument is held in a fixed position, and the current flow past that point is measured (usually by a rotary current meter attached to a mooring). Increasingly, current velocities are being measured using Acoustic Doppler Current Profilers.

15   Circulation patterns can also be inferred from the distribution of properties of the water, such as temperature or chlorophyll content. Satellite images can provide near-instantaneous (synoptic) pictures of the sea-surface, and avoid the problems that arise from interpolating between widely spaced measurements, perhaps also taken over a long period of time.

16   Satellite radar altimetry may be used to determine variations in the height of the sea-surface (i.e. departures from the geoid), and hence effectively provide a direct measure of the dynamic topography of the sea-surface.

17   The eastern boundary currents of the subtropical gyres are associated with coastal upwelling which occurs in response to equatorward longshore winds. Areas of divergence and upwelling are characterized by cooler than normal surface waters, a raised thermocline and, because nutrients are continually being supplied to the photic zone, high primary productivity.

18   The most important factor affecting wintertime climatic conditions over the northern Atlantic Ocean and the Nordic Seas is the state of the North Atlantic Oscillation (NAO). This is a continual oscillation in the difference in atmospheric pressure between the Iceland Low and the Azores High. A strong NAO (positive NAO Index) leads to strong westerlies, and warm wet winters in north-west Europe. In the ocean, the variability associated with the NAO has a roughly decadal time-scale.

Now try the following questions to consolidate your understanding of this Chapter.

QUESTION 4.14   Both Franklin and Rennell identified the Gulf Stream to be what we know today as a geostrophic current. True or false?

QUESTION 4.15   Bearing in mind that the Gulf Stream is an intense frontal region, can you suggest why it is often delineated by accumulations of *Sargassum*?

QUESTION 4.16   It is thought that the meanders and eddies that characterize the Gulf Stream downstream of Cape Hatteras may be related to a chain of seamounts extending out from the continental shelf (Figure 4.20(b)). Bearing in mind information given in Section 4.2.1, can you suggest why changes in water depth over the seamounts might lead to meanders and eddies?

QUESTION 4.17   We have said that flow in the Gulf Stream is in approximate geostrophic equilibrium. Why would it not be strictly valid to regard the meandering flow in the eddy-generating region downstream of Cape Hatteras as in geostrophic equilibrium? (You might find the beginning of Section 3.2, on inertia currents, helpful in answering this.)

QUESTION 4.18   Which of the following statements (a)–(d) concerning Lagrangian and Eulerian methods of current measurement are true, and which are false?

(a) Average current velocities determined by Lagrangian methods are likely to be *minimum* estimates.

(b) Eulerian methods require the cooperation of people other than those involved in the research programme, whereas Lagrangian methods do not.

(c) The two methods of current measurement that make use of Doppler shift in frequency are both Lagrangian methods.

(d) Calculation of surface currents using ship's drift (dead-reckoning) is an Eulerian technique.

CHAPTER 5 OTHER MAJOR CURRENT SYSTEMS

In the previous Chapter we considered the subtropical gyres; here we will turn our attention to both higher and lower latitudes. We will look at the equatorial current systems, and the contrasting patterns of ocean circulation in northern and southern polar/subpolar regions. We will also see how the natural oscillation of the North Atlantic atmosphere and ocean appears quite subdued when compared with that of the more closely coupled atmosphere and ocean of the tropical Pacific – the El Niño–Southern Oscillation phenomenon (sometimes referred to as ENSO).

5.1 EQUATORIAL CURRENT SYSTEMS

The first thing to notice about the equatorial current systems (Figures 5.1 and 3.1) is that they are not symmetrically disposed with respect to the Equator, but are shifted several degrees of latitude to the north of it.

By reference to Figure 2.3, can you suggest why this is?

As discussed in Section 2.1, the Intertropical Convergence Zone, the zone along which the Trade Winds of the two hemispheres meet, is distorted towards land in the hemisphere experiencing summer. There is a much greater proportion of land in the Northern Hemisphere than in the Southern and so the mean position of the ITCZ is shifted north of the Equator, most notably so in the Pacific.

In order to understand current flow in low latitudes, it is important to remember the following:

1   The Coriolis force is zero on the Equator.

2   Nevertheless, by latitudes of about 0.5° it is large enough to have a significant effect on flowing water.

Figure 5.1 shows schematically the winds, and the resulting surface currents and Ekman transports, in the region of the Equator. These simplified pictures represent what actually happens, to varying extents, in each of the three oceans. Look first at Figure 5.1(a). The short blue arrows represent the Ekman transport – the net transport in the wind-driven layer (Section 3.1.2); these transports are to the left of the wind in the Southern Hemisphere, to the right of the wind in the Northern Hemisphere, and *in the direction of the wind on the Equator.* The South-East Trade Winds *blow across* the Equator, with the result that there is a *divergence* to the south of the Equator; this is known as the **Equatorial Divergence**.

Remember that the ITCZ is characterized by light winds (the Doldrums) and so there is no significant Ekman transport in the region *between* the two Trade Wind zones. Water moving across the Equator in response to the South-East Trade Winds therefore converges with water in this region (at about 4° N). On the other hand, there is a *divergence* at about 10° N as a result of water moving away from the Doldrum region in response to the North-East Trade Winds.

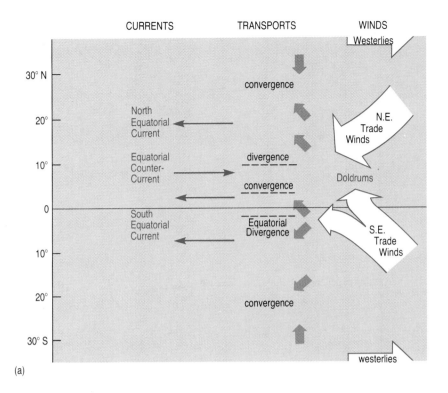

(a)

Figure 5.1 (a)   The relationships between the wind direction, the surface current and the Ekman transport (the total transport in the wind-driven layer, shown by short blue arrows), in equatorial latitudes. Note the Doldrum belt between about 5° and 10° N.

(b)   North–south diagrammatic section showing the vertical and meridional circulation in equatorial latitudes, and the shape of the sea-surface and thermocline. Regions of eastward and westward flow are indicated by the letters E and W. The darker blue region (in which geostrophic current is assumed to be zero, cf. Figure 3.20(a)) is the deep water below the thermocline. The blue oval at about 100 m depth at the Equator represents the Equatorial Undercurrent (see Section 5.1.1). (Note that the vertical scale is greatly exaggerated.)

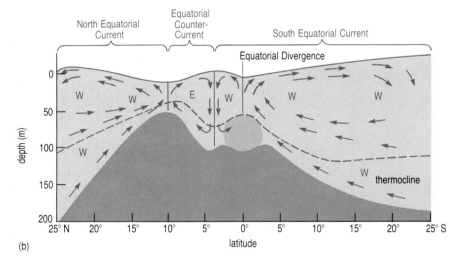

(b)

Figure 5.1(b) is a diagrammatic north–south section across the Equator, showing the vertical and meridional circulation in the mixed surface layer above the thermocline/pycnocline. There is little horizontal variation in density in the upper part of this layer, particularly in the top few tens of metres, and surfaces of constant pressure are more or less parallel to surfaces of constant density, and both are parallel to the sea-surface; in other words, conditions here may be regarded as barotropic. By the depth of the thermocline, however, there are significant lateral variations in density and conditions are baroclinic. The slopes in the thermocline are therefore contrary to those of the sea-surface.

Figure 5.1(b) is most easily interpreted in terms of easterly and westerly geostrophic flow if we use the simplified approach illustrated in Figure 3.20(a), and assume that all of the water above the bottom of the thermocline behaves as one homogeneous layer (shown lighter blue). This is a justified assumption in low latitudes, because here the density gradient, or pycnocline, between the mixed surface layer and deeper, colder waters is relatively sharp (cf. the dashed profile in Figure 3.19(a)).

QUESTION 5.1   Bearing this in mind, outline how the three regions of westward flow in Figure 5.1(b) – i.e. (1) the **North Equatorial Current** between 25° and 10° N, (2) the **South Equatorial Current** between about 4° N and about 0°, and (3) the South Equatorial Current between 0° and about 20° S – may each be explained in terms of a geostrophic balance between horizontal pressure gradient forces and the Coriolis force.

Westward flow in surface waters is also a direct result of the Trade Winds. These blow at about 45° to the Equator and surface flow in the South and North Equatorial Currents is 45° *cum sole* of this, i.e. to the west in both cases.

The easterly **Equatorial Counter-Current** between about 4° and 10° N may be explained as follows. The overall effect of the Trade Winds is to drive water towards the west, but the flow is blocked by the landmasses along the western boundaries. As a result, in equatorial regions the sea-surface slopes up towards the west, causing an eastward horizontal pressure gradient force. Because winds are light in the Doldrums, water is able to flow 'down' the horizontal pressure gradient in a current that is contrary (i.e. 'counter') to the prevailing wind direction. Furthermore, even this close to the Equator, there is some deflection by the Coriolis force. The deflection is to the right and therefore towards the Equator, and so contributes to the convergence at about 4° N (along with the transport across the Equator resulting from the South-East Trades). The sea-surface therefore slopes up from about 10° N to about 4° N, giving rise to a northward horizontal pressure gradient force which drives a geostrophic current towards the east.

This general pattern of westward-flowing North and South Equatorial Currents and an eastward-flowing Counter-Current may be observed in all three oceans. The South Equatorial Current is, on average, the broadest and strongest. Often in the Atlantic and occasionally in the Pacific, it shows two cores of maximum velocity, one at latitude about 2° N and the other on the southern side of the Equator at about 3–5° S (cf. the two parts of the current in Figure 5.l(b)). The Equatorial Counter-Current is best developed in the Pacific, where it reaches its maximum speed between 5° and 10° N, some distance below the surface. In the Atlantic, the Equatorial Counter-Current is present throughout the year only in the eastern part of the ocean, between 5° and 10° N, where it is known as the Guinea Current. In both the Pacific and the Atlantic, there is also a weak South Equatorial Counter-Current which can be distinguished in the western and central ocean between about 5° and 10° S (not shown in Figure 5.1). For this reason, its counterpart to the north is often referred to as the North Equatorial Counter-Current.

As oceanographic instrumentation and techniques have advanced, it has been possible to distinguish finer details in the patterns of ocean circulation. Figure 5.2 shows a schematic section across the Equator in the central Pacific at 170° W, based on measurements made in the late 1970s and mid-1980s. This shows not only the equatorial currents mentioned above, plus the Equatorial Undercurrent (to which we will turn shortly), but also, beneath the Equatorial Undercurrent, North and South Subsurface Counter-Currents flowing eastwards either side of a strong westward-flowing Equatorial Intermediate Current, which is effectively an intensification of the South Equatorial Current. The values for volume transport given on this diagram (for different years and different longitudes) illustrate how variable the equatorial current systems can be.

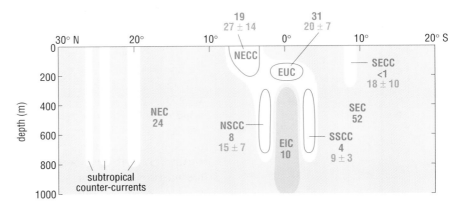

Figure 5.2 Schematic diagram to show the structure of the equatorial current system in the central Pacific, at 170° W, down to a depth of 1000 m (i.e. much deeper than in Figure 5.1(b)). Westward flow in the North and South Equatorial Currents (NEC and SEC) is shaded pale blue; strong westward flow (the Equatorial Intermediate Current, EIC) is darker blue. Eastward flow (including the South Equatorial Counter-Current, SECC) is unshaded; areas of strong eastward flow in the Equatorial Undercurrent (EUC), the North Equatorial Counter-Current (NECC) and the North and South Subsurface Counter-Currents (NSCC and SSCC) are outlined in blue. Note that a banded pattern is also shown for the northern subtropics, where counter-currents have been observed. The numbers are volume transports in sverdrups (10^6 m^3 s^{-1}). Those in red are for 155° W, based on data for April 1979–March 1980, and those in green are for 165° W, based on data for January 1984–June 1986.

The equatorial current system of the Atlantic does not have the generally simple east–west flows of the Pacific (Figure 3.1). This is partly because of the relative narrowness of the ocean basin (it is about one-third the width of the Pacific) combined with the influence of the shape of the African and American coastlines.

QUESTION 5.2   There is another reason, also connected with the shape of the continental landmasses, why surface currents in the tropical Atlantic are not simply east–west. Can you say what it is by looking at Figure 2.3?

5.1.1   THE EQUATORIAL UNDERCURRENT

The **Equatorial Undercurrent** (Figures 5.1(b) and 5.2) is a major feature of equatorial circulation. Such an undercurrent occurs in all three oceans, although it is only a seasonal feature in the Indian Ocean.

Equatorial Undercurrents flow from west to east, below the direct influence of the wind, yet they *are* wind-driven. How can this be?

The effect of the wind is transmitted downwards to deeper layers via turbulence (eddy viscosity) and is mainly confined to the mixed surface layer above the thermocline/pycnocline (Section 3.1.1). At high latitudes, winter cooling of surface waters causes them to become denser, destabilizing the upper part of the water column, so that it may more easily be mixed by wind and waves. At low latitudes, there is no winter cooling and the mixed surface layer is thin – as shown in Figure 5.1(b), in the vicinity of the Equator it may only be 50–100 m thick. In addition, as mentioned earlier, the pycnocline is a *sharper* boundary in low latitudes than in other regions; this is partly because surface heating is most intense there.

Remembering information given in Figure 2.2(b) and Section 2.3.1, can you give the other reason for the low density of surface waters in low latitudes?

The cumulonimbus activity associated with the ITCZ means that the equatorial zone is characterized by heavy rainfall, which significantly lowers the density of surface waters. Indeed, salinity has a greater control over the density distribution in the equatorial zone than in most other regions of the ocean.

Now study Figure 5.3 which shows a schematic east–west section across the equatorial ocean. The Trade Winds blow from east to west and drive a westward flow in a fairly well-defined mixed surface layer. As a result of this westward transport (and despite some return flow in the Equatorial Counter-Current(s)), water piles up against the western boundary of the ocean; there is therefore a sea-surface slope up towards the west, and a concomitant adjustment in the thermocline so that it slopes down to the west. Because the mixed layer is thin, the horizontal pressure gradient which results from the sea-surface slope extends to greater depths than the effect of the wind. Hence, although current flow 'down' the horizontal pressure gradient is opposed by the wind-driven current in the mixed surface layer, this is not the case *below* the mixed layer. As a result, in the thermocline there is an eastward jet-like current, which accelerates until mixing between it and surrounding water causes enough friction (i.e. eddy viscosity) for a balance, and hence a steady speed, to be attained. This jet-like current is the Equatorial Undercurrent (EUC).

The most powerful Equatorial Undercurrent is the one in the Pacific (see Figures 5.4 and 5.5). Its existence had been suspected as early as 1886, but it was not investigated properly until 1951 when it was fortuitously rediscovered by Townsend Cromwell and Ray Montgomery, who were on an expedition to study tuna. It is often referred to as the Cromwell Current.

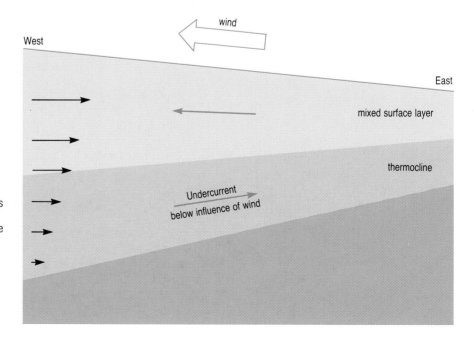

Figure 5.3 Schematic east–west section across the upper few hundred metres of the equatorial ocean, showing how the slope in the sea-surface caused by the Trade Winds, and the resulting horizontal pressure gradient, leads to the generation of the Equatorial Undercurrent. The lengths of the black arrows indicate the relative magnitude of the zonal horizontal pressure gradient. The vertical scale is greatly exaggerated.

Figure 5.4 shows the relationship between the mean wind stress and the resulting east–west slopes in the sea-surface and thermocline along the Equator in the Pacific (cf. the schematic diagram in Figure 5.3). In Figure 5.4(c), the position of the core of the Undercurrent (the region of highest velocity) is indicated by the blue crosses; note that in the eastern Pacific, where the thermocline is especially shallow, and upwelling brings cooler water to the surface, the flow in the Undercurrent may extend to the surface.

According to Figure 5.4, is the depth of the EUC shown in Figure 5.2 consistent with the location of the section?

Yes, it is. Figure 5.4 shows that at 170° W the thermocline is still fairly deep, so the core of the EUC is at about 200 m.

The Equatorial Undercurrent is very fast, especially in the eastern Pacific; velocities in the core are typically of the order of 1.5 m s^{-1}. Indeed, the Equatorial Undercurrent is the fastest component of the equatorial current system. In the context of its contribution to ocean circulation, however, it is not the speed of a current so much as its volume transport that is important. We can make an estimate of the volume transport of the Equatorial Undercurrent in the Pacific by reference to Figure 5.5(a), a meridional section showing the distribution of velocity in the vicinity of the Equator. Note that the vertical scale is greatly exaggerated so that the actual shape of the Undercurrent – that of a horizontal 'ribbon' of water – is not immediately apparent.

Figure 5.4 (a)   Mean westward wind stress along the Equator in the Pacific (negative values correspond to eastward wind stress), and (b) the dynamic height of the sea-surface, assuming no horizontal pressure gradient at 1000 m; remember that dynamic metres are numerically very similar to geometric metres. (c) Vertical distribution of temperature (°C) along the Equator between ~ 160° E and 94° E. The blue crosses indicate the position of the core of the Cromwell Current (cf. Figure 5.5(d)).

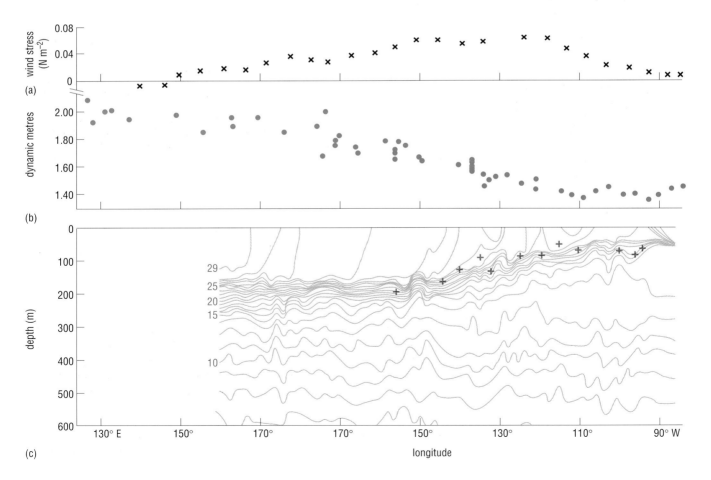

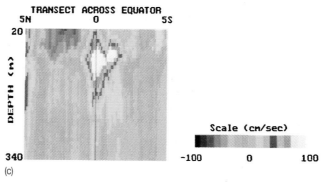

(c)

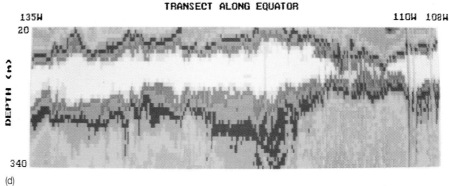

(d)

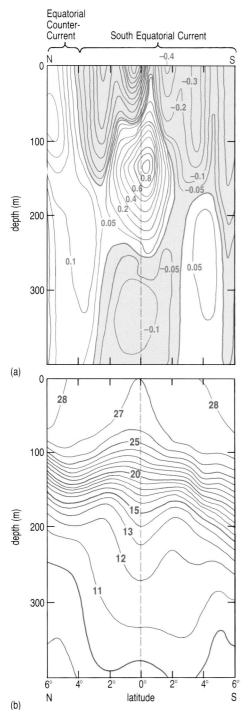

(a)

(b)

Figure 5.5 *Left:* Section across the equatorial Pacific along 150° W between about 6° N and 6° S, showing the distribution of (a) velocity (m s⁻¹; the blue area corresponds to westward flow) and (b) temperature (°C). Note that the core of the Cromwell Current is in the thermocline (cf. Figure 5.1(b)) and that in the region of the core, vigorous mixing causes the isotherms to spread apart (relative to their separation to the north and south), thus weakening the thermocline. The vertical scale is greatly exaggerated.
Above: Acoustic Doppler Current Profiler sections (c) across the Cromwell Current in the eastern Pacific at 110° W, and (d) along it, between 135° and 108° W. The core (pale grey area, corresponding to speeds of ~ 1 m s⁻¹) clearly shoals from west to east (cf. Figure 5.4(c)). Note that the tops of the sections do *not* correspond to the sea-surface but to a depth of 20 m.

QUESTION 5.3

(a) Use Figure 5.5(a) to estimate the approximate vertical extent of the Cromwell Current, and its width, given that 1° of latitude is about 110 km. How many times wider is this 'ribbon of water' than it is deep?

(b) Now estimate (i) the cross-sectional area of the Cromwell Current and (ii) its average velocity, and use this information to calculate an approximate volume transport in m³ s⁻¹. How does your estimated volume transport compare with the values given in Figure 5.2?

Figure 5.5(c) and (d) show transverse and longitudinal sections across the Cromwell Current, obtained by a shipborne Acoustic Doppler Profiler (cf. Section 4.3.7). A comparison of Figure 5.5(d) with the part of Figure 5.4(c) between 135° W and 108° W illustrates vividly how, as the resolution of observational techniques improves, so the flow patterns in the ocean are seen to be more and more complex.

How well do the areas of easterly and westerly current flow shown in Figure 5.5(a) and (c) tie up with those shown schematically in Figure 5.2?

They tie up pretty well. Eastward-flowing water is shown in blue in Figure 5.5(a) (as it is in Figure 5.2) and in various shades of blue in the ADCP section. Both the sections in Figure 5.5(a) and (c) show not only an Equatorial Undercurrent and North Equatorial Counter-Current, but also at greater depths, a South Subsurface Counter-Current and (merging with the North Equatorial Counter-Current) a North Subsurface Counter-Current. Both also suggest an Equatorial Intermediate Current below the Equatorial Undercurrent. The main differences seem to be in the extent to which the different eastward flows have merged or are clearly separate. (Rest assured that you will not be expected to remember all the details of the equatorial current system.)

Perhaps surprisingly, almost all the water in the Cromwell Current originates in the Southern Hemisphere. Much of it comes from the South Equatorial Current (which extends to significant depths – Figure 5.2) via a current that flows at depth along the northern part of the Great Barrier Reef, and then along the northern coast of New Guinea. In the eastern Pacific, some of the water leaving the Cromwell Current flows north, and some flows south, so contributing to both the North Equatorial Current and the South Equatorial Current. These flows are shown schematically in Figure 5.6.

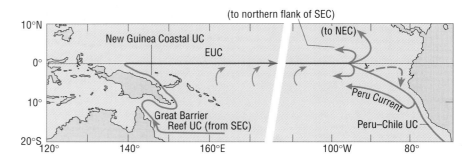

Figure 5.6   Schematic map to show how the Equatorial Undercurrent in the Pacific (the Cromwell Current) is fed at its western end by the New Guinea Coastal Undercurrent (which contains water from the South Equatorial Current) and itself feeds both the South Equatorial Current and the North Equatorial Current at its eastern end. Apart from the Peru Current, all the flows are subsurface (UC = Undercurrent). The dashed arrow represents a current that only flows from August to December.

In the Atlantic, the transport in the Equatorial Undercurrent is about one-third of that in the Cromwell Current. The Atlantic Equatorial Undercurrent was first observed in 1886 by John Buchanan, who had earlier participated in the *Challenger* Expedition. The Undercurrent was then forgotten about until it was rediscovered in 1959 – it is sometimes called the Lomonosov Current, after the vessel from which it was observed. This rediscovery was not accidental. Oceanographers were keen to know whether the Pacific Equatorial Undercurrent had a counterpart in the Atlantic: if none could be found, that would indicate that the Cromwell Current was caused by some peculiarity of the Pacific; if one *was* found, it would seem much more likely that an Undercurrent was a consequence of some fundamental aspect of equatorial circulation. It is now known that an Equatorial Undercurrent is indeed an intrinsic component of the equatorial circulation pattern but the explanation of its generation given above (Figure 5.3) probably represents only part of the story. As in the Pacific, the system of eastward equatorial currents in the Atlantic has two narrow, swift flows, either side of the main undercurrent in the thermocline and somewhat deeper – a current pattern too complex to be explained by the mechanism illustrated in Figure 5.3.

The Equatorial Undercurrent is generally aligned along the Equator even though the Trade Winds that drive it blow over a fairly wide latitude band. One of the reasons for this is that, on both sides of the Equator, the geostrophic flow resulting from the wind-induced eastward horizontal pressure gradient (Figure 5.3) is *towards* the Equator; once on the Equator the water can flow directly 'down' the pressure gradient.

Nevertheless, Equatorial Undercurrents do not always flow along the Equator. Their cores have been detected as much as 150 km away, and in all three oceans there have been observations indicating that the current as a whole may oscillate about the Equator, with wavelengths of the order of 1000 km. The initial displacement of the current may be caused by northerly or southerly winds, or by the presence of islands, but because the current is flowing east, when it strays from the Equator it is always deflected back by the Coriolis force: if it strays to the north, the Coriolis force acts to turn it to the south; if it strays to the south, the Coriolis force acts to turn it to the north. This is in contrast to the action of the Coriolis force on westward-flowing currents, which are always deflected polewards (cf. Figure 1.2(b)).

Before leaving this discussion of the main equatorial currents, look at Figure 5.7, a temperature section across the eastern Atlantic made during August and September 1963. For convenience, the 17 °C isotherm may be taken to correspond to the bottom of the thermocline.

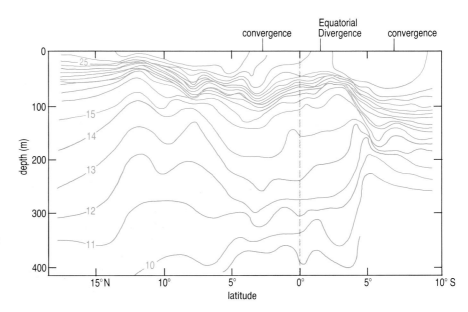

Figure 5.7   The vertical temperature distribution (in °C) across the Equator in the Atlantic at 20° W, plotted from measurements made in August–September 1963 from the research vessel *John Pillsbury*.

Compare Figure 5.7 with Figure 5.1(b), and use the slopes in the thermocline (which are in the opposite sense to the slopes in the sea-surface) to try to identify (1) the North Equatorial Current, (2) the North Equatorial Counter-Current and (3) the South Equatorial Current (ignoring for the moment their precise positions relative to the Equator).

The two regions of westward flow that make up the South Equatorial Current may be seen in the region above the thermocline between about 8° S and the Equatorial Divergence (thermocline sloping up towards the north), and between the Equatorial Divergence and about 3° N (thermocline sloping down towards the north). The change in the direction of slope of the thermocline occurs a few degrees to the south of the Equator – this probably means that when the observations were made, the South Equatorial Current had recently shifted with respect to the Equator and geostrophic equilibrium had yet to be re-established. The region of westward flow corresponding to the North Equatorial Current may be seen between about 12° and 18° N (thermocline sloping downwards to the north).

Note that two counter-currents are visible: the North Equatorial Counter-Current flows between the North and South Equatorial Currents in roughly the position shown in Figure 5.1(b) (thermocline sloping down towards the south), and the South Equatorial Counter-Current (upper part of the thermocline sloping up to the south) may be seen south of about 8° S.

Is it possible to distinguish the Equatorial Undercurrent?

No obvious spreading of the isotherms can be seen in the upper part of the thermocline in the region of the Equator, but the spreading of the 13–17 °C isotherms *is* probably related to vigorous mixing associated with the Equatorial Undercurrent. Such well-mixed regions where temperature and density vary little with depth are sometimes referred to as thermostads or **pycnostads**. A similar thermostad/pycnostad may be seen on Figure 5.5(b) in the region of the Equator, between about 200 m and 400 m depth.

Another feature visible in Figure 5.7 is the region of 'doming' isotherms at about 12° N. This is known as the 'Guinea Dome', and will be discussed in the next Section.

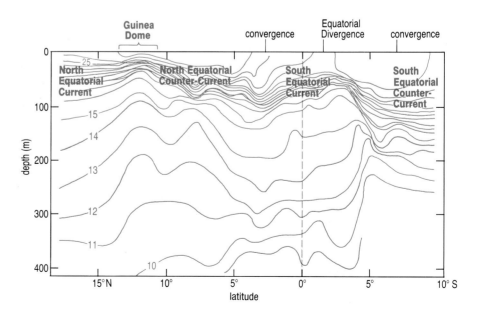

Figure 5.8   As for Figure 5.7, but with the regions corresponding to the various currents identified. (The doming of isotherms at about 12° N is known as the Guinea Dome, and will be discussed in Section 5.1.2.)

Figure 5.8 repeats Figure 5.7, but here the various regions of current flow have been labelled. Do not worry if you found it hard to interpret Figure 5.7 – its rather 'messy' contour pattern illustrates the extent to which diagrams such as Figures 5.1(b) and 5.2 are simplifications of the complex distributions observed in reality.

By now it will be clear that although the equatorial current systems of the Atlantic and Pacific Oceans share certain recognizable features, there are also differences between them. Furthermore, at any one time and place the actual patterns may look very different from the cross-sections shown here. Changes in the extent, speed and even the direction of the currents occur in response to changes in the overlying wind pattern, especially changes in the position of the ITCZ. This is particularly true in the western Pacific Ocean and (as you will see in Section 5.2) the Indian Ocean, where seasonally reversing winds known as Monsoons have dramatic effects on current speeds and directions.

5.1.2 UPWELLING IN LOW LATITUDES

Figure 5.9 is a schematic map showing the different upwelling regions of the eastern tropical Atlantic. A comparable map could be drawn for the eastern tropical Pacific.

Figure 5.9 The positions of the various upwelling areas in the tropical Atlantic, in relation to the eastward subsurface flow in the equatorial current system (the westward-flowing North and South Equatorial Currents have been omitted for the sake of clarity). The upper part of the map may be compared with Figure 4.38. Note that to the north of the Cape Blanc zone of year-round upwelling there is another zone of seasonal upwelling (off the map).

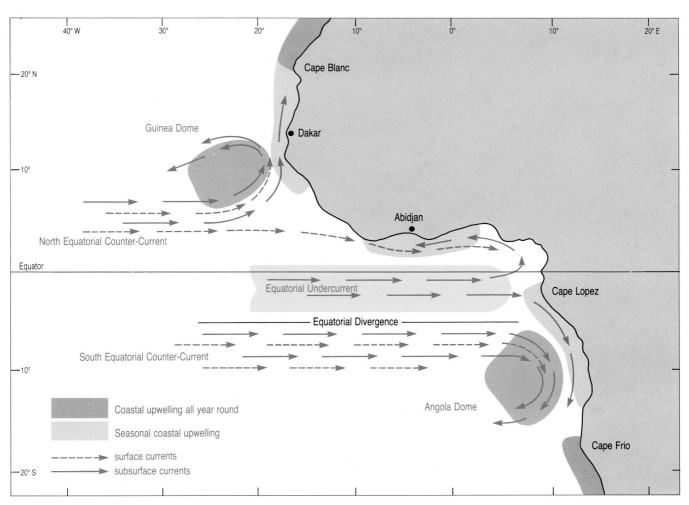

QUESTION 5.4

(a) Why does coastal upwelling occur all year round to the north of Cape Blanc and to the south of Cape Frio, but only seasonally between these latitudes?

(b) By reference to Figure 5.1, explain why there is upwelling in the region indicated by the band of pale blue shading aligned along the Equator.

Mid-ocean equatorial upwelling, like coastal upwelling, is subject to seasonal variations. The main period of mid-ocean upwelling in the eastern equatorial Atlantic is from July to September. This upwelling maximum is *partly* the result of increased divergence of surface water in response to the seasonal increase in the strength of the South-East Trade Winds (Figure 5.1(a)). However, the main reason is probably that this increase in wind strength increases the flow in the South Equatorial Current, so that the north–south slope in the thermocline becomes *so* steep that it intersects the surface (cf. Figure 5.1(b)), allowing deeper water to come to the surface.

Remember also that in low latitudes the thermocline slopes down towards the western coastal boundary, in response to the generally westward wind stress of the Trades (Figures 5.3 and 5.4). Because the mixed surface layer is therefore thinner in the eastern ocean, it is more easily affected by the wind, and sub-thermocline water may more readily be brought up to the surface. This is the main reason why mid-ocean upwelling is a feature of the *eastern* equatorial oceans in particular.

In certain localities of the eastern tropical Atlantic and Pacific, the isotherms bow up into a dome-like shape. Figure 5.9 shows the position of the Guinea Dome (also visible on Figure 5.8) and the Angola Dome; a similar dome in the tropical Pacific is named the Costa Rica Dome. These *thermal domes* seem to have a number of causes. Their existence is certainly linked to strong subsurface cyclonic flow patterns: Figure 5.9 shows how the positions of the Guinea Dome and the Angola Dome relate to subsurface flow in the North and South Equatorial Counter-Currents, while Figure 5.8 shows how the Guinea Dome corresponds to the northern edge of the region of sloping isotherms associated with the North Equatorial Counter-Current.

The degree of 'doming' varies seasonally. During the northern summer when the ITCZ is in its most northerly position, the Guinea Dome protrudes up into the thermocline, but the Angola Dome is weak. In the southern summer, the situation is reversed: the Guinea Dome no longer distorts the thermocline, and the Angola Dome is at its most prominent (although it is never as strong a feature as the Guinea Dome).

QUESTION 5.5   Winds are generally light and variable in the vicinity of the ITCZ. In some circumstances, however, they do significantly affect surface waters, particularly in regions where the atmospheric pressure is especially low.

(a) Will winds be cyclonic or anticyclonic around such regions?

(b) How does this help to explain the fact that the Guinea Dome 'protrudes' into the surface waters when the ITCZ is in its most northerly position?

As in other areas of upwelling, productivity of phytoplankton and hence of other organisms is enhanced in the surface waters above thermal domes.

To round off this Section, look at Figure 5.10 which shows the distribution of sea-surface temperature in the tropical Atlantic and, for comparison, the variation in the depth of the 21 °C isotherm (which corresponds roughly to the middle of the thermocline) at the same time of year.

Does Figure 5.10 show the situation in the northern summer (July to September) or the southern summer (January to March)?

Figure 5.10(a) shows that the surface ocean is generally warmer to the north of the Equator than to the south. Furthermore, Figure 5.10(b) shows that the thermocline is very shallow along the Equator in the eastern Atlantic; more importantly, it shows that the thermocline is also shallow along the coast of the Gulf of Guinea and that the Guinea Dome (at ~10° N) is well developed. These upwelling-related features may also be seen to some extent in Figure 5.10(a). The maps must therefore show the situation in July to September, rather than January to March.

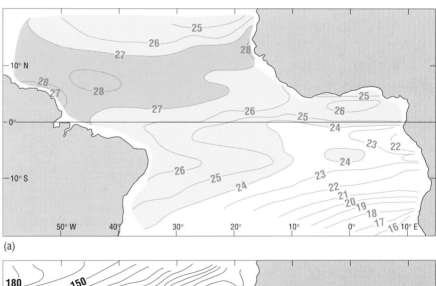

(a)

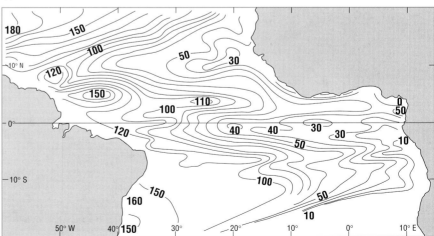

Figure 5.10 (a)   The mean sea-surface temperature (°C) in the tropical Atlantic at one season of the year.
(b)   The average depth (in metres) of the 21 °C isotherm (which corresponds approximately to the middle of the thermocline) in the tropical Atlantic at the same season of the year as (a).

(b)

5.2   MONSOONAL CIRCULATION

The word 'monsoon' is derived from an Arab word meaning 'winds that change seasonally'.

From Figure 2.3, which regions of the ocean are affected by monsoons?

The most immediately obvious region is the Indian Ocean, north of about 15° S. Also affected by seasonally reversing winds are the westernmost part of the Pacific Ocean, including the regions of the Malaysian and Indonesian archipelagos and, to a lesser extent, the west coast of Africa and the Caribbean. Here, we will concentrate on the monsoonal reversals over the Indian Ocean.

5.2.1   MONSOON WINDS OVER THE INDIAN OCEAN

As discussed in Section 2.1, the winds over the Indian Ocean change dramatically with the seasons. In the northern winter (Figure 2.3(b)), the air over southern Asia is cooler and denser than air over the ocean, and so the surface atmospheric pressure is greater over the continent than over the ocean. The resulting pressure gradient leads to a low-level northerly or north-easterly flow of air from the Asian landmass to south of the Equator. This flow of air is the North-East Monsoon. After crossing the Equator, the flow is turned to the left by the Coriolis force and converges with the South-East Trades at about 10–20° S. As the year progresses, the Asian landmass heats up and the high pressure over southern Asia weakens. By May/June, a low has developed; suddenly the wind direction changes, and a southerly or south-westerly wind blows across the region until September. This is the South-West Monsoon, the stronger of the two monsoons.

During which season of the year, i.e. during which monsoon, does the wind pattern over the Indian Ocean most resemble that in the other two oceans?

Figure 2.3(a) and (b) show that the wind pattern over the Indian Ocean most resembles that over the other two oceans in the southern summer/northern winter. At this time of year, the North-East Monsoon blows across the ocean, and the wind field resembles that associated with the North-East Trade Winds.

The different types of weather characteristic of the two monsoons are well known. In January and February, during the North-East Monsoon, the winds bring dry cool air to India from the Asian landmass. In May and June, during the South-West Monsoon, the winds cross the Arabian Sea and bring humid maritime air to India. The moisture that provides the heavy monsoon rains is partly a direct result of evaporation from the warmed surface of the Arabian Sea, and partly the result of upward convection of warm moist air above the Arabian Sea, which leads to the formation of cyclonic vortices which draw in more moisture-laden air from adjacent regions (cf. Section 2.3.1).

The monsoonal nature of the winds may also be thought of as a manifestation of the seasonal shift in the position of the ITCZ, from about 20° S in January to about 25° N, over Asia, in July. The summer convection over Asia is driven not only by direct heating by the warm continental mass below, but also by the condensation of moisture from air rising over high land (particularly the southern Himalayas), to form the monsoon rains. This releases latent heat originally taken up from the ocean as latent heat of evaporation, warming the air, causing it to rise still further, and intensifying the convection.

A particularly interesting aspect of the South-West Monsoon is the appearance over the western side of the ocean of an intense, southerly low-level atmospheric jet (Figure 5.11). This resembles an oceanic western boundary current, although here it is the high tablelands of eastern Africa that form the western boundary.

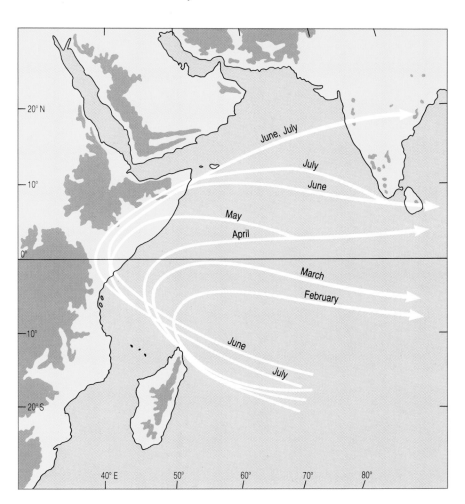

Figure 5.11 Monthly positions of the core of the low-level (1–2 km) atmospheric jet over the Indian Ocean. Brown shading represents land higher than 1 km above sea-level.

5.2.2 THE CURRENT SYSTEM OF THE INDIAN OCEAN

As you might expect, the surface circulation of the northern Indian Ocean changes seasonally in response to the monsoons, but most resembles that of the other two oceans in the northern winter, during the North-East Monsoon (cf. Section 5.2.1). At this time of year, both a North and a South Equatorial Current are present, as well as an Equatorial Counter-Current (Figure 5.12(a),(b)). In the northern summer, by contrast, the flow in the North Equatorial Current reverses and combines with a weakened Equatorial Counter-Current to form the South-West Monsoon Current (Figure 5.12(d),(e)). The South Equatorial Current is still present, although its flow is not as strong as during the North-East Monsoon.

158

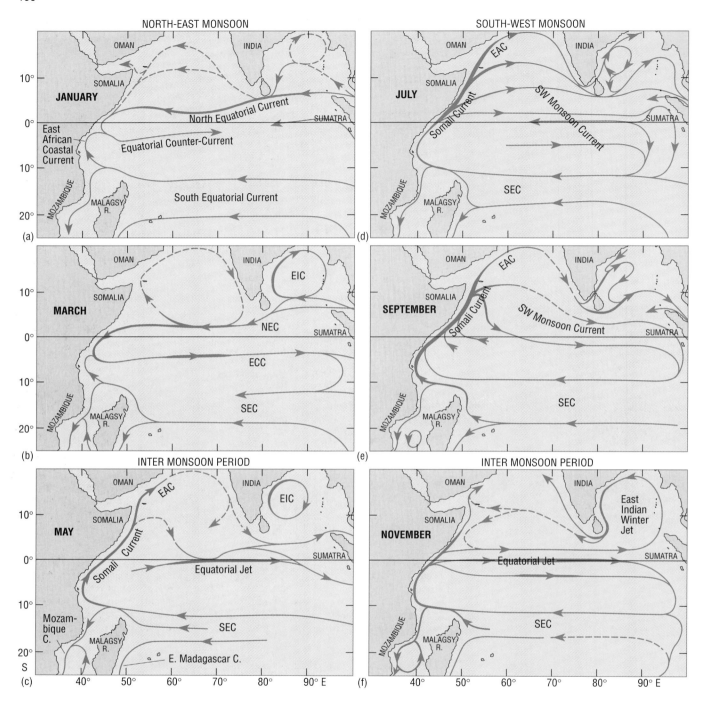

Figure 5.12 Surface currents in the northern Indian Ocean, as deduced from ships' drift data. The North-East Monsoon is most fully established from January to March ((a),(b)), and the South-West Monsoon is most fully established during July–September ((d),(e)). The thicknesses of the lines indicate the relative intensities of the flows. For example, current speeds in the Equatorial Jet ((c),(f)) may reach 1.0–1.3 m s^{-1}, but are mostly 0.3–0.7 m s^{-1}. The South Equatorial Current (SEC) is spread over the area between the two flow lines. The westward flow along the Equator within the South-West Monsoon Current, shown in (d), is discussed in Section 5.3.1. EAC is the East Arabian Current.

The most spectacular seasonal change is the reversal of the Somali Current, off east Africa. This reversal has been known about for centuries. In the ninth century, Ibn Khordazbeh noted (in sailing instructions for masters of dhows carrying slaves from east Africa to Oman) that 'the sea flows during the summer months to the north-east', and 'during the winter months to the south-west'. So, during the North-East Monsoon the Somali Current flows to the south-west, but for the rest of the year it flows to the north-east. During the South-West Monsoon it becomes a major western boundary current (Figure 5.12(d),(e)) – its surface velocity may reach 3.7 m s^{-1}, and the volume transport in the upper 200 m of its flow is about 60×10^6 m^3 s^{-1}.

The low-level atmospheric jet mentioned in the previous Section (Figure 5.11) is thought to play an important role in the generation of the intense coastal upwelling that occurs off Somalia during the South-West Monsoon. At this time of year, upwelling also occurs off Arabia, both along the coast where the north-easterly current diverges from it, and offshore in response to local cyclonic winds. The Somali and Arabian regions of upwelling are the most vigorous in the Indian Ocean (Figure 5.13).

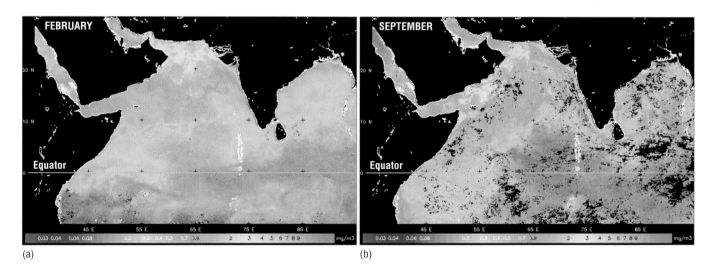

(a) (b)

Figure 5.13 Satellite images showing chlorophyll concentrations in surface waters of the northern Indian Ocean in (a) February and (b) September. Note the high chlorophyll concentrations (orange/yellow) off Somalia and Oman in September, resulting from upwelling associated with the South-West Monsoon. (The high chlorophyll concentrations close to the coasts of India and Bangladesh are a result of nutrient-rich river outflows.)

Would you expect open-ocean upwelling to occur in the region of the Equator, as it does in the Pacific and Atlantic?

No, because these regions of upwelling occur as a result of the South-East Trade Winds blowing across the Equator and causing a surface divergence just to the south of it (Figure 5.1(a)). Figure 2.3 shows that these wind conditions are not typical of either season of the year.

Would you expect there to be an Equatorial Undercurrent in the Indian Ocean?

Only for part of the year, not as a permanent feature. The existence of an Undercurrent depends on there being wind stress towards the west, to drive a westward surface current and hence cause a sea-surface slope up to the west; this would provide an eastward horizontal pressure gradient force to drive an eastward current below the wind-driven layer. In the equatorial Indian Ocean, the direction of the wind varies seasonally, but best fulfils the necessary conditions for an Undercurrent during the first few months of the year. The first direct observations of the Undercurrent were made during the International Indian Ocean Expedition (1962–65) at the end of the North-East Monsoon, when it was seen to be an ocean-wide feature; but this may not have been typical. In general, in the Indian Ocean the Undercurrent seems to be a stronger and more persistent flow in the western part of the ocean than in the central or eastern parts.

The changing pattern of surface currents in Figure 5.12 was deduced from ships' drift data collected by the UK Meteorological Office from log books of merchant vessels. Such information can provide reasonably accurate estimates of surface flow velocities in the Indian Ocean because, for most of the year, currents here are generally stronger than those in the Pacific and Indian Oceans.

What intense surface current feature appears *between* the monsoons (Figure 5.12(c) and (f))?

It is an *eastward Equatorial Jet*, which is driven by westerly winds over the central equatorial ocean. Although the jet is detected in ships' drift data between April and June, and in October/November, it is possible that it is in fact a brief event, lasting perhaps only a month at a time.

Generally speaking, how would you expect the direction of the slopes of the sea-surface and of the thermocline to change in response to the changing directions of the wind and surface current along the Equator?

When winds and currents along the Equator become westward, the sea-surface will (eventually) slope up to the west, and the thermocline slope down to the west; when winds and currents along the Equator are eastward, the sea-surface will (eventually) slope up to the east, and the thermocline slope down to the east.

The complexity of the surface circulation of the Indian Ocean, which contrasts with the relative simplicity of the gyral systems of the Pacific and Atlantic Oceans, is a result of the frequency and rapidity with which the overlying wind system changes. Wind speeds and directions change so fast that there is not always time for the upper ocean to adjust so that it is in equilibrium with the wind – as a result, during the relatively short inter-monsoon period, while the prevailing wind directions are changing dramatically, there is an increase in the large number of eddies, both cyclonic and anticyclonic and ranging in size from 100 to 1000 km across. The ocean's response time – or, looked at another way, its 'memory' – is many times longer than that of the atmosphere, and one of the most interesting questions that can be asked about the ocean circulation is: How is it possible for the ocean to react as fast as it does?

We will address this question in the next Section, but we cannot leave the Indian Ocean without considering a western boundary current which is even more powerful than the Somali Current during the South-West Monsoon. This is the Agulhas Current, second only to the Gulf Stream in its volume transport. Flowing polewards along the coast of Africa from south of 20° S and the island of Madagascar (the Malagasy Republic), the Agulhas Current seems mostly to derive *not* from the Mozambique Current (as might be expected), but from the East Madagascar Current, which is itself quite a strong western boundary current (both are labelled on Figure 5.12(c)). Unlike other western boundary currents, the Agulhas Current shows little seasonal variation, averaging speeds of 1.6 m s^{-1} throughout the year, and exceeding 2.5 m s^{-1} in most months. Near 31° S, transport in the current is estimated to be 70×10^6 m^3 s^{-1}, but it increases by approximately 6×10^6 m^3 s^{-1} for every 100 km travelled, so by the time it approaches the shallow Agulhas Bank at about 35° S its volume transport has grown to between 95×10^6 m^3 s^{-1} and 135×10^6 m^3 s^{-1}.

Off the tip of South Africa the Agulhas Current loops back on itself, forming a feature known as the 'Agulhas retroflection'. As shown in Figure 5.14, the retroflection is a source of 'ring-like' eddies (cf. Section 4.3.6), which form when the westerly part of the loop 'pinches off'. Most of the eddies, which are rings of Agulhas Current water encircling Indian Ocean water, are injected into the Benguela Current and are carried away north-westwards into the Atlantic. They are highly energetic and are thought to have life-spans of many years.

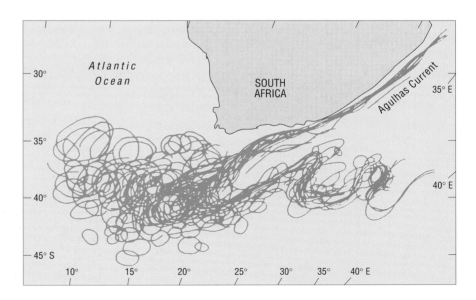

Figure 5.14   Map showing the changing path of the Agulhas Current and its current loop or retroflection off South Africa. The blue line in fact corresponds to different positions of the temperature front on the coastal side of the (warm) current over a 12-month period during 1984–85. The rings are eddies which form from the loop when it extends westwards towards the Atlantic; when an eddy pinches off, the loop retreats to its most easterly position, and the cycle repeats.

5.3 THE ROLE OF LONG WAVES IN OCEAN CIRCULATION

The changes in the slopes of the sea-surface and thermocline along the Equator in the Indian Ocean, mentioned in Section 5.2.2, occur surprisingly fast. Exactly how fast the ocean can respond to seasonal changes in the wind has been studied in the simpler and steadier Atlantic, where the stress of the South-East Trade Winds across the equatorial ocean causes the sea-surface to slope up towards the west, and the thermocline to slope down (cf. Figure 5.3).

In the Atlantic, the South-East Trades are at their weakest during March–April and at their strongest during August–September. Figure 5.15 shows how the slopes of the sea-surface and thermocline across the Atlantic basin vary over the course of the year.

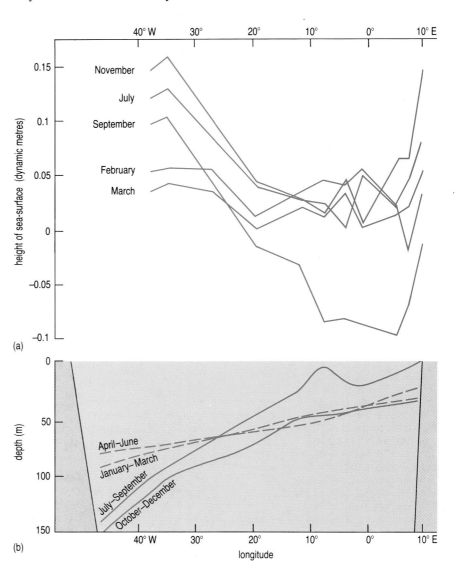

Figure 5.15 (a) Seasonal variations in the height of the sea-surface across the equatorial Atlantic. (The curves appear jagged because they are based on mean values at particular longitudes.)
(b) Seasonal variations in the depth of the 23 °C isotherm across the equatorial Atlantic.

Do the minimum and maximum slopes of the sea-surface and thermocline correspond to the periods of minimum and maximum wind stress, respectively?

Yes, they do, as far as the monthly mean values in Figure 5.15 allow us to tell. The important point about this is not the correlation itself but the fact that it indicates that the upper waters of the Atlantic Ocean *respond very quickly indeed* to changes in the overlying wind field.

This fast response time cannot be explained simply in terms of water being transported across the equatorial Atlantic and 'piling up' at the western boundary: the upper ocean as a whole must somehow adjust to the overlying wind and to the fact that there is a boundary along the western side of the ocean. In other words, the ocean in the central and eastern parts of the basin must in some sense 'know about' or 'have felt' the western boundary. The way in which the mid-ocean receives information about the existence of a boundary is through perturbations or disturbances that travel through the ocean as pulses or waves. This is analogous to the way in which *we* receive information concerning the world about us – through light waves or sound waves.

These wave-like disturbances not only enable water in mid-ocean to respond to the existence of coastal boundaries, they also transmit the effects of changes in the overlying wind field from one region of ocean to another, and do so much faster than would be possible simply through transportation of water in wind-driven currents. It is believed that the speed of the reversal of the Somali Current is one example of such an effect: as described in Section 5.2.2, the surface waters off Somalia flow fairly fast south-westwards during the North-East Monsoon, but can nevertheless be moving *very* fast in the opposite direction only a few months later. This dramatic 'switch' is thought to be possible because the upper ocean off eastern Africa 'feels' the effect of winds blowing in the central region, as well as being directly driven by local winds.

Various different types of waves may be generated in the ocean. We are all familiar with the wind waves which occur at the surface of the ocean, in which particles of water are displaced from their 'normal' or equilibrium position in a vertical direction and return under the influence of gravity. The types of gravity waves that are of interest as far as ocean circulation is concerned have wavelengths which may be anything from tens to thousands of kilometres, periods ranging from days up to months, or even years, and vertical displacements which vary from centimetres up to tens of metres. These vertical displacements may be more or less constant with depth (Figure 5.16(a)); alternatively, they may be greatest where there is a strong vertical density gradient – in the thermocline, for example (Figure 5.16(b)). In the first case, the vertical density distribution is not affected by the passage of the wave, and so if isobaric and isopycnic surfaces are initially parallel to one another, they will remain so (Figure 5.16(a)). For this reason, such waves are sometimes referred to as 'barotropic waves'; we will refer to them as 'surface waves' although it should be remembered that motion associated with them extends to significant depths, because their wavelengths are much greater than the depth of the ocean. In the second case, the vertical density distribution is affected by the passage of the waves, so that density surfaces are caused to intersect isobaric surfaces (Figure 5.16(b)); such waves are therefore referred to as 'baroclinic' waves. Baroclinic waves generally have much larger amplitudes than surface (barotropic) waves.

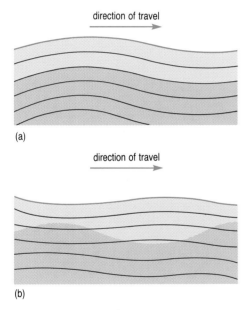

direction of travel

(a)

direction of travel

(b)

Figure 5.16   Examples of (a) a 'surface' long wave and (b) a long wave in the thermocline, here shown as a sharp boundary between less dense surface water (light greenish-blue) and more dense deeper water (darker blue). In (a), the surface ocean as a whole moves up and down, and isobaric and isopycnic surfaces remain parallel. Such waves are therefore described as 'barotropic'. In (b), the passage of the wave changes the vertical density distribution, so that isopycnic surfaces are alternately compressed and separated. In addition, there are pressure variations over the surface of the density interface (the light greenish-blue/dark blue boundary) so that isobaric and isopycnic surfaces intersect; such waves are therefore described as 'baroclinic'.

Because these disturbances have very long wavelengths and periods, they are significantly affected by the Coriolis force. As a result, motion occurs in a *horizontal* as well as a vertical direction. This may be seen most clearly in the flow patterns associated with **Kelvin waves** and planetary or **Rossby waves**, the two classes of waves that are most important as far as ocean circulation is concerned.

5.3.1 OCEANIC WAVE GUIDES AND KELVIN WAVES

A general feature of all wave motions, through water or any other medium, is that where the physical characteristics of the medium change with position, waves seeking to cross the line of change may be reflected, or in some other way deflected, so that they become trapped within a **wave guide**. Common examples of wave guides employed in communications technology are optical fibres and coaxial cables, both of which are used to carry information along a specified path. An example of a wave guide in the ocean is the sound channel, a depth zone where the velocity of sound in seawater is relatively low, and within which sound waves may be trapped by refraction (Section 4.3.3). Sounds emitted in the sound channel – by, for example, Sofar floats – are transmitted over distances of thousands of kilometres with relatively little attenuation.

Now imagine a parcel of water in the Northern Hemisphere, moving northwards with a coastal boundary on its right. The Coriolis force continually tends to deflect the parcel to the right, but because the coastal boundary is in the way, only limited deflection is possible. Water piles up against the boundary, giving rise to an offshore horizontal pressure gradient force; this keeps the parcel of water moving parallel to the coast, in a geostrophic current. Consequently, a coastal boundary constrains the way in which water can move in response to the forces acting on it. As a result, coasts may act as wave guides to the perturbations known as Kelvin waves (Figure 5.17(a)), which can travel as surface (barotropic) waves or as baroclinic waves.

In a Kelvin wave, perturbations of the sea-surface or of the thermocline propagate parallel to and close to the coast, as though unaffected by the Earth's rotation, because the Coriolis force directed towards the coast is opposed by a horizontal pressure gradient force that results from the slope of the sea-surface. Thus, a necessary condition for the propagation of Kelvin waves is that the horizontal pressure gradient force and Coriolis force act in opposition.

QUESTION 5.6 Given that necessary condition, is it possible for a Kelvin wave in the Northern Hemisphere to propagate with the coastal boundary on the left? Or for a Kelvin wave in the Southern Hemisphere to propagate with the coastal boundary on the right?

Kelvin waves are similar to surface wind waves in that the principal maintaining force is gravity. Particle movement within the wave is such that the amplitude of the vertical displacement is greatest at the coast and decreases exponentially away from it, so that at any point and any time the Coriolis force balances the pressure gradient resulting from the slope of the sea-surface (or thermocline) (Figure 5.17(b)).

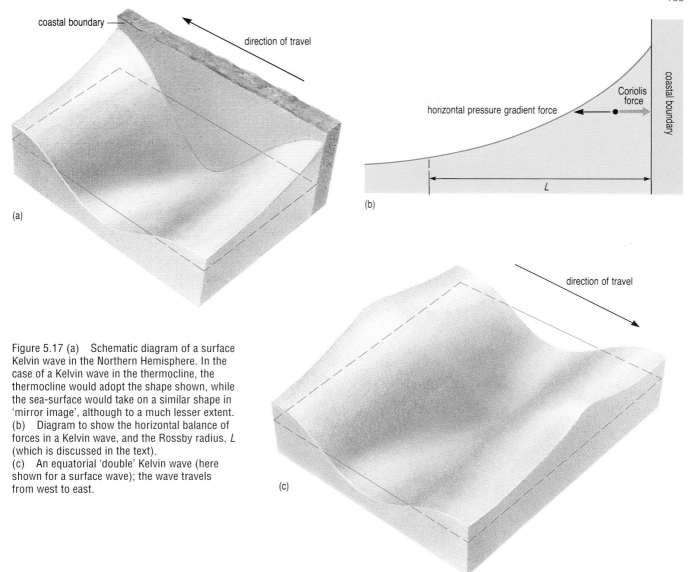

Figure 5.17 (a)   Schematic diagram of a surface Kelvin wave in the Northern Hemisphere. In the case of a Kelvin wave in the thermocline, the thermocline would adopt the shape shown, while the sea-surface would take on a similar shape in 'mirror image', although to a much lesser extent.
(b)   Diagram to show the horizontal balance of forces in a Kelvin wave, and the Rossby radius, *L* (which is discussed in the text).
(c)   An equatorial 'double' Kelvin wave (here shown for a surface wave); the wave travels from west to east.

A Kelvin wave may be regarded as being 'trapped' within a certain distance of the coast, because by that distance its amplitude has significantly decayed away. This distance is known as the **Rossby radius of deformation** (L) and can be calculated from $L = c/f$ where f is the Coriolis parameter and c is the wave speed. In mid-latitudes, the Rossby radius for a Kelvin wave with its maximum displacement in the thermocline is generally about 25 km.

QUESTION 5.7   What is the Rossby radius L for a Kelvin wave in the thermocline at 10° N, if its speed is 2 m s^{-1}? (You will need to use the fact that $\Omega = 7.29 \times 10^{-5}$ s^{-1}.)

Thus, in low latitudes, coastal Kelvin waves are not as closely trapped to the coast as they are in mid-latitudes.

The Rossby radius of deformation is not simply a measure of the degree to which Kelvin waves are trapped. More generally, it is the distance that a wave with speed c can travel in time $1/f$, and it therefore provides a guide to the distance that a wave (i.e. a disturbance) can travel before being significantly affected by the Coriolis force. Because f is zero at the Equator and a maximum at the poles, Rossby radii decrease from infinity at the Equator to a minimum at the poles. The tendency for disturbances in current patterns to take on a curved or gyral character therefore *increases* with increasing distance from the Equator, and as a result, the higher the latitude, the smaller ('tighter') eddies, or other wave-like disturbances, tend to be. (This effect can be thought of as a manifestation of the increase in planetary vorticity with latitude, as illustrated in Figure 4.7(a).)

Almost everybody has experienced the effects of a Kelvin wave at first hand. The twice-daily rise and fall of sea-level corresponding to high and low tide occurs in the form of coastal Kelvin waves, which progress anticlockwise round ocean basins (i.e. with the coast on the right) in the **amphidromic systems** of the Northern Hemisphere, and clockwise round basins in the Southern Hemisphere. (*Note*: Tides and wind-generated waves are discussed elsewhere in this Series.)

We have seen that coastal Kelvin waves may travel along coastal boundaries because the Coriolis force cannot play the same part in the balance of forces as it usually does. Along the Equator, the Coriolis force actually *is* zero and a similar effect arises, leading to the existence of an **equatorial wave guide**. An equatorial Kelvin wave is like two parallel coastal Kelvin waves (one in each hemisphere) joining at the Equator, which they 'feel' as a boundary (Figure 5.17(c)). Like coastal Kelvin waves, equatorial Kelvin waves propagate with the 'boundary' on the right in the Northern Hemisphere and on the left in the Southern Hemisphere. As a result, Kelvin waves can only propagate *eastwards* along the equatorial wave guide.

Surface equatorial Kelvin waves travel very fast, at about $200 \, \mathrm{m \, s^{-1}}$. Their Rossby radius of deformation is about 2000 km, and so they can hardly be regarded as 'trapped' at all. However, this is not the case for Kelvin waves in the thermocline. They travel much more slowly, with c typically between 0.5 and $3.0 \, \mathrm{m \, s^{-1}}$, and have Rossby radii of 100–250 km. They may be detectable at the surface, as sea-level is slightly raised above regions where the thermocline is depressed, and slightly depressed above regions where the thermocline is raised.

Kelvin waves in the thermocline can have dramatic effects, particularly in low latitudes where the mixed surface layer is thin. They may be generated by an abrupt change in the overlying wind field, as occurs for instance in the western Atlantic when the ITCZ moves northwards over the region and it comes under the influence of the South-East Trades. This causes a disturbance in the upper ocean (cf. Figure 5.18(a)), which travels eastwards along the equatorial wave guide as a double Kelvin wave in the thermocline (this takes about 4–6 weeks) and, on reaching the coast, splits into two coastal Kelvin waves, each travelling away from the Equator (Figure 5.18(b)). In the region of the disturbance where the thermocline bulges upward, colder, deeper water comes nearer to the surface. By the time the wave reaches the coasts of Ghana and Ivory Coast (north of the Equator) and Gabon (south of the Equator), this cooler, sub-thermocline water is detectable at the surface, contributing to the 'seasonal coastal upwelling' off Abidjan and south of Cape Lopez, shown in light green in Figure 5.9.

Why is this 'pulse' not detected earlier, as it moves across the western and central equatorial Atlantic?

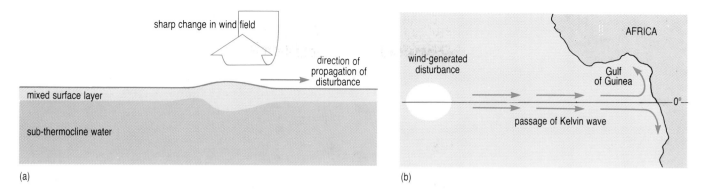

(a)

(b)

Figure 5.18 (a) Schematic diagram showing a disturbance of the upper ocean caused by an abrupt change in the overlying wind field. (b) Such a disturbance may be generated in the western Atlantic and travel eastwards as an equatorial Kelvin wave; at the eastern boundary, this splits into two coastal Kelvin waves, which contribute to seasonal upwelling in the Gulf of Guinea (cf. Figure 5.9).

As discussed in Section 5.1.1, the equatorial thermocline slopes up from the west and is nearest to the surface at the eastern coast. The passage of the Kelvin wave only brings cooler, sub-thermocline water to the surface where the thermocline is already fairly shallow.

In the Indian Ocean, it is the equatorial wave guide that permits such swift responses to changes in wind direction. At the start of the South-West Monsoon, westerly winds in the western part of the ocean cause water to move away from the African coast, lifting the thermocline. The resulting upward bulge in the thermocline travels east as a Kelvin wave (Figure 5.18), accompanied by a strong eastward flow of surface water, until it reaches Sumatra, thus bringing about a sea-surface slope up to the east, and a slope in the thermocline down to the east. Because eastward flow does not start simultaneously all the way along the Equator, but propagates along the Equator as a wave, at certain times of year there are areas of both eastward *and* westward flow on the Equator (Figure 5.12(a) and (d)).

When the winds over the Indian Ocean change from westerlies to northerlies/north-easterlies at the end of the South-West Monsoon, the upper ocean begins to 'rearrange itself' and in less than a week the sea-surface slope up to the east is beginning to decline. This change is being brought about not by means of Kelvin waves, but by Rossby or planetary waves.

5.3.2 ROSSBY WAVES

While Kelvin waves carry information across ocean basins from west to east, Rossby or planetary waves, which are slower, carry information from east to west. They propagate zonally along lines of latitude, and arise from the need for potential vorticity to be conserved (Section 4.2.1). Imagine a parcel of water at some given latitude ϕ in the Northern Hemisphere and assume that, initially, it has no relative vorticity – i.e. no rotational movement in relation to the Earth, and no current shear. If the parcel of water is displaced polewards, it enters a region of higher positive planetary vorticity, f (Figure 5.19(a)(i)). Its potential vorticity $(f + \zeta)/D$ must remain constant, and so to compensate for its gain in f the water parcel will acquire negative relative vorticity ζ and will tend to rotate clockwise, causing adjacent water also to move clockwise. If it moves southwards and overshoots its original latitude, its loss in planetary vorticity will be compensated for by a gain in positive relative vorticity, and it will rotate anticlockwise (Figure 5.19(a)(ii)). A water parcel may swing back and forth about its original latitude, alternately gaining planetary vorticity while losing relative vorticity, and losing planetary vorticity while gaining relative vorticity.

Now imagine a row of such water parcels in a current or airstream flowing along a line of latitude. If the flow is displaced polewards or equatorwards, horizontal oscillations of the kind described above may occur, so that the flow undulates about the original line of latitude (Figure 5.19(b)); these undulations are Rossby or planetary waves. In the ocean, the scale of such undulations is of the order of hundreds of kilometres; in the atmosphere, it varies from about 5000 to 20 000 km.

From your reading of Chapter 2, can you give an example of Rossby waves in the atmosphere?

The large-scale undulations in the jet stream of the upper westerlies, shown in Figure 2.7, are atmospheric Rossby waves.

In both the atmosphere and the ocean, the overall effect of the clockwise and anticlockwise rotations associated with Rossby waves is to cause the *wave-form* – i.e. the undulations – to move *westwards* relative to the flow. This is true even if the flow is itself moving eastwards, as is the case with the jet stream in the upper westerlies. However, flow velocities in the

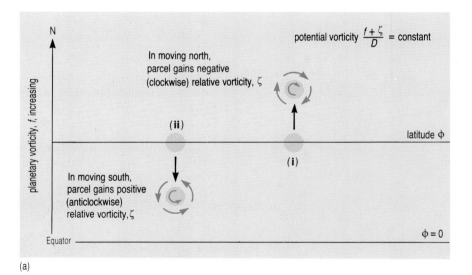

(a)

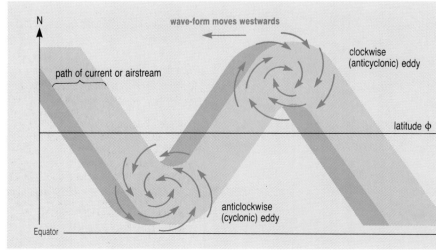

(b)

Figure 5.19 (a) Diagram to show how in a Rossby wave the need to conserve potential vorticity $(f + \zeta)/D$ leads to a parcel of water oscillating about a line of latitude ϕ while alternately gaining and losing relative vorticity ζ. For details, see text.
(b) The path taken by a current or airstream affected by a Rossby wave. Note that the flow pattern is characterized by anticyclonic and cyclonic eddies, and that the wave-form moves *westward* relative to the current or airstream.

atmosphere may reach 100 m s^{-1} and so Rossby waves in an airstream may move eastward *relative to the Earth,* while still moving westward relative to the airstream. If the eastward motion of air in the airstream is approximately equal to the westward motion of the wave-form, stationary Rossby waves result. In the ocean, flow velocities rarely reach 1 m s^{-1} and so even in eastward-flowing currents, Rossby waves nearly always move westward relative to the Earth. Indeed, the Antarctic Circumpolar Current is the only current in which Rossby waves are carried eastward.

The way in which Kelvin and Rossby waves affect ocean circulation depends on the latitude. At middle and high latitudes, information about a change in the wind stress propagates mainly westwards, by means of Rossby waves, so the ocean near western boundaries is affected by events in mid-ocean to a much greater extent than the ocean near eastern boundaries. By contrast, at low latitudes information can travel westwards by Rossby waves or eastwards by Kelvin waves in the equatorial wave guide. In addition, because of the equatorial wave guide, the upper ocean in low latitudes can respond to changing winds much faster than is possible away from the Equator. This is partly because the equatorial wave guide supports both Rossby and Kelvin waves, and partly because Rossby waves travel fastest there. For example, a Rossby wave can take as little as three months to travel west across the equatorial Pacific, whereas it could take ten years to cross the Pacific at $30° \text{ N}$ or $30° \text{ S}$.

It would not be appropriate to go further into the details of either Rossby or Kelvin waves in this Volume. However, one of the most intriguing aspects of these waves is that when an equatorial Kelvin wave reaches the eastern boundary, it not only splits and travel polewards along the coast (as described in Section 5.3.1, for the tropical Atlantic), but may also be partially reflected as a Rossby wave. This can be seen in the computer-generated diagrams shown in Figure 5.20.

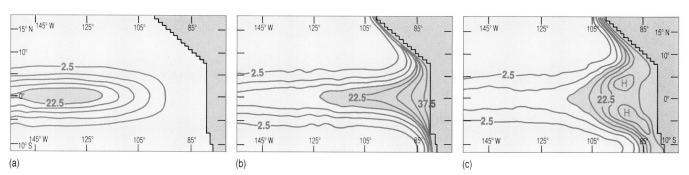

(a) (b) (c)

Figure 5.20 Computer-generated diagrams showing the progress from mid-Pacific to the South American coast, of an internal equatorial Kelvin wave. The contour numbers may be regarded as either the depression of the thermocline in metres or the accompanying rise in sea-level in centimetres. The diagrams show the situation at successive monthly intervals. In (c), the equatorial Kelvin wave has split into two poleward-travelling coastal Kelvin waves. Note that the coastal boundary has the effect of increasing the amplitude of the disturbance. The equatorial Kelvin wave has also just been partially reflected as an equatorial Rossby wave, as can be seen by the circular contours which result from the rotatory motion associated with the wave. Because the two eddies are on *either side of the Equator*, both are anticyclonic and lead to topographic highs (H), although the northerly one is clockwise and the southerly one anticlockwise (cf. Figure 5.19(b)). (In (b) and (c), the small-scale waves in the contours are artefacts of the model.)

5.4 EL NIÑO–SOUTHERN OSCILLATION

An El Niño event is a period of anomalous climatic conditions, centred in the tropical Pacific, that occurs at intervals of 2–7 years. The most obvious sign that an El Niño event is underway is the appearance of unusually warm surface water off the coast of Ecuador and Peru. This generally occurs within a few months of Christmas, and the name El Niño – meaning the Christ Child or, literally, 'The Boy' – was originally the local name for a seasonal warm current that flows south along the coast of Peru, particularly from Christmas time onwards.

Interest in the phenomenon of El Niño goes back to the mid-19th century, but it was the El Niño event of 1972–73 that stimulated large-scale research into climatic fluctuations, which began to be seen as a result of the interaction between atmosphere and ocean. At the time, 1972 was known as 'the year of climate anomalies' – there were devastating droughts in Australia, India, Africa and the (then) Soviet Union, and excessive rainfall around the Mediterranean and along the Pacific coast of South America. In addition, a number of important fisheries had greatly reduced catches. The most catastrophic and long-term collapse was that of the Peruvian anchoveta fishery, where the situation was exacerbated by overfishing. The extent to which these events were directly related to the El Niño event was not clear, but confidence in the world's ability to feed itself was badly shaken. This concern initiated political support for the large-scale climate research programmes that now form a significant part of scientific effort world-wide.

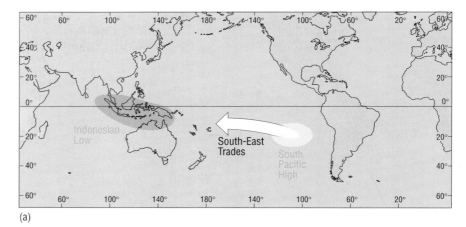

(a)

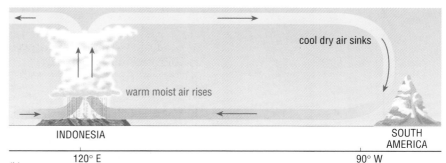

(b)

Figure 5.21 (a)   Schematic map showing the positions of the Indonesian Low and the South Pacific High. (This map should be studied in conjunction with Figure 2.3, showing the global wind pattern.)
(b)   The zonal atmospheric circulation between the Indonesian Low and the South Pacific High.

El Niño events are perturbations of the *ocean–atmosphere system*. It is not known whether the perturbations originate in the atmosphere or the ocean, but for convenience we will start by considering what happens in the atmosphere during an El Niño event. The prevailing winds over the equatorial Pacific are the South-East Trades. Their strength depends on the difference in surface atmospheric pressure between the subtropical high pressure region in the eastern South Pacific – where cool, dry air converges and subsides – and the low pressure region over Indonesia – where warm, moist air rises, producing cumulonimbus clouds and heavy rainfall (Figure 5.21). During an El Niño event, the Indonesian Low has anomalously high pressure (i.e. is a weaker low than usual) and moves eastwards into the central Pacific, while the South Pacific High becomes anomalously low. The South-East Trades weaken, and there are bursts of westerlies in the western Pacific.

By reference to Figure 5.4, can you suggest what effect a relaxation of the South-East Trades will have on the upper waters of the equatorial Pacific?

The sea-surface slope will 'collapse', so that both it and the thermocline become near-horizontal, enabling a considerable volume of warm mixed-layer water to move eastwards across the ocean. In the western Pacific, the collapse in the Trade Winds occurs abruptly, and so the resulting change in the upper ocean – a depression in the thermocline accompanied by a slight rise in sea-level (cf. Figure 5.18(a)) – propagates eastwards along the Equator as a pulse, or series of pulses, of Kelvin waves. At the eastern boundary, the equatorial Kelvin waves split into northward- and southward-travelling coastal Kelvin waves (cf. Figure 5.18(b)), as well as being partially reflected as Rossby waves (cf. Figure 5.20). The speed of these Kelvin waves has been calculated to be about 2.5 m s⁻¹, but the bulge travels slightly faster than this because it is carried forward by flow in the Equatorial Undercurrent.

Figures 5.22 and 5.23 (overleaf) summarize the main differences between normal conditions in the Pacific basin, and conditions during an El Niño event. As shown in part (a) of these Figures, the highest sea-surface temperatures of 28–29 °C are normally found in the western ocean; during an El Niño event, this area of exceptionally warm water moves into the central ocean, along with the vigorous convection of moist air usually associated with the Indonesian Low, and the Intertropical Convergence Zone shifts southwards and eastwards (Figure 5.23(b)).

Why could this eastward shift of the region of vigorous convection, and in the position of the ITCZ, be the result of an eastward movement of exceptionally warm surface water, rather than the indirect cause of it?

A warm sea-surface leads to increased upward convection of moist air and, as discussed in Section 2.3.1, the increase in convection is particularly marked when sea-surface temperatures exceed ~28 °C. The eastward movement across the Pacific of the Indonesian Low, the shift in the position of the ITCZ, and the exceptionally warm surface water are therefore all intimately linked together. Thus, although the general eastward movement of warm mixed-layer water may be explained in terms of the slackening of the Trade Winds in response to a change in the strength and position of the Indonesian Low, this is clearly only part of the story.

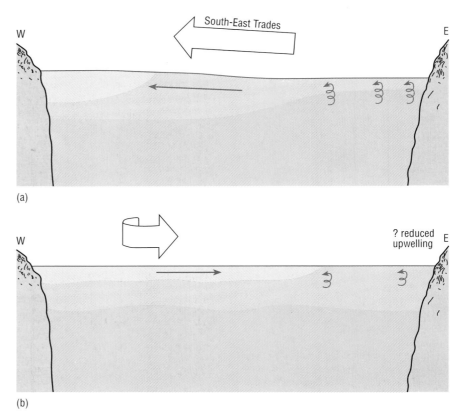

South-East Trades

W

E

(a)

W

E

? reduced
upwelling

(b)

Figure 5.22 Cross-section along the Equator in the Pacific (a) in a normal year, and (b) during an El Niño event. The pink area represents the pool of very warm water usually located in the western Pacific. The curly arrows represent upwelling. Note that although upwelling may not be entirely suppressed during an El Niño event, the upwelled water now comes from within the warm mixed layer and is not nutrient-rich.

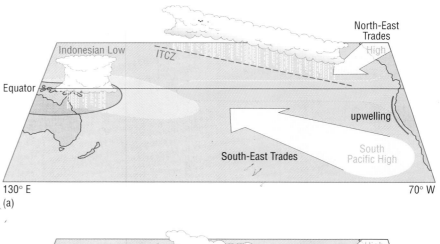

Indonesian Low ITCZ North-East Trades High

Equator

upwelling

South-East Trades South Pacific High

130° E 70° W
(a)

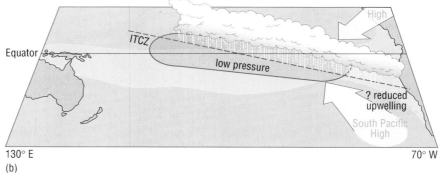

Figure 5.23 Schematic perspective plan-view diagrams showing conditions in the Pacific (a) in a normal year, and (b) during an El Niño event. Note that although no El Niño can be labelled 'typical', the features shown in the diagram seem to occur in most El Niños. The pink tone indicates regions where the sea-surface temperature is higher than about 28 °C; the orange tone in (a) indicates regions that are normally dry.

High

Equator ITCZ

low pressure

? reduced upwelling

South Pacific High

130° E 70° W
(b)

The general disruption of ecosystems and fisheries for which El Niño events are notorious is largely the result of the usually nutrient-rich surface waters of the eastern tropical Pacific being replaced by warm, nutrient-poor waters from the west. The upwelling that usually brings nutrient-rich water up from below the thermocline sometimes stops altogether. Even if it does not, the lowering of the thermocline in the eastern tropical Pacific means that any water that *is* upwelled comes from within the already nutrient-depleted surface layer (Figures 5.22(b) and 5.23(b)).

El Niño events are now often referred to as ENSO events, where the SO stands for 'Southern Oscillation'. 'Southern Oscillation' is the term used for the continual rise and fall of the atmospheric pressure difference at sea-level, between the Indonesian Low and the South Pacific High (cf. the North Atlantic Oscillation, Section 4.5). The two meteorological stations generally used are those at Darwin, northern Australia (for the Indonesian Low), and Tahiti (for the South Pacific High). When the pressure difference between the two is larger than average, the SO Index is described as positive; when it is smaller than average, it is described as negative. As you might expect, El Niño events occur when the SO Index is large and negative (Figure 5.24(a)). Figure 5.24(b) shows the variation of the so-called Multivariate ENSO Index for the last 50 years of the twentieth century.

Figure 5.24 (a)   Monthly averages of the Southern Oscillation Index during the period 1970–1990, with El Niño events (sometimes referred to as 'warm events') indicated by red bars, and La Niña events ('cold events') indicated by blue bars.
(b)   The variation of the Multivariate ENSO Index during the period 1950–2000. This index is computed by the NOAA–CIRES Climate Diagnostics Center, at the University of Colorado, using weighted means for six atmospheric and oceanic parameters observed over the tropical Pacific Ocean (see text). The plot has been inverted from its usual orientation, so that it matches the Southern Oscillation Index plot in (a).

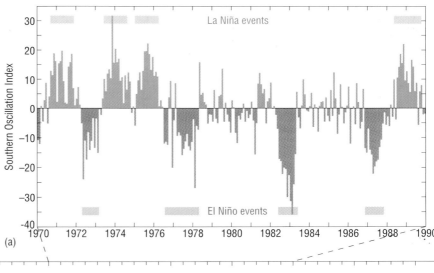

(a)

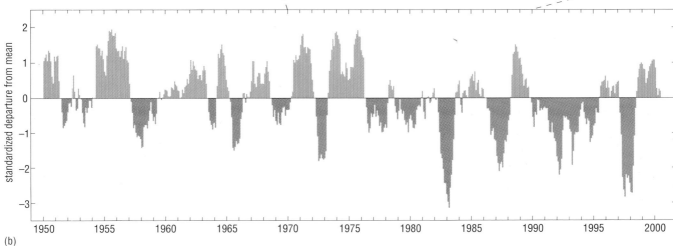

(b)

This Index reflects the variation in six important oceanic and atmospheric variables relating to conditions in the tropical Pacific, namely atmospheric pressure at sea-level, zonal and meridional winds, sea-surface temperature, surface air temperature, and the overall cloudiness. The correspondences between the plots in (a) and (b) show that while the changes in the Southern Oscillation Index (i.e. changes in the atmosphere) are clearly a fundamental part of the driving mechanism, it is the tropical atmosphere–ocean system as a whole that is 'oscillating'.

Looking at Figure 5.24, it is easy to see why many climate researchers no longer view El Niño events as anomalous, but rather as extreme cases of one possible state of the atmosphere–ocean system in the tropical Pacific. This view has led to extreme cases of the other ('normal') state (i.e. an unusually large pressure difference between Darwin and Tahiti, strong upwelling along the coast of South America, etc.) being referred to as La Niña (the Girl). The fact that the tropical Pacific was in or near El Niño mode for much of the 1990s only tends to support the view that El Niño should not be thought of as the anomalous condition.

The recent increase in the frequency of El Niño events is thought by many researchers to be a consequence of global warming, and therefore a trend that is likely to continue. There is now a network of meteorological stations around the Pacific, and the changing patterns of climatic variables such as those employed to calculate the Multivariate ENSO Index are continually monitored. Even if we do not yet fully understand the atmosphere–ocean feedback loops which lead to extreme conditions, we can monitor the 'symptoms' and so predict and hence hopefully mitigate any catastrophes. An important part of the climate researchers' arsenal is the information provided by satellites about, for example, wind speeds (e.g. Figure 2.3(c)) and outgoing long-wave radiation (for sea-surface temperatures and atmospheric water vapour content). There is also, of course, the continually changing shape of the sea-surface itself (e.g. Figure 4.33). Radar altimeters aboard the *TOPEX–Poseidon* satellite provide dramatic images of the 'warm water bulge' that travels across the equatorial Pacific, and polewards along the coasts of North and South America, during an El Niño event (Figure 5.25).

The anomalous climatic conditions associated with the ENSO phenomenon cause havoc over a large part of the globe. Regions of the Pacific basin that are normally wet experience drought, and regions that are normally arid experience destructive torrential rain, which may lead to floods, mudslides and epidemics. Cyclones are more frequent and occur eastward of their normal limit (Section 2.3.1); during the unusually long and severe El Niño event of 1982–83, French Polynesia was struck by five cyclones although it is normally unaffected by them. As in 1972, anomalous climatic conditions often occur outside the tropical Pacific basin: extremely cold winters in North America and Eurasia are also in some way linked with El Niño events, perhaps because the anomalous pattern of surface atmospheric pressure has repercussions for (or is linked to) the paths of the northern polar jet stream (Figures 2.2(b) and 2.7(a)) and the subtropical jet, another high-level airstream which, like the polar jet stream, may undulate, affecting storm tracks and the distribution of warm and cold air.

It is now thought that the paths of the jet streams are also affected by another, longer-scale cyclical change in the pattern of sea-surface temperature in the Pacific. The phenomenon is known as the Pacific Decadal Oscillation (although it may have two periodicities, of 15–20 years and ~ 70 years) and it affects the whole ocean from the Aleutians to the southern Pacific.

FEBRUARY 1997

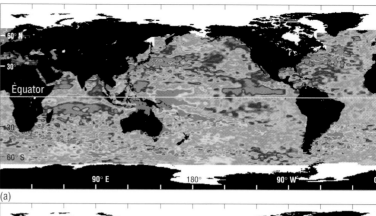

(a)

APRIL 1997

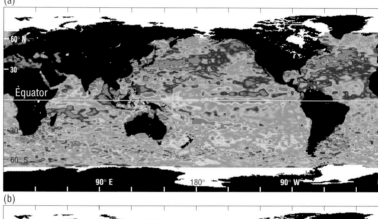

(b)

JULY 1997

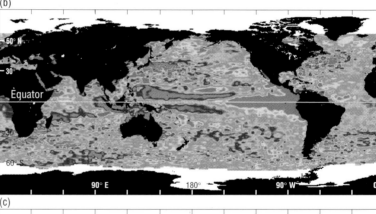

(c)

JANUARY 1997

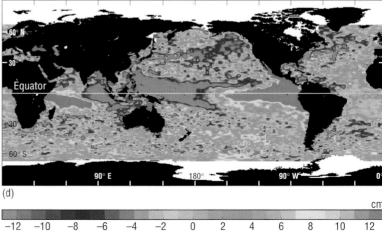

(d)

Figure 5.25   Stages in the development of the 1997–98 ENSO event, as shown by anomalies in sea-surface height recorded by altimeters aboard the *TOPEX–Poseidon* satellite. The anomaly is the difference between the sea-surface height and the mean for the ocean as a whole; reds and yellows indicate sea-level higher than average (shown green); blues and purples indicate sea-level lower than average. (a) is the normal situation, with higher sea-level in the western Pacific, associated with the pool of very warm water. In (b) the warm pool is beginning to move across the equatorial Pacific; (c) when it reaches the coast of America, the Kelvin wave splits, so that warm water spreads north and south. (d) shows another pulse arriving at the eastern boundary six months later. This 'double peak' of the 1997–98 ENSO event may also be seen in Figure 5.24(b).

cm

−12 −10 −8 −6 −4 −2 0 2 4 6 8 10 12

Climatic oscillations in the Indian Ocean

Since at least the 1870s, there have been attempts to predict the Indian monsoon rainfall, with the aim of minimizing the catastrophic effects of droughts. A good starting point would seem to be to look for correlations with changes in the tropical Pacific, but analysis of climatic variables suggests that the ENSO cycle has only a small effect on precipitation patterns around the Indian Ocean. Furthermore, unlike the Atlantic and the Pacific, the Indian Ocean has no readily identifiable oscillations – no equivalent of ENSO or the NAO (Section 4.5).

As far as climate is concerned, the tropical Indian Ocean has significant differences from the Atlantic and Pacific. The low-latitude area of high precipitation is in the east (over Indonesia), while in the Pacific and the Atlantic it is in the west (over Indonesia and Amazonia, respectively), and for five months of the year, surface current flow is towards the east (in the South-West Monsoon Current, Figure 5.12(d),(e)). In all three oceans the areas of most vigorous upwelling and lowest sea-surface temperatures occur where (and in the case of the Indian Ocean, when) the thermocline is shallowest. In the tropical Indian Ocean, these areas are found in the west (during the South-West Monsoon, Figure 5.13(b)), while in the Atlantic and Pacific they occur in the east (e.g. Figure 5.9).

However, in some years, in late summer and autumn, anomalously warm water and reduced upwelling are observed in the western Indian Ocean, accompanied by heavy or even catastrophic rains over eastern Africa; at the same time there are unusually low sea-surface temperatures off Sumatra in the east and reduced rainfall or drought over Indonesia. This anomalous 'mode' of the Indian Ocean and overlying atmosphere does not have a simple link to the strength of the monsoons. Neither does it have a clear link with ENSO, although the sea-surface temperature in the Indian Ocean is (not surprisingly) affected by conditions in the western Pacific, and the flow of Pacific water through the Indonesian archipelago is less during an El Niño event. As in the Pacific, the interacting factors which bring about the anomalous conditions are reversing winds and changes in the distribution of sea-surface temperature; and the switch to the anomalous mode is transmitted across the ocean basin by Kelvin waves and Rossby waves.

5.5   CIRCULATION IN HIGH LATITUDES

Between about 50° and 70° of latitude, surface winds have a cyclonic tendency, associated with the subpolar low pressure regions (Figures 2.2(a) and 2.3). In the oceans at northern high latitudes, where current flow is modified by the presence of landmasses, cyclonic surface gyres result. The subpolar gyre in the North Pacific consists of the Oyashio and Alaska Current; gyral flow in the North Atlantic is disrupted by the Greenland landmass (Figure 3.1) and there are cyclonic gyres in the northernmost Atlantic and in the Norwegian and Greenland Seas. In southern high latitudes, there are no landmasses to impede flow and the westerly winds drive the (eastward-flowing) **Antarctic Circumpolar Current** (formerly known as the West Wind Drift). Easterly winds blowing off the ice-covered continent drive the Antarctic Polar Current which flows west in a narrow zone around most of Antarctica.

5.5.1 THE ARCTIC SEA

The Arctic Sea is largely surrounded by continental masses; it is linked to the Pacific via the Bering Straits – which are only about 45 m deep – and to the Atlantic mainly via the Norwegian and Greenland Seas (Figure 5.26). Deep Atlantic water is prevented from entering these seas, and hence from entering the Arctic Sea, by shallow submarine plateaux which extend from Greenland to Scotland, via Iceland and the Faeroes. The significance of this submarine barrier for the deep circulation of the ocean will become clear in Chapter 6.

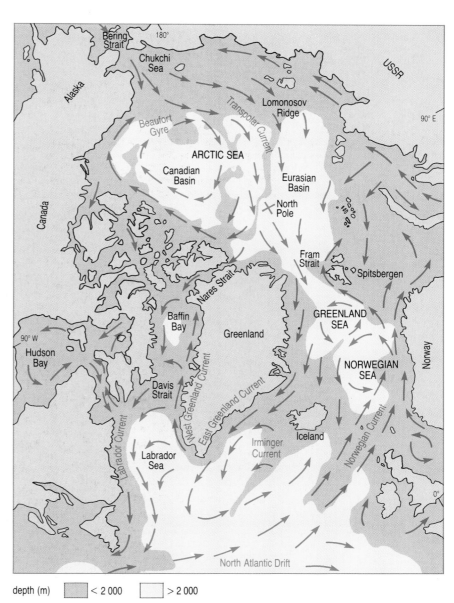

depth (m)   < 2 000   > 2 000

Figure 5.26   The bathymetry and surface currents of the Arctic Sea and the adjacent seas of the North Atlantic. (The North Atlantic Drift is the old name for the North Atlantic Current, the downstream continuation of the Gulf Stream.)

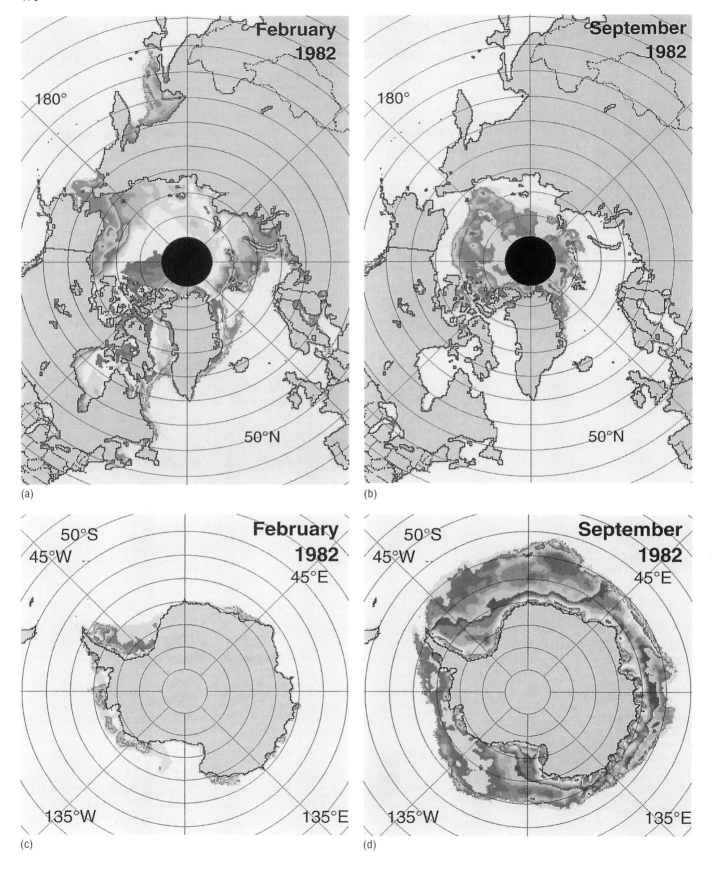

(a) February 1982   180°   50°N

(b) September 1982   180°   50°N

(c) February 1982   50°S 45°W 45°E   135°W 135°E

(d) September 1982   50°S 45°W 45°E   135°W 135°E

The enclosed nature of the Arctic region also greatly affects its ice cover. Figure 5.27 shows the seasonal variation in the ice cover of both the Arctic and the Antarctic.

Which of the two regions shows the lesser variability in ice cover?

The Arctic. Much of the Arctic ice remains 'locked' in the sea all year round; only about 10% leaves annually, via the Fram Straits between Greenland and Spitsbergen (see Figure 5.26). Much of the Arctic Sea is permanently covered with ice (Figure 5.27(a)), and most Arctic pack ice is several years old. This situation contrasts markedly with that in the Southern Ocean where the limits of ice cover shift over about 20° of latitude during the course of a year (Figure 5.27(c) and (d)), and most of the ice cover is renewed annually.

Even in those parts of the Arctic Sea where the ice cover is permanent, the pack ice does not form a solid mass. Under the influence of winds and currents it continually cracks and shifts, so that layers of ice raft over one another; in addition, pressure ridges form, locally increasing the ice thickness from 3–4 m to 40–50 m.

If so much of the Arctic Sea is permanently covered by a thick layer of ice, how has the general surface circulation shown in Figure 5.26 been determined?

The general circulation in the upper layers of the Arctic Sea was initially deduced from the average motion of the ice, as revealed by the movement of ice-bound ships and camps on the ice; more recently, buoys fixed in the ice have been tracked by satellite. There are difficulties in separating pack-ice motion caused by local winds from that related to general current patterns, but it seems that the shorter-period fluctuations are largely related to local winds, and longer-period motion to the surface circulation. In addition to direct current measurements, geostrophic calculations have been made using temperature and salinity data collected from oceanographic stations based on ships or on ice-islands and ice-floes.

The circulation that has been deduced is a clockwise (anticyclonic) gyre centred over the Canadian Basin, with the main surface outflow being the East Greenland Current (Figure 5.26). This current carries southwards not only the pack ice but also icebergs which have calved from glaciers reaching the east coast of Greenland. Off the southern tip of Greenland, the East Greenland Current converges with the warm Irminger Current and most of the ice melts. Some, however, may be carried around to the west coast of Greenland where it is supplemented by large numbers of icebergs from glaciers reaching the west Greenland coast. The ice circulates in Baffin Bay and the Labrador Sea and eventually travels southwards in the Labrador Current. Off the Newfoundland Grand Banks, the Labrador Current converges with the Gulf Stream (Figure 4.31) and here even the largest icebergs gradually break up and melt.

Because it is thought that the effects of global warming should show up first at high northern latitudes, the extent and thickness of Arctic sea-ice has been closely monitored over recent decades. There is indeed evidence that Arctic sea-ice is thinning and penetrating less far south in winter, but it is still hard to be sure that this is part of a long-term trend rather than (at least in part) a climatic oscillation or anomaly.

Figure 5.27 (opposite)   Seasonal changes in ice cover in northern and southern high latitudes, as determined using microwave measurements obtained from the *Nimbus* satellite programme during February and September, 1982. The different colours represent the percentage of sea-surface covered by ice: a purple tone indicates 100% coverage, while a light blue tone represents 20% or less.
(a) and (b)   Near-maximum and near-minimum ice cover in the Arctic region.
(c) and (d)   Near-minimum and near-maximum ice cover in the Antarctic region.

One of the most extreme variations in 'ocean climate' observed during the last century was the event now known as the **Great Salinity Anomaly**. The low salinities were first noticed in the waters to the west of Scotland during 1973–79, with minimum values being recorded in 1975 (cf. Figure 5.28) but it was subsequently discovered that this was part of a larger and longer-lasting phenomenon. A pulse of low salinity water, which had been off eastern Greenland in 1968, travelled round to the Labrador Sea, where it was observed in 1972 (Figure 5.29); it then travelled cyclonically round to the north-east Atlantic and the Norwegian Sea, and was again detected off eastern Greenland in 1981–82. The cause of the event is still not known for sure – one possibility is that it was triggered by anomalous northerly winds over eastern Greenland in the 1960s; another that it may have begun with an unusually large export of ice from the Arctic Sea. Whatever the cause, for more than a decade, the layer of unusually low salinity water occupying the uppermost 500–800 m of the water column had a marked effect on the deep circulation of the North Atlantic, for reasons that will become clear in Chapter 6.

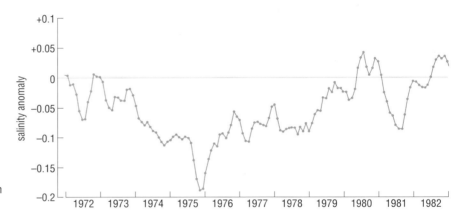

Figure 5.28   Monthly salinity anomalies (i.e. departures from monthly means) for the central Rockall Channel (also known as the Rockall Trough) between January 1972 and December 1982. The pulse of low salinity water is clearly seen to have passed through the Channel in autumn 1975. Salinity values are given in parts per thousand (‰), now omitted, by convention. The minimum salinity of 0.2 below the monthly mean may not seem remarkable, but as you will see in Chapter 6, this is a significant decrease (it is about five times the normal variability in this area). (Similar, though less marked, salinity anomalies have been detected in data for the North Atlantic for the 1950s, 1980s and early 1990s.)

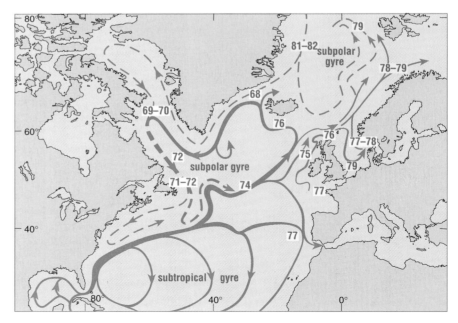

Figure 5.29   The circulation of the Great Salinity Anomaly around the northern North Atlantic between 1968 and 1982, shown in relation to the current pattern in the northern North Atlantic (cf. Figure 4.20(a)). Currents shown with dashed arrows are cold.

5.5.2 THE SOUTHERN OCEAN

The more or less zonal distribution of the ice cover around Antarctica, shown in Figure 5.27(d), reflects the predominantly zonal current flow. As shown in Figure 5.30, there are cyclonic subpolar gyres in the Weddell and Ross Seas but the predominant feature of the circulation of this region is the Antarctic Circumpolar Current – the only current to flow around the globe without encountering any continuous land barrier.

Given that the overlying winds are essentially westerly, in which direction would you expect the sea-surface to slope in the region of the Antarctic Circumpolar Current?

As this is in the Southern Hemisphere, Ekman transport is to the left of the wind, and the sea-surface slopes down towards the Antarctic continent (see Frontispiece and Figure 3.21). This sea-surface slope generates a geostrophic slope current to the east, i.e. flow in the same direction as the wind but extending to greater depths than the surface wind-driven layer (cf. Figure 3.25 for a similar situation in the subtropical gyres). Below the wind-driven layer, the density distribution is such that, in general, the horizontal pressure gradient force and the Coriolis force balance, and geostrophic equilibrium is maintained.

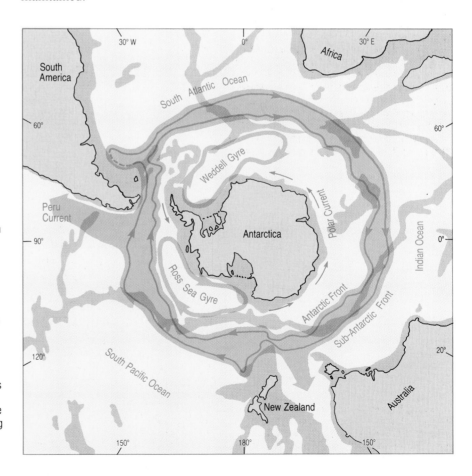

Figure 5.30   Schematic map showing the mean path of the Antarctic Circumpolar Current (blue tone); the two dark blue lines represent the average positions of the Antarctic Front and the Sub-Antarctic Front, and the jets which flow along them (discussed in the text). Note that a significant part of the current branches northward and flows up the west coast of South America as the Peru Current; there is also a branch of the current flowing northwards below the surface between Australia and New Zealand. The approximate positions of the gyres in the Weddell Sea and Ross Sea are also shown, as is the path of the Polar Current. The Antarctic Divergence is between the Polar Current and the Antarctic Circumpolar Current. Blue–grey shading indicates water depths less than 3000 m.

In surface layers, the direct effect of the wind stress, combined with the Coriolis force, leads to a northward component of flow, and a region of convergences forms within the strongest part of the Antarctic Circumpolar Current. This region of convergences was originally thought to be a single convergence, and was named the **Antarctic Convergence**. It is now known to consist of a series of convergences, or fronts, and has been renamed the **Antarctic Polar Frontal Zone (APFZ)**. The fronts in the APFZ are associated with strong zonal current jets, with velocities reaching $0.5–1.0\,\mathrm{m\,s^{-1}}$. Two major jets occur in association with the northern and southern boundaries of the APFZ – the Sub-Antarctic Front and the Antarctic Front (also known as the Polar Front); average positions of these jets are shown in Figure 5.30.

Despite its great length – about 24 000 km – the Antarctic Circumpolar Current has remarkably consistent characteristics wherever it is observed; furthermore, the Sub-Antarctic Front and the Antarctic Front persist throughout the extent of the current, although the distance between them is very variable (Figure 5.30). Figure 5.31 shows temperature and salinity sections across the Antarctic Circumpolar Current in the Drake Passage, between South America and the islands that lie to the north of the Antarctic Peninsula (cf. Figure 5.30). The blue bands indicate the positions of the fronts. The narrowest front is a boundary between oceanic waters and colder, fresher water that originated in the Weddell Sea; the other two fronts are the Antarctic Front and the Sub-Antarctic Front, mentioned above.

Figure 5.31   Sections of (a) temperature (°C) and (b) salinity across the Drake Passage. The isotherms and isohalines sloping up to the south from a depth of about 3000 m delineate a 'wedge' of warmer, less saline water flowing over colder, more saline water. The sections were made during the southern summer. The blue toned regions are fronts. At the surface, the sharpest changes in salinity and, particularly, temperature occur at the Antarctic Front; for this reason, the Antarctic Front was the first to be observed and was identified as the Antarctic Convergence.

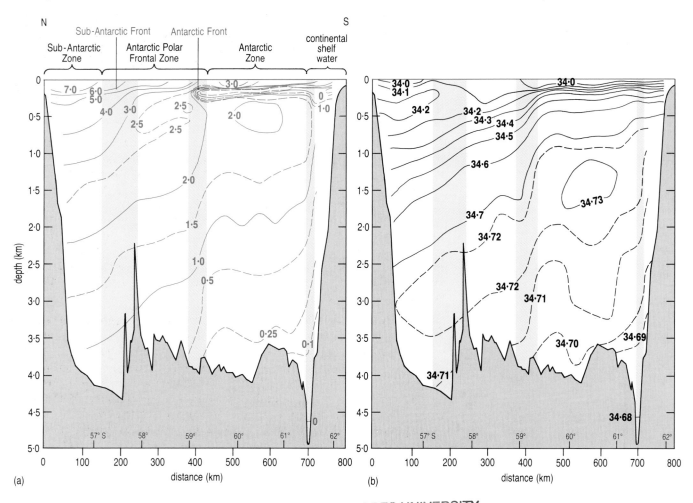

Why do the fronts indicated on Figure 5.31 imply locally stronger current flow (the 'jets' referred to earlier)?

As mentioned above, flow in the Antarctic Circumpolar Current is generally in geostrophic equilibrium. The isopycnic surfaces slope up towards the south, and the steeper their slope the greater the velocity of the eastward geostrophic current. The fronts on Figure 5.31 are characterized by steeper slopes in the isotherms and isohalines, and hence in the isopycnals; they are therefore also characterized by faster geostrophic currents (Section 4.3.2). Figure 5.32 is a more detailed picture of flow through the Drake Passage. This shows not only three eastward current jets (as in Figure 5.31) but also two weaker eastward filaments separated by areas of westerly flow.

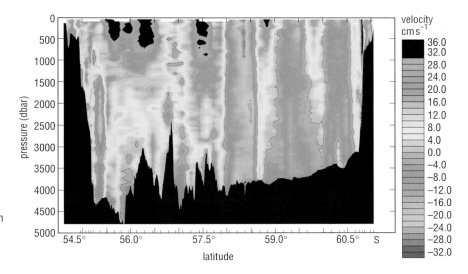

Figure 5.32   Section across the Drake Passage showing current velocity: black, purple, red and orange correspond to eastward flow; yellow, green and light blue to westward flow. The section was obtained using an LADCP ('lowered ADCP') deployed from a research vessel at the locations indicated by the ticks on the bottom axis.

Generally, the path of the Antarctic Circumpolar Current does not depart greatly from the average path shown in Figure 5.30, but in the central and western South Pacific, it can shift by more than $10°$ of latitude between summer and winter. Flow within the APFZ is complex, with many eddies and meanders. The current jet associated with the Antarctic (Polar) Front has been observed to meander southwards and capture colder water, and there are numerous eddies within the frontal zone, both cold-core and warm-core. You may remember from Sections 3.5.2 and 4.3.6 that such eddies are a very important heat-transfer mechanism, in this case enabling cold water to move north across the Antarctic Circumpolar Current, and warm water to move south.

Although it shows up clearly on maps of dynamic topography because it is so broad, away from the frontal jets the Antarctic Circumpolar Current is not particularly fast: south of the Antarctic Front, surface speeds are about $0.04 \, \text{m s}^{-1}$, while in the faster region, to the north of the Antarctic Front, they may reach $0.15-0.2 \, \text{m s}^{-1}$. However, the current is very deep, extending to the sea-floor at about 4000 m depth, and its volume transport is therefore enormous. Estimates suggest that the average transport through the Drake Passage is about $130 \times 10^6 \, \text{m}^3 \, \text{s}^{-1}$, so in terms of volume transport, the Antarctic Circumpolar Current is certainly the mightiest current in the oceans.

For a long time, oceanographers were puzzled that the Antarctic Circumpolar Current was not faster than it is, given that it travels uninterrupted around the globe under the cumulative influence of the westerlies. The answer seems to be that the eastward wind stress is more or less balanced by frictional forces resulting from interaction of the flow with the sea-floor topography (which, as indicated by Figure 5.30, may also account for some of the north–south variation in the path of the current). Friction generated by turbulence adjacent to and within the Antarctic Circumpolar Current itself may also play a part, along with that acting on the current as it flows through the restricted Drake Passage.

So far, we have only been considering horizontal motion. However, as discussed above, the APFZ is a region of convergence of surface waters; and between the eastward-flowing Antarctic Circumpolar Current and the westward-flowing Polar Current is a divergence of surface water – the **Antarctic Divergence**.

What significance do such divergences and convergences have for the three-dimensional circulation of the ocean?

They must lead to vertical motion: upwelling at the divergences and sinking at the convergences. The Antarctic Divergence is biologically one of the most productive regions of open ocean in the world. The nutrient-rich water upwelled there leads to high primary productivity which supports large populations of zooplankton, and bigger organisms ranging from krill to whales.

The convergences in the Antarctic Polar Frontal Zone are an important source of cold deep sub-thermocline water for the world ocean. Even colder 'bottom water' is formed off the Antarctic continent. These deep and bottom waters will be discussed further in Chapter 6.

The Antarctic Circumpolar Wave
By now, you will be becoming familiar with the idea that the ocean and atmosphere are not separate but closely coupled one to another, and that the ocean–atmosphere system has natural oscillations with a wide variety of periods. It is clear that the oscillations in the different oceans are often (but not always) inter-related, generally via the atmosphere, particularly by upper level winds. There is, however, another way in which the three oceans are linked, namely by the Southern Ocean, which encircles the globe, and this is the setting for one of the most fascinating of the natural oscillations. This wave-like disturbance involves not only the ocean and atmosphere, but also the cryosphere – the ice-cover.

The **Antarctic Circumpolar Wave** has been detected through analyses of data relating to atmospheric pressure at sea-level, meridional wind stress, sea-surface temperature (monthly averages) and sea-ice extent, collected over 13 years. When the normal seasonal changes in these four parameters are subtracted from the variability actually observed, a wave-like pattern of highs and lows is revealed. The wave has two wavelengths end-to-end, so there are two high-pressure anomalies, and two low-pressure anomalies, two maxima and two minima in meridional (north–south) wind stress, two areas of anomalously high sea-surface temperature and two areas of anomalously low sea-surface temperature, and two 'bulges' in ice-extent, all travelling eastward around the Antarctic continent (Figure 5.33).

The wave has a period of 4–5 years, and therefore takes 8–10 years to travel all the way round the Antarctic continent. The maxima and minima of the different components do not all coincide (i.e. they are out of phase with one another), but they all travel around the continent with the same speed, suggesting that they are closely linked together by feedback loops acting over the time-scales in question. The speed falls within the range of current velocities found in the Antarctic Circumpolar Current, which is perhaps not surprising given that the ACC can transport unusually warm or unusually cold surface waters around the globe.

The wave is not a purely Antarctic phenomenon. It is believed to be generated by ENSO signals propagating from the equatorial Pacific, probably via the atmosphere. The anomalies in sea-surface temperature travel northwards in the Pacific with the Peru (Humboldt) Current and northwards in the Atlantic with the Benguela Current, and generally spread equatorwards in all three oceans, probably as a result of Ekman transport in response to the westerly winds over the Southern Ocean.

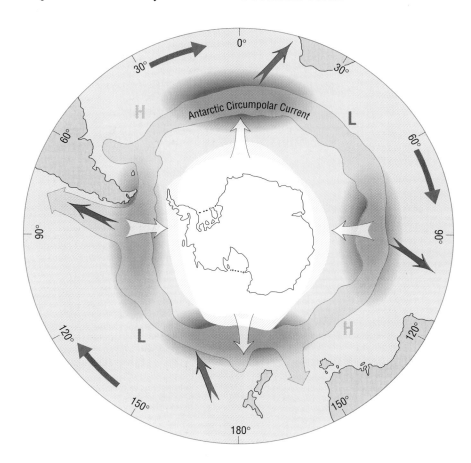

Figure 5.33 Simplified schematic map of the Antarctic region and the mean path of the Antarctic Circumpolar Current, to show the characteristics of the Antarctic Circumpolar Wave, which involves sea-surface temperature (red, warm; blue, cold), atmospheric pressure at sea-level (H = anomalously high pressure, L = anomalously low pressure), meridional wind stress (open arrows), and sea-ice extent. The heavy arrows indicate the general eastward motion of the Wave, and the blue and red arrows indicate the direction of travel of anomalous conditions within the ocean, southward from the western Pacific, eastward around the Antarctic continent (with the ACC), and northward into the eastern Pacific, Atlantic and Indian Oceans.

In what other ways might the wave contribute to the variability of global climate, i.e. variability in the global atmosphere–ocean system?

Interannual variability in atmospheric pressure and winds in the Antarctic region would directly affect global climate via the atmospheric and oceanic circulation, and variability in sea-surface temperature would contribute to this by altering the Equator-to-Pole temperature gradient. Furthermore, these changes would be likely to affect climate on longer time-scales via their effect on the rates of formation of cold deep and bottom water masses, which are discussed in the final Chapter.

5.6 SUMMARY OF CHAPTER 5

1   The major components of equatorial current systems are westward-flowing North and South Equatorial Currents, one or more eastward-flowing Counter-Currents (surface and subsurface), and an eastward-flowing Equatorial Undercurrent, which is generally centred on the Equator. Flow in the North and South Equatorial Currents is partly directly driven by the Trade Winds and is partly geostrophic flow.

2   The equatorial current system is best developed in the Pacific Ocean, where the surface waters are under the cumulative influence of the prevailing Trade Winds over the greatest distances. In the Atlantic, the equatorial circulation is affected by the shape of the ocean basin and, indirectly, by the effect of the continental masses on the ITCZ. In the Indian Ocean, the circulation is monsoonal, most resembling that in the other tropical oceans in the northern winter.

3   The Intertropical Convergence Zone is generally displaced north of the Equator so that the South-East Trade Winds blow across it. As a result, divergence of surface waters, and upwelling, occur just south of the Equator. There is a convergence of surface waters at about 4° N.

4   The prevailing easterly winds over the tropical ocean cause the sea-surface to slope up (and the thermocline to slope down) towards the west. As a result, there is an eastward horizontal pressure gradient force and the Equatorial Counter-Current(s) flow(s) down this gradient towards the east in zones of small westward wind stress (the Doldrums).

5   This eastward horizontal pressure gradient force also drives the Equatorial Undercurrent, which flows in the thermocline below the mixed surface layer. The Equatorial Undercurrent is a ribbon of fast-flowing water, many hundred times wider than it is thick. It is generally aligned along the Equator, although it may have long-wavelength undulations; if it is diverted away from the Equator, the Coriolis force turns it equatorwards again. The Equatorial Undercurrent has a significant volume transport, particularly in the Pacific.

6   In the Pacific and the Atlantic, extensive areas of upwelling occur just south of the Equator, in association with the South Equatorial Current. There is also coastal upwelling along the eastern boundaries – either year-round or seasonal – as a result of the Trade Winds blowing along the shore.

7   Surface divergence and upwelling may occur below the ITCZ because it is a region of low pressure and cyclonic winds. When the ITCZ is over certain regions of doming isotherms (apparently associated with flow in

subsurface counter-currents), the doming intensifies and 'protrudes' into the thermocline. These thermal domes seem to be a feature of the eastern sides of oceans. In the Pacific and the Atlantic, all types of upwelling occur most readily on the eastern side of the ocean, because there the thermocline is at its shallowest, and the mixed layer at its thinnest.

8   The winds over the Indian Ocean change seasonally as a result of the differential heating of the ocean and the Asian landmass. During the North-East Monsoon (northern winter), the winds are from Asia and are dry and cool; during the stronger South-West Monsoon, the winds carry moisture from the Arabian Sea to the Indian subcontinent. Because the winds over the equatorial zone change over the course of the year, so does the direction of the sea-surface slope along the Equator. As a result, in the Indian Ocean the Equatorial Undercurrent is only a seasonal feature of the circulation.

9   The most dramatic seasonal change in the surface circulation of the Indian Ocean is the reversal of the Somali Current which flows south-westwards during the North-East Monsoon but is a major western boundary current during the South-West Monsoon. At that time of the year, the North Equatorial Current reverses and becomes the South-West Monsoon Current. During the South-West Monsoon, there are regions of intense upwelling on the western side of the ocean, off Somalia and Oman.

10   The Agulhas Current is the next most powerful western boundary current, second only to the Gulf Stream. Its retroflection off the tip of southern Africa is a source of eddies, many of which are carried into the Atlantic.

11   The ocean can respond to the winds in distant places by means of large-scale disturbances that travel as waves. These waves may propagate along the surface (barotropic waves) or along a region of sharp density gradient such as the thermocline (baroclinic waves); surface waves, in particular, travel very fast. Two of the most important types of waves are Kelvin waves and Rossby (or planetary) waves. Rossby waves result from the need for potential vorticity to be conserved and, relative to the flow, only travel westwards. Kelvin waves may travel eastwards along the Equator (as a double wave) or along coasts (with the coast to the right in the Northern Hemisphere and to the left in the Southern Hemisphere). In these cases, the Equator and the coast, respectively, are acting as wave guides. Because of the equatorial wave guide, the ocean in low latitudes can respond much more rapidly to changes in the overlying wind than can the ocean at higher latitudes.

12   El Niño or ENSO events are climatic fluctuations centred in the tropical Pacific, in which the east–west slopes in the sea-surface and thermocline collapse, and warm water spreads across the tropical Pacific, along with areas of vigorous convection and heavy rainfall. During El Niño events, the difference in pressure between the South Pacific High and the Indonesian Low is less than usual (i.e. the Southern Oscillation Index is large and negative), and the Trade Winds are weaker than usual. When the Southern Oscillation Index is large and positive (i.e. conditions are an extreme version of the 'normal' situation), there is said to be a La Niña.

13   Unlike the tropical Pacific and Atlantic, the Indian Ocean does not have well defined climatic oscillations, but it does have an anomalous mode in which conditions along the Equator become more like those in the other two oceans (warm in the west, cool in the east).

14   As a result of the contrasting distributions of land and sea in northern and southern high latitudes, both the type of ice cover and the current pattern of the two regions are very different. A large proportion of Arctic pack ice is several years old, while most Antarctic ice is renewed yearly. The main circulatory pattern in the Arctic Sea is an anticyclonic gyre with cross-basin flow between the Bering Straits and the Fram Strait, where the outflow becomes the East Greenland Current.

15   The Great Salinity Anomaly was a pulse of low salinity water which, between 1968 and 1981–82, travelled westwards round Greenland, around the Labrador Sea, and the subpolar gyre, and then back to the north-east Atlantic and the Norwegian and Greenland Seas.

16   The major current feature of the Southern Ocean is the Antarctic Circumpolar Current (ACC) which, by virtue of its great depth, has an enormous volume transport. The cumulative influence of the westerly wind stress acting on the ACC is balanced mainly by frictional forces generated by the interaction of the ACC with the sea-floor topography. The strongest currents in the ACC flow along fronts in the Antarctic Polar Frontal Zone, and these current jets often form meanders and eddies. The Antarctic Polar Frontal Zone is a region where surface water converges and sinks; the Antarctic Divergence, between the Antarctic Circumpolar Current and the Antarctic Polar Current, is a region of upwelling.

17   The Antarctic Circumpolar Wave is a wave-like progression of maxima and minima of atmospheric pressure, meridional wind stress, sea-surface temperature and sea-ice extent, which travels eastwards around the Antarctic continent. The wave has two wavelengths end-to-end and, as it has a period of 4–5 years, it takes 8–10 years to travel all the way round the Antarctic continent. Its importance for the global climate is not yet understood.

Now try the following questions to consolidate your understanding of this Chapter.

QUESTION 5.8   A number of ALACE floats (Section 4.3.4) were deployed in the South Atlantic in 1990. Figure 5.34 (opposite) shows their paths up until mid-1997. Study the paths and on the basis of your study so far (mainly Chapter 5, but also Figure 3.1) make a list of the current features that you can recognize. Which floats had been travelling the fastest?

QUESTION 5.9   Monsoon winds are sometimes described as land breezes and sea breezes, on a very large scale. Would you say that this is a fair description of them?

QUESTION 5.10   At the beginning of Section 5.2.2, we say: 'As you might expect, the surface circulation of the northern Indian Ocean … most resembles that of the other two oceans in the northern winter'. However, there is a significant feature of the circulation in the northern *summer* that is shared by the other two oceans. What is it?

QUESTION 5.11   The Rossby radius of an equatorial Kelvin wave is defined mathematically as the distance from the Equator by which the wave's amplitude has fallen to $1/e$ of its maximum value (e, the base of natural logarithms is ~2.72, so $1/e$ is about 0.4). Use this information to estimate the Rossby radius of deformation for the (computer-generated) equatorial Kelvin wave shown in Figure 5.20. Use diagram (a), and note that the contours are 5 m (or cm) apart; 1° of latitude is about 110 km.

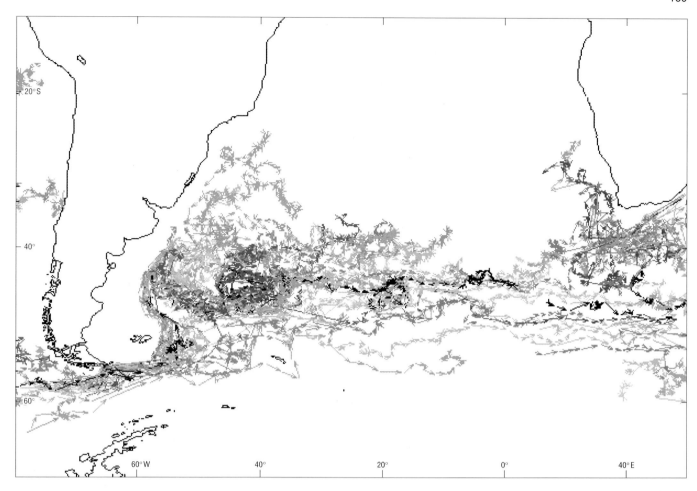

Figure 5.34 The tracks of ALACE floats deployed in the South Atlantic in January 1990. Each colour corresponds to a different float; this map shows their paths up until June 1997. (For use with Question 5.8.)

QUESTION 5.12 As you saw in Section 5.3.2, Rossby waves result from the need for potential vorticity $(f + \zeta)/D$ to be conserved. Bearing that in mind, can you suggest why similar waves (known as 'shelf waves') occur over regions of relatively sharp depth change between the continental shelf and the deep sea-floor? Begin by imagining a current flowing above (say) the continental slope, parallel to a particular depth contour. What happens if, for some reason, the flow is displaced either towards or away from the coast (assume that distances involved are sufficiently small for changes in latitude to be neglected)? (You do not need to go into details.)

QUESTION 5.13 The vessel *Fram* took three years to drift in the ice from north of the Bering Straits to Spitsbergen (Figure 5.26).

(a) With which current did the *Fram* travel?

(b) Given that the distance involved is roughly 4000 km, approximately what is the implied average speed of this current?

CHAPTER 6

GLOBAL FLUXES AND THE DEEP CIRCULATION

In Chapters 1 and 2, we saw that the ocean–atmosphere system is a huge heat engine, for which solar radiation is the power source. Figure 6.1 is a schematic summary diagram, showing how the net radiative gain at low latitudes and net radiative loss at high latitudes (cf. Figure 1.4) is compensated for by the atmosphere–ocean heat redistribution system (cf. Figure 1.5), which has three interacting components – winds, surface currents, and deep flow in the thermohaline circulation.

We begin this Chapter about the three-dimensional circulation of the oceans by considering conditions at the sea-surface, because it is there that the cold, dense waters of the deep ocean originate.

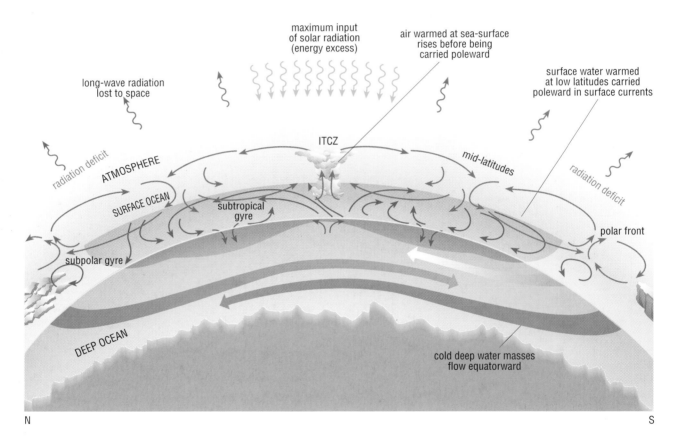

Figure 6.1   Schematic diagram of the Earth's heat-redistribution system (not to scale), consisting of three interacting components: the wind system in the atmosphere (Chapter 2), the surface current system (including the subtropical and subpolar gyres; Chapters 3 to 5); and the density-driven thermohaline circulation in the deep ocean, which will be discussed in Section 6.3.

6.1 THE OCEANIC HEAT BUDGET

Figure 1.4 summarized the heat gained and lost by the Earth–atmosphere–ocean system as a whole; here we will consider the heat gained and lost by the ocean.

6.1.1 SOLAR RADIATION

Figure 6.2 shows how the amount of solar radiation received annually varies over the Earth's surface. Intuitively, you might expect that the contours would be parallel to lines of latitude. This is clearly not the case.

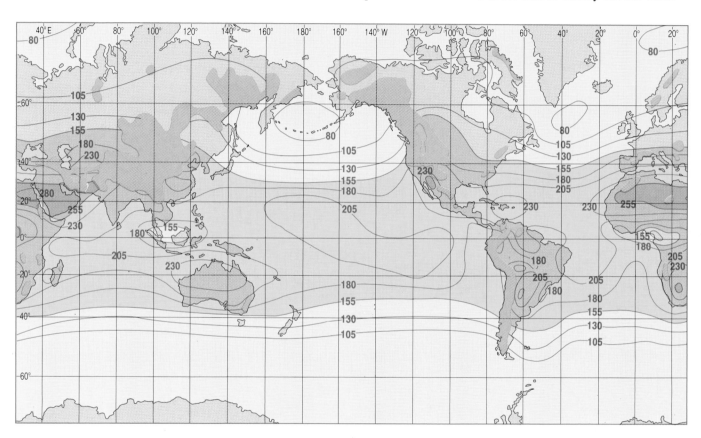

Figure 6.2 The amount of solar radiation received at the surface of the Earth, in W m^{-2}, averaged over the course of a year. (Contours over high ground have been omitted.)

In general, at a given latitude, are contour values greater over the oceans or over the continents? Why might this be?

Insolation – the amount of incoming solar radiation reaching the Earth's surface – is generally greater for continental areas than for the oceans. As discussed in Chapter 2, the atmosphere over the oceans contains a large amount of water, particularly in low latitudes. Along with gases such as CO_2 and SO_2, water vapour and clouds absorb on average about 20% of incoming solar radiation (cf. Figure 6.3). Clouds especially have a marked effect as they not only absorb incoming radiation but also reflect it back to space. Over land, the atmosphere tends to be less cloudy and drier, particularly in tropical and subtropical areas.

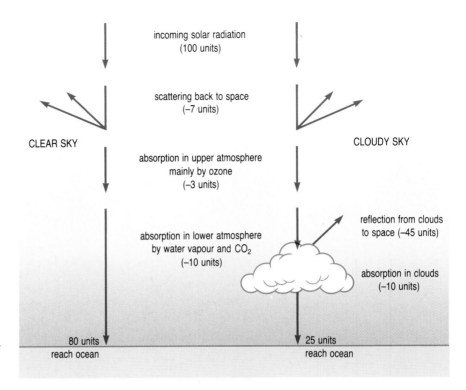

incoming solar radiation
(100 units)

scattering back to space
(−7 units)

CLEAR SKY

CLOUDY SKY

absorption in upper atmosphere
mainly by ozone
(−3 units)

absorption in lower atmosphere
by water vapour and CO₂
(−10 units)

reflection from clouds
to space (−45 units)

absorption in clouds
(−10 units)

80 units
reach ocean

25 units
reach ocean

Figure 6.3   Diagram to illustrate the effect of the atmosphere on the radiation budget of the ocean. The values are very approximate and represent average conditions; for example, in low latitudes as much as 40% of incoming solar radiation may be absorbed by the atmosphere over the oceans.

QUESTION 6.1

(a) Why would you expect the effect of water vapour over the ocean to be greater at lower latitudes?

(b) Now look at Figure 6.2. Thinking back to Section 5.4, can you suggest (1) the reason for the contour pattern in the vicinity of Indonesia, and (2) why the high insolation region indicated by the 205 W m⁻² contour in the tropical Pacific has a bulge towards the south-east?

The supply of water vapour to the atmosphere over Indonesia is not simply a result of the pressure distribution and the wind system. The Indonesian rainforest, itself a product of these climatic influences and a reservoir of terrestrial water, draws in moisture by actively transpiring water to the overlying atmosphere, intensifying atmospheric convection and cloud formation. Other low latitude regions where water vapour and cloud reduce incoming solar radiation are the Malaysian archipelago, equatorial Africa and Brazil.

Water vapour and clouds are not the only factors that affect incoming solar radiation, although they are by far the most important influences at the present time. The amount of solar radiation received at the Earth's surface may also be reduced by the presence in the atmosphere of ash, smoke, dust and gases from volcanoes and industrial complexes, as well as dust from arid regions.

6.1.2   THE HEAT-BUDGET EQUATION

As discussed in Section 1.1, the Earth as a whole not only receives solar radiation, which is largely short-wave, but also re-emits long-wavelength radiation. This is because all bodies with a temperature above absolute zero emit radiation: the higher the temperature of the body concerned, the greater the total amount of radiant energy emitted. In fact, the intensity (I) of the radiation emitted increases in proportion to the fourth power of the absolute temperature (T); i.e. $I = \sigma T^4$. This is known as Stefan's Law and the constant σ is known as Stefan's constant. Furthermore, the higher the temperature of the body concerned, the more the radiation spectrum is shifted towards shorter wavelengths. Thus, the surfaces of the oceans and continents not only absorb and reflect the incoming short-wave solar radiation that has penetrated the atmosphere but also re-emit radiation which is mostly of a much longer wavelength, because of their relatively low temperatures. This longer-wavelength radiation is either lost to space or is *absorbed* by clouds, water vapour and other gases – especially carbon dioxide and ozone – all of which re-emit long-wave radiant energy in all directions. In calculating the amount of radiant energy absorbed by the oceans, we therefore have to consider not only the incoming short-wave ($< 4\,\mu m$) radiation (given the symbol Q_s) but also the *net* emission of long-wave radiation (also known as back-radiation, and so given the symbol Q_b). For all latitudes, $Q_s - Q_b$ is generally positive, i.e. the oceans absorb more radiant energy than they emit (Figure 6.4), although at higher latitudes the value of $Q_s - Q_b$ varies significantly with the time of year.

Of the total amount of energy received from the Sun by the world's oceans, about 41% is lost to the atmosphere and, indirectly, to space, as long-wave radiation, and about 54% is lost as latent heat through evaporation from the sea-surface. A relatively small amount – about 5% – is lost to the overlying atmosphere by conduction. Heat loss by evaporation is generally given the symbol Q_e, and heat loss by conduction, the symbol Q_h.

The temperature of a body is a measure of the thermal energy it possesses. If the average temperature of the oceans is to remain constant, the gains and losses of heat must even out over a period. In other words, the **heat budget** must balance.

Figure 6.4 The radiation balance ($Q_s - Q_b$) at the Earth's surface, in W m^{-2}, averaged over the course of a year. Values have been converted from non-SI units; and contours have been omitted over high ground. The white area shows the approximate winter limit of sea-ice cover.

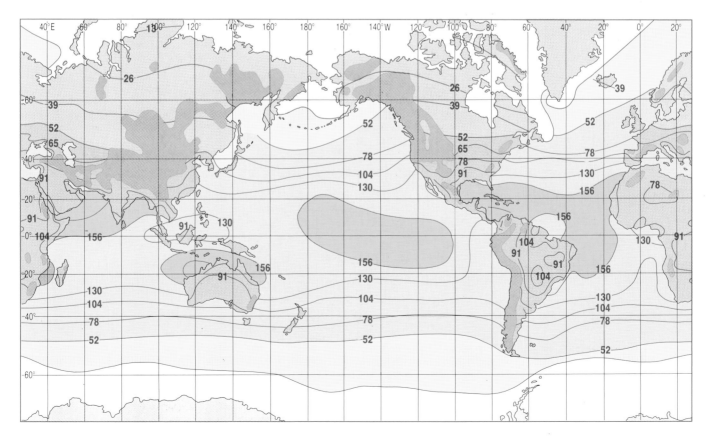

Heat is not only being continuously gained and lost from the oceans, but also redistributed within them, by currents and mixing.

Figure 6.5(a) and (b) show the global distributions of sea-surface temperature in July and January. Would you say that these temperature distributions reflect the influence of ocean currents (Figure 3.1)?

Yes. In particular, surface water on the western sides of oceans is generally warmer than the water on the eastern side, particularly in the hemisphere experiencing summer. This is a result of flow around the subtropical gyres, with the western boundary currents carrying warm water from lower latitudes – the effect of the Gulf Stream in transporting relatively warm water across the Atlantic can also be clearly seen, especially in Figure 6.5(a), for the northern summer. By contrast, the eastern boundary currents carrying cold water from higher latitudes cause temperatures on the eastern sides of oceans to be somewhat lower than they would otherwise be. Low temperatures in eastern boundary currents are also a result of upwelling – the effect of the Benguela upwelling is particularly evident, especially in the southern winter (Figure 6.5(a)); so, too is upwelling at the northern edge of the South Equatorial Current in the Pacific and the Atlantic Oceans.

Heat brought into a region of ocean by currents and mixing, i.e. by advection, is given the symbol Q_v. The term 'advection' (cf. Section 1.1) is normally taken to relate to horizontal transport of water into an area, but water carried to the surface in upwelling currents, or carried away from it in downwelling currents, also contributes to Q_v. Indeed, heat transport – like current flow – has areas of convergence and divergence (cf. Figure 3.27). However, while converging surface water tends to sink, a region of convergence of heat (Q_v positive) can result in a rise in temperature leading to heat being lost upwards (as Q_b, Q_e and Q_h) as well as being mixed downwards.

In summary, the heat-budget equation for any part of the ocean should include the following terms:

Q_s – solar energy, received by the ocean as short-wave radiation;

Q_b – the net loss of energy from the surface of the ocean as long-wave (back-) radiation;

Q_e – the heat lost by evaporation from the surface, less any heat gained by condensation at the surface;

Q_h – the net amount of heat transferred to the atmosphere by conduction across the air–sea interface (but see later);

Q_t – the amount of surplus heat actually available to increase the temperature of the water; when there is a heat deficit, this term will be negative, and there will be a fall in the temperature of the water;

Q_v – the net amount of heat gained from adjacent parts of the ocean by advection (including upwelling or sinking of water) and mixing; when heat is lost by advection, this term will be negative. (For the ocean as a whole, Q_v is of course zero as it refers to the redistribution of heat *within* the ocean.)

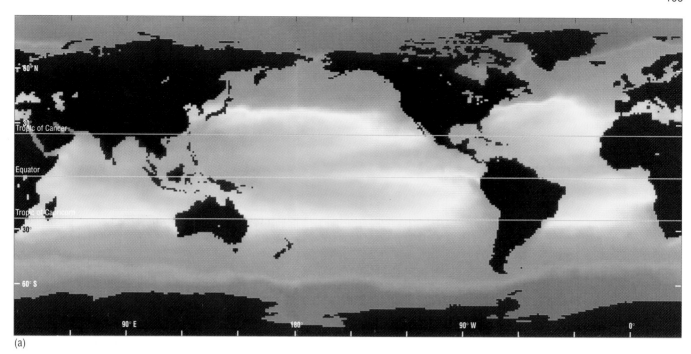

(a)

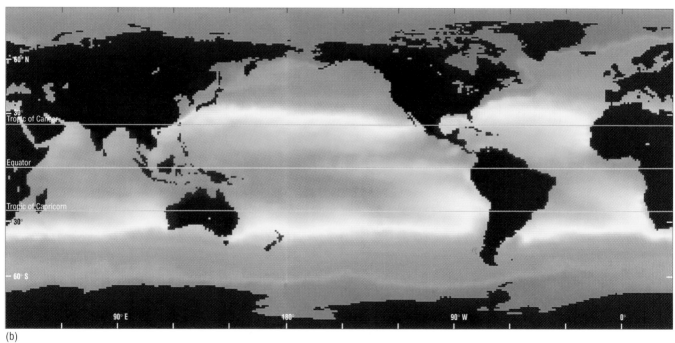

(b)

Figure 6.5   The global distribution of
sea-surface temperature (°C) (a) in July,
(b) in January.
Note: Ignore differences in *intensity* of colour
between (a) and (b).

The full heat-budget equation for a part of the oceans is therefore:

$$Q_s + Q_v = Q_b + Q_h + Q_e + Q_t \qquad (6.1)$$

Assessing the relative sizes of the quantities in Equation 6.1 has been an ongoing challenge for oceanographers. However, we can review the principles behind the methods for estimating them, and look briefly at some of the problems encountered in practice.

Values for radiation gain (Q_s) can be estimated from knowledge of incoming solar radiation (cf. Figures 1.4(a) and 6.2). Values for radiation loss (Q_b) can be estimated using the temperature of the surface skin of the ocean. This is often determined by measuring the temperature at a few metres depth, and then applying a correction; this method can give good results as long as a shallow diurnal (i.e. daytime) thermocline has not developed. Sea-surface temperature is also needed to estimate Q_h and Q_e.

The most accurate results are obtained by using a combination of *in situ* and satellite measurements. Global scale temperature data are now available through the use of satellite-borne radiometers (cf. Figure 4.31(a)). These measure the radiation intensity, at different wavelengths, at the top of the atmosphere; the sea-surface temperature itself is determined using assumptions about the effects of cloud cover and about atmospheric concentrations of those variable constituents that absorb and re-emit radiation, in particular water vapour and aerosol droplets.

For many decades, meteorological and oceanographic data have been collected regularly, not only from research vessels, but also from weather ships (now mostly decommissioned), buoys, and enormous numbers of commercial and Admiralty vessels and other 'ships of opportunity', now referred to as Voluntary Observing Ships. Despite the observation code (Section 4.1.1), observational practice and positioning of instruments vary from ship to ship, so care has to be taken in comparing or combining data. There are also problems regarding the quantity and, in particular, the distribution of data. Of necessity, observations are concentrated along shipping lanes, and until recently, some regions – especially in the Pacific – were completely devoid of reliable data.

Nevertheless, useful empirical relationships have been derived using those variables that are regularly determined, especially mean cloudiness, relative humidity immediately above the surface of the water, and the surface temperature. Figure 6.6 illustrates the empirically determined relationship between these last two variables and Q_b.

Given that the intensity of radiation (i.e. the amount of energy) emitted by a body increases as its temperature increases, is the variation of Q_b shown in Figure 6.6 what we might expect?

At first sight, no: according to Figure 6.6, for a given relative humidity Q_b decreases with temperature. The explanation for this apparent anomaly lies in the fact that the warmer the air over the oceans, the more water it can hold before becoming saturated. Thus, a given relative humidity value at high temperature corresponds to a greater atmospheric water vapour content than the same relative humidity at a lower temperature. The more water vapour there is in the atmosphere, the more long-wave radiation is absorbed by it, and the more is radiated *back* to the sea-surface, thus decreasing the *net* loss of long-wave energy *from* the sea-surface. Global patterns of outgoing long-wave radiation (often referred to as OLR), compiled from satellite observations, are extremely useful climatological tools.

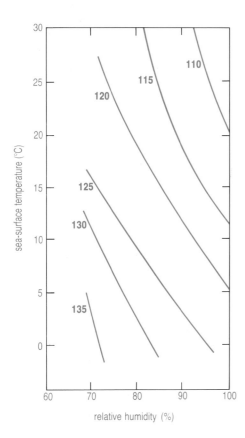

Figure 6.6   Curves showing how the net back-radiation, Q_b (in $W\,m^{-2}$), from the sea-surface to a clear sky, varies as a function of the sea-surface temperature and the relative humidity at an altitude of a few metres. Relative humidity is a measure of the degree of saturation of the air. It is defined as: (actual water vapour pressure)/(saturation water vapour pressure at the ambient temperature), expressed as a percentage.

Earlier, we defined Q_h as the amount of heat removed from the sea by conduction across the air–sea interface. However, if *only* conduction were involved, Q_h would be very small. As discussed in Section 2.2.2, over most of the oceans, especially in winter and/or in windy conditions, the atmosphere is unstable and subject to turbulent convection. Thus, air warmed by the underlying sea is swiftly removed, allowing more cool air to come into contact with the sea-surface. The same argument may be applied to Q_e: air above the sea-surface may become saturated with water vapour as the result of evaporation, but turbulent convection causes it to be quickly replaced by new, drier air.

Because Q_h and Q_e depend critically on the degree of turbulence in the atmosphere above the sea-surface, and fluctuate rapidly with the movement of eddying parcels of air, they are very hard to estimate. The most reliable method for estimating Q_h and Q_e involves determining the degree of correlation between the rapidly changing vertical upward velocity and the rapidly changing temperature (in the case of Q_h) or relative humidity (for Q_e). In each case, the higher the correlation, the higher the flux (of sensible heat on the one hand and latent heat on the other). In practice, this 'eddy correlation' method is very time-consuming but it is useful because it can be used to calibrate the results from simpler techniques, which can be undertaken from merchant vessels or other Voluntary Observing Ships.

These simpler methods depend upon calculating the vertical gradients of temperature (for Q_h) and atmospheric water vapour content (for Q_e) immediately above the sea-surface, and multiplying by a wind-dependent 'transfer coefficient' to give what are often described as 'bulk formulae'. The resulting estimates can be quite crude, not least because of problems in determining the wind speed, which may be done using an anemometer on a vessel or buoy and/or on the basis of the sea-state, which can be related to the Beaufort Wind Scale. Furthermore, the conditions under which the measurements are made can be less than ideal (e.g. because of the effect of heat from the ship itself). However, with the availability of the eddy correlation method mentioned above, and access to global surface wind data from satellites (e.g. Figure 2.3(c)), estimates of Q_h and Q_e are gradually improving.

In most areas of the ocean, the sea-surface is warmer than the overlying air, heat is lost from the sea by conduction and convection, and Q_h (defined as a loss) will be positive. Similarly, over most of the ocean, the water content of the atmosphere decreases with increasing distance from the surface, evaporation takes place, and Q_e (defined as a loss) will be positive. One of the few regions of the ocean where Q_e is negative is the Grand Banks off Newfoundland. Here, the sea-surface temperature (which is affected by the Labrador Current, cf. Figure 4.31) is generally lower than the air temperature, and the relative humidity is such that water vapour condenses onto the sea-surface, leading to a gain of (latent) heat by the sea, and formation of the fog for which the Grand Banks are famous. This type of fog, resulting from warm, moist air coming into contact with a cold sea-surface, is known as **advection fog**.

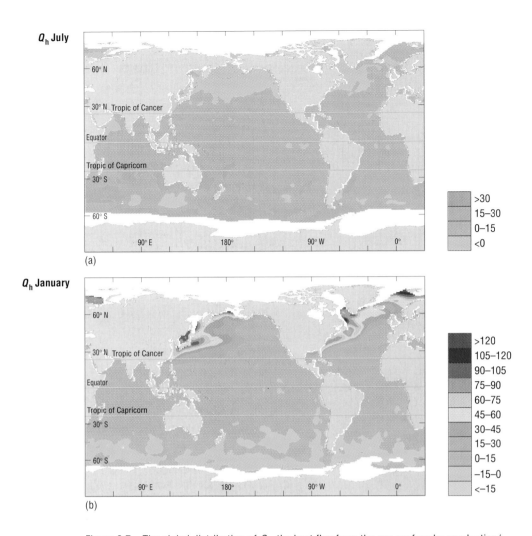

Q_h July

>30
15–30
0–15
<0

(a)

Q_h January

>120
105–120
90–105
75–90
60–75
45–60
30–45
15–30
0–15
–15–0
<–15

(b)

Figure 6.7   The global distribution of Q_h, the heat flux from the sea-surface by conduction/convection, in W m^{-2}, for (a) July and (b) January. Negative values (orange) indicate a gain of heat by the sea.

QUESTION 6.3   Figures 6.7 and 6.8 show the global distributions of annual mean values of Q_h and Q_e; negative values indicate a gain of heat by the sea and positive values indicate a loss of heat from the sea. To what phenomena do you ascribe the following?

(a) The relatively high positive values of both Q_h and Q_e in the western ocean off Japan and off the eastern United States, in January.

(b) Areas of negative Q_h, and relatively low positive values of Q_e, in the eastern parts of the equatorial Pacific and the equatorial Atlantic.

In extremely calm conditions, processes occurring at the molecular level become relatively more important, but this rarely occurs over the ocean. In turbulent conditions, eddying parcels of air carry heat and moisture away from the sea-surface, and Q_h (relating to 'sensible' heat, i.e. heat that gives rise to an increase in temperature that can be detected or 'sensed') and Q_e (relating to latent heat) vary together. In other words, generally speaking, under particular weather and sea conditions, the ratio Q_h/Q_e remains more or less constant. This ratio is known as Bowen's ratio, R.

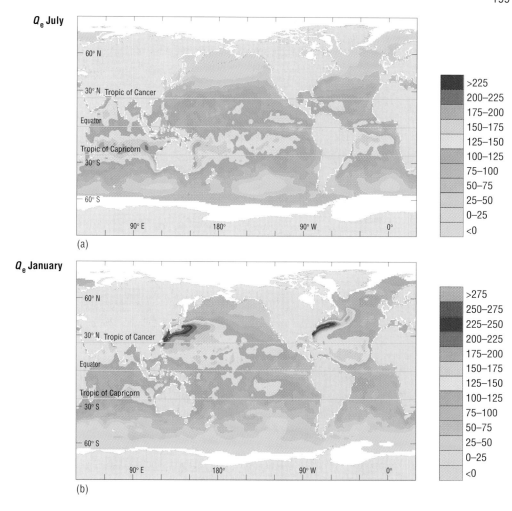

Q_e July

(a)

Q_e January

(b)

Figure 6.8   The global distribution of Q_e, the heat flux from the sea-surface by evaporation, in W m^{-2}, for (a) July and (b) January. Negative values (pink) indicate a gain of heat by the sea.

Figure 6.9 (overleaf) shows the variation with latitude of various contributions to the oceanic heat budget in July and January. Part (a) relates to the radiative terms, Q_s and Q_b, and part (b) to the 'turbulent fluxes', Q_h and Q_e.

According to Figure 6.9(b), what is the approximate value for Bowen's ratio (1) at the Equator, (2) at about 70° N in winter?

According to Figure 6.9(b), at the Equator, Bowen's ratio is about 8/80 or about 0.1; at about 70° N in winter it is about 60/90 or ~ 0.7. Put another way, Q_h ranges from about one-tenth of Q_e in low latitudes, up to about two-thirds of Q_e (or more) at high latitudes.

Part (c) of Figure 6.9 shows the net result of the gains and losses of heat in (a) and (b). Note that within 10–15° of the Equator there is a net heat gain all year, but outside these latitudes the sea-surface in the winter hemisphere experiences a net loss. This is shown even more dramatically in Figure 6.10, which also demonstrates how different the patterns of winter heat loss are in the two hemispheres.

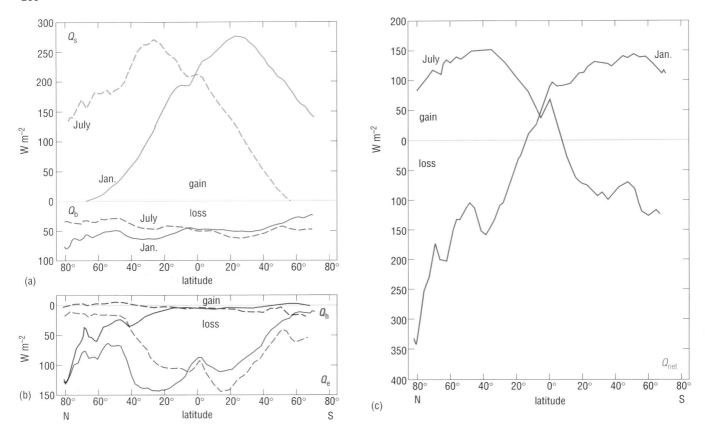

Figure 6.9 The variation with latitude of the mean values (per unit area) for heat-budget terms relating to heat transfer across the air–sea interface. In each case, the full curve is for January and the broken curve is for July.
(a) The terms for radiative gain and loss (Q_s and Q_b);
(b) the terms relating to the 'turbulent fluxes', Q_h (sensible heat, brick-red curve) and Q_e (latent heat, blue curve);
(c) the net heat loss/gain, Q_{net}.

By reference to Figures 6.7 to 6.10, what *are* the most marked differences in the patterns of heat loss in the two hemispheres?

The first factor is that in the Northern Hemisphere large amounts of heat are lost as Q_e from the warm western boundary currents (cf. Question 6.3) (this warm water moving polewards also accounts for high values of Q_b in the same regions). The second marked difference is that Q_h plays a much greater role in winter cooling of the ocean in the Northern Hemisphere than in the Southern Hemisphere (cf. Figures 6.7 and 6.9(b)) – this is a consequence of outbreaks of extremely cold air from the continents passing over relatively warm surface waters.

So far in this discussion of the oceanic heat budget, we have been ignoring Q_t on the assumption that for any given latitude band, the heat gained is equal to the heat lost – i.e. on average, $Q_t = 0$ – at least over periods of several years.

How *can* the average value of Q_t for any given location be zero if, as Figure 6.9(c) shows, there is always a net gain of heat at low latitudes?

Figure 6.9 also does not include the Q_v term for heat transport within the oceans. As discussed in earlier Chapters, ocean currents transport heat from low to high latitudes; the advective loss of heat from low-latitude regions ensures they do not continually heat up, while the advective gain of heat by high-latitude regions ensures they do not continually cool down. For the

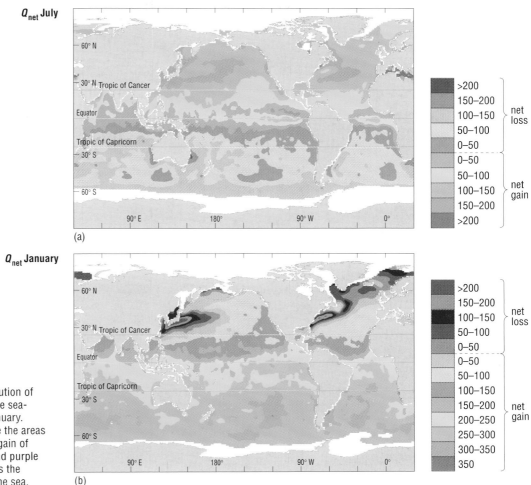

Q_{net} July

(a)

Q_{net} January

(b)

>200	
150–200	net
100–150	loss
50–100	
0–50	
0–50	
50–100	net
100–150	gain
150–200	
>200	

>200	
150–200	net
100–150	loss
50–100	
0–50	
0–50	
50–100	
100–150	
150–200	net
200–250	gain
250–300	
300–350	
350	

Figure 6.10 The global distribution of Q_{net}, the net heat loss/gain at the sea-surface, for (a) July and (b) January. Red, orange and yellow indicate the areas where there is the greatest net gain of heat by the sea; darker blues and purple indicate the areas where there is the greatest net loss of heat from the sea.

purposes of heat-budget calculations, it may therefore be assumed that at any location, over periods of a few years (so global warming may be ignored), the mean temperature of the water remains constant and Q_t is zero.

As a result of the poleward transport of heat in warm currents, the surface of the sea is generally above the freezing point of seawater (~ −1.9 °C) except at very high latitudes. If sea-ice *does* form, however, the radiation balance is changed dramatically. The **albedo** of the surface – i.e. the percentage of incoming radiation that is reflected – increases, perhaps to as much as 80–90%. Thus Q_s is greatly reduced; however, Q_b for ice is much the same as it is for water, so $Q_s - Q_b$ is also significantly decreased. Once ice has formed, therefore, it tends to be maintained. On the other hand, it has been estimated that the balance in the Arctic Sea is fairly fine, so that if the ice cover in a particular area were to *melt*, the resulting increase in $Q_s - Q_b$ might well *keep* the sea ice-free. Because of the positive feedback loop between decreasing ice/snow cover (on both land and sea) and decreasing albedo, climatologists believe that the effects of global warming will develop more rapidly in the Arctic than elsewhere. In this context, we have space only to note that over the last decade or so, pack ice in the Arctic Sea has become thinner, and open water has been observed at the North Pole in summer.

However, it is not a simple task to work out what effect local ice-melting or ice-formation will have on the overall heat budget. For example, when ice cover increases, heat losses by conduction/convection (Q_h) and by evaporation (Q_e) are reduced, but the temperature of surface waters is still likely to be lowered until a new heat balance is attained. During this period, the input of heat from adjacent parts of the ocean (Q_v) is likely to increase substantially. A small initial decrease in surface temperature in regions that are already close to freezing point can therefore have a considerable effect on the heat budget, not only of the overlying atmosphere but also of a very much wider area of ocean. In Section 6.3, you will see that the interaction between atmosphere, ocean and ice is further complicated by the important role played by the ocean's salt content.

6.2 CONSERVATION OF SALT

In considering the heat budget of the ocean, we have assumed that, over periods of several years at least, the Earth's heat supply – incoming solar radiation – remains constant, and that as a result the ocean is neither heating up nor cooling down. We have in effect been applying the principle of conservation of energy. In Section 4.2.3, we introduced the principle of continuity, which is another way of expressing the principle of conservation of mass which, because seawater is nearly incompressible, in turn approximates to the conservation of volume. Another conservation principle that is very important in oceanography is the principle of conservation of salt.

The principle that the amount of salt in the oceans remains constant may, at first sight, appear to be seriously flawed, because salt is being continually added by rivers, at a rate of about 3.6×10^9 tonnes a year. However, there is a consensus among marine scientists that the rates of input of dissolved substances to the oceans are balanced by their rates of removal to the sediments, so that the oceans are in a **steady state**. The salt content of the oceans, and hence the average salinity of seawater, therefore change little with time. So, for all practical purposes, the principle of conservation of salt is a valid one.

Of course, at a given location in the ocean, the salinity may be changed. Within the oceans, this occurs through the mixing together of waters with different salinities to produce water with an intermediate salinity. At the surface of the ocean, salinity is increased by evaporation, and decreased as a result of dilution by rain and snow and, occasionally, by condensation on the sea-surface. Figure 6.11(a) shows the distribution of the mean surface salinity and Figure 6.11(b) demonstrates the correlation between surface salinity and the balance between evaporation and precipitation ($E - P$).

QUESTION 6.4

(a) In general, are surface salinities higher in the Pacific Ocean or the Atlantic Ocean?

(b) In what way does Figure 6.11(b) demonstrate the effect of the Intertropical Convergence Zone?

(c) The surface salinities shown are annual averages, and locally there are marked seasonal variations. What feature would you expect to see in surface salinities in the Bay of Bengal during the northern summer?

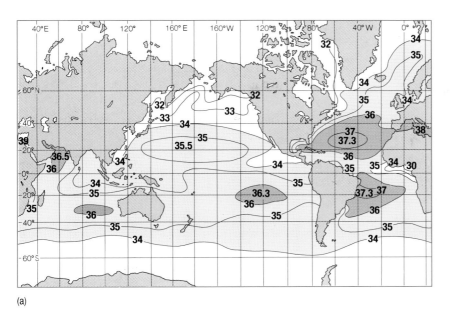

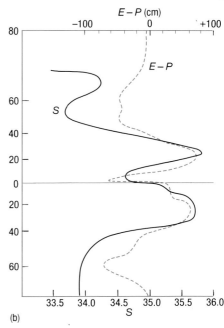

(a)

(b)

Figure 6.11 (a)   The mean annual distribution of surface salinity. Note that although they are effectively parts per thousand by weight, salinity values have no units because the salinity of a water sample is determined as the ratio of the electrical conductivity of the sample to the electrical conductivity of a standard. These salinity values are sometimes quoted as 'p.s.u.' or practical salinity units.
(b)   Average values of salinity, S (black line), and the difference between average annual evaporation and precipitation ($E - P$) (blue line), plotted against latitude.

The difference in surface salinities between the Pacific and the Atlantic is reflected in the marked difference between the average salinities of the two oceans as a whole: about 34.9 for the Atlantic and about 34.6 for the Pacific.

Awareness of the global variations of such factors as sea-surface salinity, local evaporation–precipitation balances – and indeed of the various heat-budget terms – is essential if we are to quantify fluxes of water (and heat) across the ocean–atmosphere boundary. The redistribution of salt and heat *within* the ocean is studied by monitoring the movement of bodies of water with characteristic combinations of temperature and salinity. These identifiable bodies of water are the subject of Section 6.3.

First, however, let us see how the principle of conservation of salt may be applied on a relatively small scale.

6.2.1   PRACTICAL APPLICATION OF THE PRINCIPLES OF CONSERVATION AND CONTINUITY

In practice, the principle of conservation of salt is most often used, together with the principle of continuity, to study the flow, or the evaporation–precipitation balance, of relatively enclosed bodies of water with limited connections with the main ocean. These might be fjords, estuaries, or semi-enclosed seas like the Mediterranean or the Baltic.

Figure 6.12 represents a channel, or some other semi-enclosed body of water, bounded by two vertical transverse sections with areas A_1 and A_2. Water enters the channel through A_1 at a rate of V_1 m^3 s^{-1} and leaves it through A_2 at a rate of V_2 m^3 s^{-1}. Between A_1 and A_2, water also enters the channel as a result of precipitation and run-off, and is removed by evaporation. The net rate at which water is being added by these processes is represented by F, which is also measured as volume per unit time (m^3 s^{-1}). If we assume that the volume of water in the portion of channel under consideration remains constant over a given period of time (and that water is incompressible), we may equate the total volume of water entering this portion of the channel with the total volume leaving, i.e.

$$V_1 + F = V_2 \qquad (6.2)$$

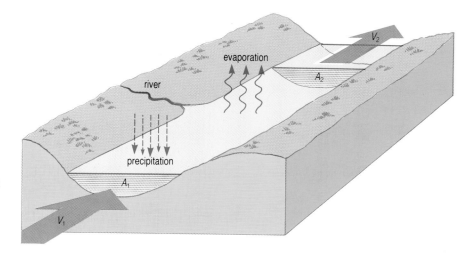

Figure 6.12   The flow of water into and out of a length of channel. Water flows into the channel through the section with area A_1 at a rate of V_1 m³ s⁻¹, and out through the section with area A_2 at a rate of V_2 m³ s⁻¹.

If we further assume that the average salinity between the sections remains constant, the amount of salt entering through A_1 must equal that leaving through A_2, because the processes of precipitation, run-off and evaporation do not involve any *net* transfer of salt. Because salinity values are effectively parts per thousand by weight (although salinity has no units – see Figure 6.11 caption), the mass of salt in a kilogram of seawater is the density (in kg m⁻³) times the salinity. Thus, the mass of salt transported across A_1 and A_2 per second must be $V_1 \rho_1 S_1$ and $V_2 \rho_2 S_2$, and these must be equal to one another, i.e.

$$V_1 \rho_1 S_1 = V_2 \rho_2 S_2$$

where ρ_1, ρ_2, and S_1, S_2 are, respectively, the mean densities and salinities of the water at A_1 and A_2. Proportional changes in density are very small compared with those of salinity – for example, a change in salinity from 30 to 36 (i.e. over almost the complete range of salinities found in the oceans) results in an increase in density of less than 0.5%. We may therefore ignore changes in density and write:

$$V_1 S_1 = V_2 S_2 \tag{6.3}$$

We now have two equations that we may solve for either V_1 or V_2. From Equation 6.3:

$$V_1 = \frac{S_2}{S_1} V_2 \tag{6.4}$$

Hence, substituting for V_1 in Equation 6.2:

$$\frac{S_2}{S_1} V_2 + F = V_2$$

$$F = V_2 \left(1 - \frac{S_2}{S_1} \right)$$

$$V_2 = \frac{F}{1 - \dfrac{S_2}{S_1}} \tag{6.5a}$$

and similarly,

$$V_1 = \dfrac{F}{\dfrac{S_2}{S_1} - 1} \tag{6.5b}$$

These equations give us a means of calculating the rates of flow across A_1 and A_2, if the mean salinity at each section and the rates of evaporation, precipitation and run-off are known. Also, the average volume of water flowing through the sections per unit time is in each case equal to the cross-sectional area $\times$ the mean current velocity, i.e. $V_1 = A_1 \bar{u}_1$ and $V_2 = A_2 \bar{u}_2$. Thus, if areas A_1 and A_2 are known, $\bar{u}_1$ and $\bar{u}_2$, the average velocities of the currents flowing through the sections, may be estimated.

In deriving Equations 6.3 to 6.5, we assumed that salt is carried into and out of the channel only by advection of the mean current; the effect of, say, an eddy of exceptionally high or low salinity could not be taken into account. Also, in estimating average values of the different parameters, any variations resulting from tidal flow would have to be allowed for. Nevertheless, this application of the principles of continuity and of conservation of salt provides an extremely useful tool in the study of semi-enclosed bodies of water. Here, we will demonstrate how the principles may be applied to flow into and out of the Mediterranean (Figure 6.13).

At the Straits of Gibraltar, Atlantic water of relatively low salinity flows into the Mediterranean Sea, while high salinity Mediterranean water flows out at depth.

According to Figure 6.13(b), what are the values of S_1, the mean salinity of water flowing into the Mediterranean, and S_2, the mean salinity of water flowing out?

Figure 6.13(b) suggests that S_1 is between 36.25 and 36.5, and that S_2 is between 37.0 and 38.0. The average values are about 36.3 for S_1 and 37.8 for S_2.

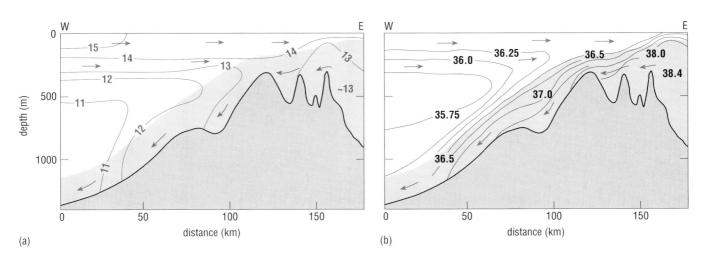

Figure 6.13   West–east sections across the Gibraltar sill of (a) temperature (°C) and (b) salinity, showing the inflow of Atlantic water in the upper layers and the outflow of Mediterranean Water (shown in blue) at depth. (The vertical exaggeration is about $\times 75$.)

QUESTION 6.5

(a) Evaporation exceeds freshwater input to the Mediterranean (i.e. input from rivers and precipitation) by $7 \times 10^4 \, \text{m}^3 \, \text{s}^{-1}$. Use this information, along with Equations 6.5(a) and (b), to calculate values for V_1 and V_2, the rates of inflow and outflow through the Straits of Gibraltar, in $\text{m}^3 \, \text{s}^{-1}$.

(b) The Mediterranean contains about $3.8 \times 10^6 \, \text{km}^3$ of water. Use your value of V_1 to estimate roughly how long it would take for all this water to be replaced once.

Direct current measurements in the upper layers of water in the Straits of Gibraltar indicate that V_1 is of the order of $1.75 \times 10^6 \, \text{m}^3 \, \text{s}^{-1}$, which suggests that application of the principles of continuity and conservation of salt can produce reasonably reliable results.

The value you calculated in Question 6.5(b) indicates that it takes about 70 years for all the water in the Mediterranean to be replaced. This is the **residence time** of water in the Mediterranean. The term 'flushing time' is also used, particularly in connection with flow in estuaries; it is useful because it is a measure of the extent to which pollutants are likely to accumulate.

As mentioned earlier, the evaporation–precipitation cycle is not the only mechanism whereby salinity may be changed. At high latitudes, the formation of ice and the addition of meltwater from glaciers and sea-ice have a significant effect on the salinity of seawater. In Section 6.3, we will see how the removal of freshwater and/or heat from surface water drives the deep thermohaline circulation.

6.3 OCEAN WATER MASSES

Most of what is known about the three-dimensional circulation of the ocean has been deduced from the study of bodies of water that are identifiable because they have particular combinations of physical and chemical properties. Such bodies of water are referred to as **water masses**, and the properties most often used to identify them are temperature and salinity.

Temperature and salinity can be used to identify water masses because (as mentioned in Section 4.2.4) they are conservative properties, that is, they are altered *only* by processes occurring at the boundaries of the ocean; *within* the ocean, changes occur only as a result of mixing with water masses having different characteristics. **Non-conservative properties**, on the other hand, are subject to alteration by physical, chemical or biological processes occurring within the oceans.

Water masses that form in semi-enclosed seas provide particularly clear examples of bodies of water with recognizable temperature and salinity characteristics. As discussed in the previous Section, deep water leaving the Mediterranean Sea through the Straits of Gibraltar is of unusually high salinity (Figure 6.13). This **Mediterranean Water** forms in winter in the north-western Mediterranean. Intense cooling and higher than normal evaporation, associated particularly with the cold, dry Mistral wind, increase the density of surface water to such an extent that there is vertical mixing, or convection, right to the sea-floor at more than 2000 m depth. The homogeneous water mass so formed has a salinity of more than 38.4 and a temperature of about 12.8 °C. As it leaves the Straits of Gibraltar at depth, intense mixing occurs at the interface with the incoming Atlantic water (Figure 6.13), and both its salinity and its temperature are somewhat reduced.

The least-mixed layer of Mediterranean Water in the adjacent Atlantic has a salinity of 36.5 and a temperature of 11 °C. Because of its relatively high density, it sinks down to about 1000 m depth where it becomes neutrally buoyant, and it spreads out at this level. Although it is being continually modified by mixing, Mediterranean Water can be recognized throughout much of the Atlantic Ocean by its distinctive signature of high temperature combined with high salinity (Figure 6.14).

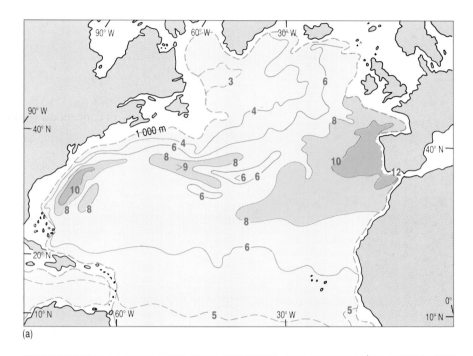

(a)

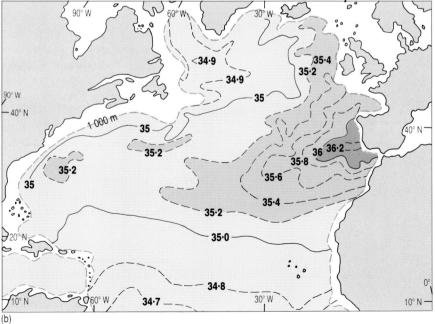

(b)

Figure 6.14   Distribution of (a) temperature (°C) and (b) salinity at 1000 m depth in the North Atlantic, showing the spread of Mediterranean Water. The broken black line is the 1000 m isobath. Note that on leaving the Straits of Gibraltar, Mediterranean Water first turns north under the influence of the Coriolis force and spreads along the coast of Portugal. It gradually mixes into the waters of the subtropical gyre and eventually spreads southwards and westwards.

Within the ocean there are a large number of distinct water masses, each characterized by temperature and salinity values reflecting a particular set of surface conditions, and generally considered to originate in a particular source region. You saw in Section 6.1 that the temperature of surface waters at any location in the ocean depends on the relative sizes of the components of the heat budget in that region; similarly, the salinity will depend upon the relative importance of the various factors discussed in Section 6.2. However, a water mass with particular temperature and salinity values will only result if a body of water is subject to specific meteorological influences over a significant period, during which it remains in the mixed surface layer. Furthermore, if the water is eventually to become isolated from the atmosphere, it must sink down from the sea-surface. These necessary conditions are satisfied in regions where surface waters converge.

QUESTION 6.6   From your reading of Chapters 4 and 5 in particular, suggest two regions of convergence of surface water, one in mid-latitudes and the other at high southern latitudes.

6.3.1   UPPER AND INTERMEDIATE WATER MASSES

If you compare Figure 6.15 with Figure 3.1, you will see that the geographical distribution of the world's upper water masses is strongly influenced by the pattern of surface currents. Upper water masses are generally considered to include both the mixed surface layer and the water corresponding to the upper part of the permanent thermocline, and they are therefore of varying thickness. If, as is the case in the region of the Equator, salinity is kept low by high precipitation and the temperature is high, the density of surface water will be low; the upper water column will therefore be stable, and only a very shallow water mass can form.

By contrast, the water masses that form in the subtropical gyres – also known as 'central waters' (cf. Figure 6.15) – are upper water masses of considerable thickness. As discussed in Section 3.4, in regions of convergence like the subtropical gyres (cf. Question 6.6) the sea-surface is raised and the thermocline depressed, leading to a thickening of the mixed surface layer (Figure 3.24(c) and (d)). Water sinks from the surface continually, but in winter, cooling of surface water leads to instability and vigorous vertical mixing occurs. As a result, water is alternately brought into contact with the surface and then carried deep down, so that a thick and fairly homogeneous water mass is formed.

The central water mass formed in the Sargasso Sea in winter (labelled Western North Atlantic Central Water on Figure 6.15) has temperatures ranging from 20.0 °C down to 7.0 °C.

What is the approximate lower depth limit of this water mass according to the temperature section across the Sargasso Sea shown in Figure 4.22?

It is about 1000–1100 m. Figure 4.22 also demonstrates two other relevant points. The first is that a large volume of water in the North Atlantic subtropical gyre has a temperature close to 18 °C. This '18 °C water' is an example of a **mode water**, that is, a volume of water within which temperature varies very little. The concept of a mode water is intimately related to the second point shown by Figure 4.22, which is that within the main body of Western North Atlantic Central Water, the isotherms are widely spaced; in other words, the waters are characterized by a thermostad or pycnostad (Section 5.1.1).

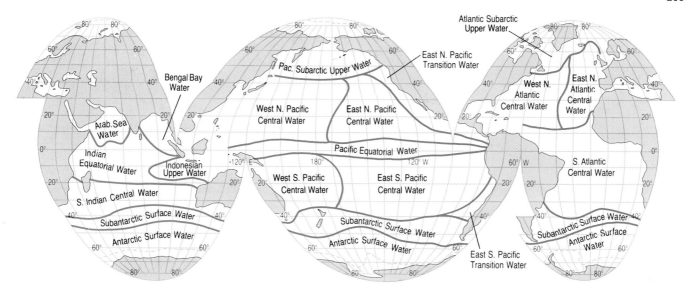

Figure 6.15 The global distribution of upper water masses. The boundaries between different water masses are not as sharp as the lines on this map might suggest. (You need not *remember* the details of this map.)

The temperature–salinity characteristics of Western and Eastern North Atlantic Central Waters are very similar. In common with central water masses in the other oceans, North Atlantic Central Waters have moderately high temperatures and above-average salinities.

Bearing in mind that the central water masses form below the anticyclonic subtropical wind systems, can you explain why their salinities are above average?

The subtropical anticyclones are regions where dry air subsides (cf. Figure 2.20), and net evaporation $(E − P)$ is high, leading to high salinities in the mixed surface layer (cf. Figure 6.11) and hence in the water mass as a whole. However, because of the differing environmental/climatic conditions on the two sides of the ocean, the salinities of Eastern North Atlantic Central Water are on average about 0.1 to 0.2 higher than those of its western counterpart. One reason for this is the stronger influence of Mediterranean Water on the eastern side of the ocean; another possibility is that because surface mixing penetrates to deeper levels on the western side of the ocean, the upper water mass is brought into close contact with the low-salinity Western Atlantic Sub-Arctic Water that underlies it.

Western Atlantic Sub-Arctic Water and Mediterranean Water are examples of *intermediate water masses*, which flow between the upper water masses and the deep and bottom water masses. Of the two, Western Atlantic Sub-Arctic Water is the more typical because, like most intermediate water masses, it forms in subpolar regions (cf. Figure 6.16) where precipitation exceeds evaporation, and its salinity is therefore low.

However, Western Atlantic Sub-Arctic Water consists largely of **Labrador Sea Water**, which – as you will see – may be found at great depths in the ocean, as well as at the intermediate depths covered by Figure 6.16. Most Labrador Sea Water forms in a cyclonic gyre on the offshore side of the Labrador Current (cf. Figure 5.26). In summer, the density of surface water in the Labrador Sea is lowered by the addition of freshwater from melting sea-ice and icebergs (Section 5.5.1). In winter, however, the surface water is cooled by the pack ice and by cold, dry Arctic air masses that have passed over northern Canada.

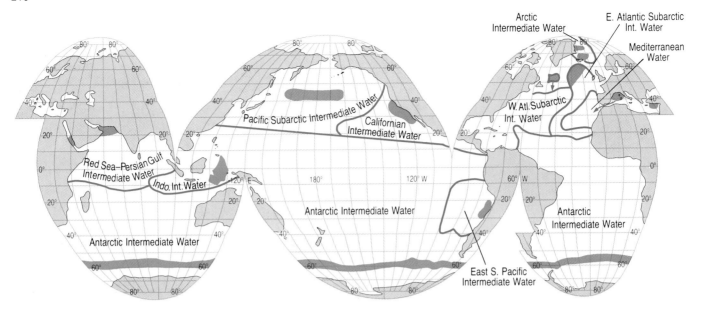

Figure 6.16 The global distribution of intermediate water masses (between about 550 and 1500 m depth). The source regions of the water masses are indicated by blue toned areas. Note that Antarctic Intermediate Water is by far the most widespread intermediate water mass. (You need not *remember* the details of this map.)

Which heat-budget loss terms are increased by the passage of these air masses?

Heat will be lost to the cold air masses by conduction and convection and, because they are also dry, by evaporation. Q_h and Q_e are therefore both increased. The density of surface water in the Labrador Sea is therefore increased through a fall in temperature and a rise in salinity. The increased salinity is still fairly low in absolute terms (~34.9) and the decreased temperature is relatively high (~3 °C) but their combined effect is sufficient to increase the density of surface water above that of the underlying water. The upper water column is therefore destabilized and vertical convection occurs, with denser surface water sinking and displacing less dense subsurface water, which rises to the surface. Surface water circulating in the gyre is thus repeatedly subjected to vertical mixing, so that eventually a water mass 1500 m thick or more may be formed. There is great variability in the depth of mixing and in the amount of Labrador Sea Water formed in any one year. In some years, there is no deep convection and no Labrador Sea Water is formed at all; Figure 6.17 shows data for a year when an enormous volume of the water mass was produced.

QUESTION 6.7

(a) Why do the cross-sections in Figure 6.17(a) and (b) indicate the presence of a large volume of Labrador Sea Water?

(b) Why are the contour patterns in Figure 6.17 consistent with cyclonic winds? And why would the effect of cyclonic winds render the uppermost waters of the Labrador Sea more susceptible to being destabilized?

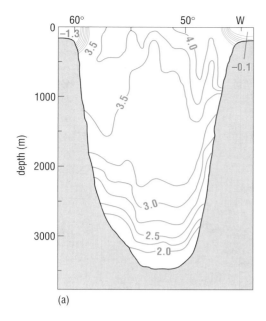

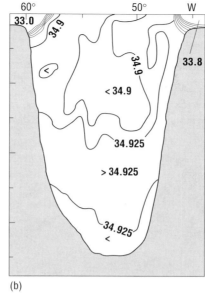

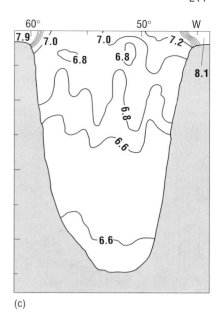

(a) (b) (c)

Figure 6.17 Section across the southern Labrador Sea at about 60° N, showing vertical distributions of (a) temperature (°C), (b) salinity and (c) oxygen (ml l⁻¹). As will be discussed in Section 6.5, atmospheric oxygen dissolves readily in cold surface water, so high concentrations of dissolved oxygen indicate water recently in contact with the atmosphere (but see the answer to Question 6.7(b)).

In the Mediterranean, too, the triggering of deep convection is aided by the fact that the cold, dry air of the Mistral blows over water circulating in a cyclonic gyre.

After sinking, Labrador Sea Water spreads out at mid-depths both eastward across the northern Atlantic, and southward in a current known as the *deep western boundary current*. As Figure 6.17 shows, Labrador Sea Water is very distinctive, with temperatures of 3–4 °C and low salinities (< 34.92), and often occupies a well-developed pycnostad. It is therefore considered to be a *subpolar mode water*.

The subpolar mode water that forms in the Antarctic Polar Frontal Zone (cf. Question 6.6) is known as **Antarctic Intermediate Water** (**AAIW**) and is the most widespread intermediate water mass in the oceans (Figure 6.16). It is characterized by temperatures of 2–4 °C and salinities of about 34.2, and is therefore significantly fresher than Western Atlantic Sub-Arctic/ Labrador Sea Water.

Can Antarctic Intermediate Water be identified on Figure 5.31?

Yes, indeed it can. In Figure 5.31(a), a 'tongue' of cold water can be seen extending down from the Antarctic Front to a depth of about 800 m; in the corresponding region in Figure 5.31(b), the near-surface isohalines dip downwards, indicating the sinking of relatively low-salinity water. The downward flow of Antarctic Intermediate Water may be seen in the slope of the isotherms and isohalines from this surface region to a depth of about 1500 m at 56° S (the northerly limit of the section).

After sinking at convergences in the Frontal Zone, Antarctic Intermediate Water spreads northwards throughout the Southern Hemisphere; in the Atlantic Ocean, its low salinity enables it to be traced to at least 20° N (Figure 6.18). The relationship between the flow of Antarctic Intermediate Water and the major deep and bottom water masses of the Atlantic Ocean is shown in Figure 6.19.

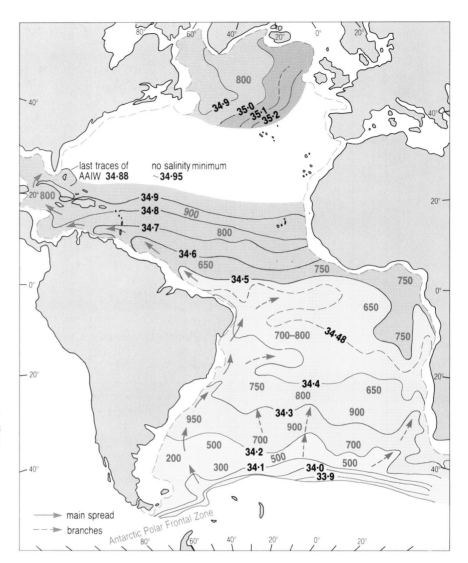

Figure 6.18 The spread of the least-mixed layer of Antarctic Intermediate Water in the Atlantic, as shown by the position of the salinity minimum at about 500–900 m depth. The numbers on the contours are salinity values; the numbers in blue give the approximate depth of the salinity minimum (in m). The thin dashed line is the 600 m isobath. The salinity contours in the north-western Atlantic show the southward spread of Labrador Sea Water (Western Atlantic Sub-Arctic Water).

Figure 6.19 Meridional cross-section of the Atlantic Ocean, showing movement of the major water masses; NADW = North Atlantic Deep Water; AAIW = Antarctic Intermediate Water; AABW = Antarctic Bottom Water. Water with salinity greater than 34.8 is shown yellow; note how the low-salinity tongue of AAIW extends northwards from the Antarctic Polar Frontal Zone, to overlie the more saline NADW. The **M** at about 35° N indicates the inflow of water from the Mediterranean. Water warmer than 10 °C is shown pink/orange, and that cooler than 0 °C (corresponding approximately to the distribution of AABW) is shown blue. The 'Subtropical Convergences' correspond to the centres of the subtropical gyres. The oxygen maxima and minima will be explained in Section 6.5.

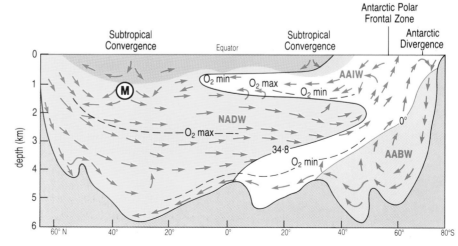

6.3.2 DEEP AND BOTTOM WATER MASSES

Deep water masses flow between the surface and intermediate water masses and the sea-bed; if the deepest water in contact with the sea-bed is distinguishable from overlying water, it is referred to as *bottom water*.

North Atlantic Deep Water

As shown by Figure 6.19, the major deep water mass of the Atlantic is **North Atlantic Deep Water** (**NADW**). The main surface source region of North Atlantic Deep Water is believed to be the subpolar gyre in the Greenland Sea. As shown in Figure 5.26, there is a fairly free passage of surface water between the Norwegian and Greenland Seas and the North Atlantic, with water flowing into the Seas mainly between Scotland and Iceland, and out mainly between Iceland and Greenland. However, the irregular plateau extending from Scotland to Greenland (and passing through the Faeroe Islands and Iceland) presents a major obstacle to flow at depths greater than about 400 m and a complete barrier at depths greater than about 850 m (Figure 6.20). Furthermore, the bottom topography isolates the Norwegian and Greenland Seas from water in the deepest basins of the Arctic Sea (Figure 5.26).

Cold polar water and ice from the Arctic enter the region via the Fram Strait, and are carried south in the East Greenland Current, which forms the western limb of the cyclonic circulation. Forming the eastern limb is the northward-flowing Norwegian Current, carrying surface water that is exceptionally warm and saline for these latitudes, with temperature and salinity values generally in excess of 8 °C and 35.25 respectively.

Can you identify the origin of this water?

This is water from the North Atlantic Current, the downstream continuation of the Gulf Stream. While circulating in the mixed surface layer of the Norwegian and Greenland Seas, this water is not only considerably cooled but also diluted, because – like the Labrador Sea – the region is generally one of excess precipitation. Furthermore, in the summer there is addition of freshwater from melting ice carried in the East Greenland Current.

In the winter, surface water in the Greenland Sea may become sufficiently dense to sink. As in the Labrador Sea, cyclonic winds lead to a surface divergence and a bowing up of the isotherms/isopycnals, circumstances which 'precondition' the surface water to be readily destabilized. Deep convection seems to occur in very short-lived events, in well-defined regions, perhaps only a few kilometres across, sometimes referred to as 'chimneys'. These form in response to particularly cold winds and/or ice-formation, but exactly how is not clear. As mentioned above, ice is carried down the western side of the gyre in the East Greenland Current. Eddies form along the ice-edge, and in some winters a 'tongue' of locally formed ice develops at about 72–76° N, sometimes extending well across the Greenland Sea (cf. Figure 5.27(b)). Development of this feature, which is known as the Odden (a Norwegian word meaning 'headland'), seems to encourage deep convection: the brine left behind when freshwater is abstracted into sea-ice increases the salinity of surface seawater, so that (given the low temperatures) it becomes sufficiently dense to sink into the depths of the Greenland Sea, from where it may spread into the Norwegian Sea (Figure 6.20).

The salinity of the dense water that accumulates and circulates in the deep basins of the Norwegian and Greenland Seas is increased through the addition of high-salinity Arctic water which enters at depth via the Fram Strait (Figure 6.20). This cold, very saline water results mainly from the seasonal formation of shelf ice around the margins of the Arctic Sea, which leaves behind dense 'brine' that sinks into the deep Arctic basins. The cold deep water that results from the high-salinity Arctic outflow and the deep water formed in the Greenland Sea is the densest in the world ocean. It has a salinity of ~34.9 and temperatures below 0 °C, and fills up the basin – which in places is 3 km deep – to the level of the submarine ridge.

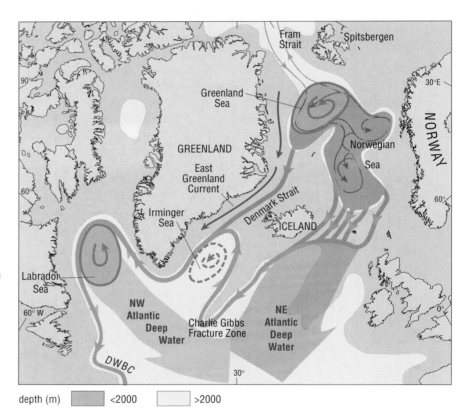

Figure 6.20 Schematic map showing the cyclonic gyres in the Greenland and Labrador Seas, which are the main areas of deep water formation in the Northern Hemisphere; the dashed cyclonic arrow is a less important region of deep-water formation, in the Irminger Sea. The dark grey arrows show the extremely saline Arctic water which enters the Greenland Sea through the Fram Strait, and which is an important component of North Atlantic Deep Water. Undiluted North Atlantic Deep Water, and the paths of the densest water that overflows the ridge between Greenland and Scotland, are shown in olive green; DWBC = deep western boundary current. The broad green arrows indicate the two types of North Atlantic Deep Water, North-East Atlantic Deep Water (darker green) and the less dense North-West Atlantic Deep Water, which overlies North-East Atlantic Deep Water in the eastern Atlantic. Note: The denser type of NADW is *also* formed from the Denmark Strait overflow.

From time to time, perhaps in response to storms overhead causing isobaric and isopycnic surfaces to tilt, the dense water accumulating in the Norwegian and Greenland basins flows over the submarine ridge. The overflows occur at various specific locations – notably through the Denmark Strait (between Iceland and Greenland), between Iceland and the Faeroe Islands, and through a narrow channel south of the Faeroe Islands (Figure 6.20). As the deep water overflows the plateau and flows down into the depths of the Atlantic, there is extensive turbulent mixing with overlying water, some of which is entrained. The characteristics of the water mass that results therefore depend not only upon the characteristics of Norwegian Sea and Greenland Sea deep waters, but also on the characteristics of the water above the various outflows and the degree of entrainment that occurs. The water mass that flows south from the overflow sites in the eastern basin of the Atlantic is sometimes known as North-East Atlantic Deep Water (or eastern North Atlantic Deep Water), and is shown as the darker of the broad green arrows in Figure 6.20. This is essentially a mixture of the original deep water and the overlying North Atlantic Central Water, and so has temperature–salinity characteristics of 2.5 °C and 35.03.

The paths followed by the densest, least-mixed overflow water are shown on Figure 6.20 in olive green. The overflow water from east of Iceland follows the topography around to the west, passes through the Charlie Gibbs Fracture Zone at ~55 °C and combines with the largest overflow from the Denmark Strait (see Figure 6.21(a)). Because of the effect of the Coriolis force, the overflow water 'hugs the topography', with the isopycnals sloping up steeply to the right of the flow (Figure 6.21(b)(i)).

Deep water may be added in the Irminger Sea off southern Greenland, where winter cooling of homogeneous Sub-Arctic Upper Water (Figure 6.15) leads to deep convection from surface to bottom. However, much more significant is the addition of dense water in the Labrador Sea, to form North-West Atlantic Deep Water, or western North Atlantic Deep Water (the lighter of the broad green arrows in Figure 6.20), which flows south in the western basin of the Atlantic and also spreads eastwards to overlie the denser North-East Atlantic Deep Water in the eastern basin.

The influence of Labrador Sea Water on North Atlantic Deep Water may be observed in northern and western parts of the North Atlantic. What other intermediate water mass has a significant influence on the characteristics of North Atlantic Deep Water, particularly on the *eastern* side of the ocean?

Mediterranean Water. As described earlier, the least-mixed layer of Mediterranean Water is generally found at a depth of about 1000 m (Figure 6.14), but its influence extends down to more than 2000 m.

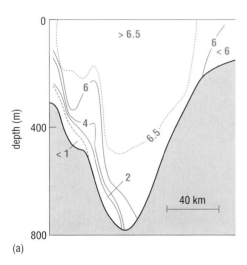

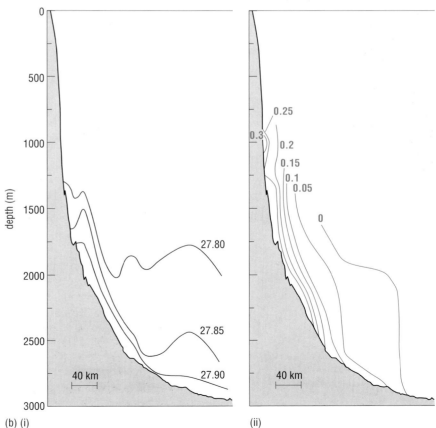

Figure 6.21 (a) Temperature section across the Denmark Strait between Greenland and Iceland, at about 65° N. The Denmark Strait is the deepest overflow channel, and so this is the largest outflow. (b) Sections across the Greenland continental slope at about 63° N, showing the overflow downstream of the Denmark Strait, as indicated by (i) isopycnals (for contour values see p. 230) and (ii) long-term mean current speed (m s⁻¹). The flow in (b) consists of the Denmark Strait overflow, plus the overflows from east of Iceland (cf. Figure 6.20). Note the different vertical and horizontal scales in (a) and (b).

Interannual variability in formation of the components of NADW

As discussed in Section 6.3.1, there is great variability from year to year in the depth of mixing in the Labrador Sea, and in the amount of Labrador Sea Water formed. Occasionally, Labrador Sea Water forms to the west of the gyre, and can enter directly into the deep western boundary current (see DWBC on Figure 6.20) without being caught up in the gyral circulation. However, it usually circulates at depth around the Labrador Sea a number of times, so that marked interannual variations in the signature of Labrador Sea Water in North-West Atlantic Deep Water are somewhat 'smoothed out'.

Nevertheless, over the last few decades, oceanographers have been noticing significant fluctuations in both the contribution of Labrador Sea Water to North-West Atlantic Deep Water and the rate of production of deep water in the Greenland Sea. Indeed, it seems that there may be a 'see-saw' relationship between convection in the two locations, in that as one increases the other decreases, and *vice versa*. These fluctuations may be part of changes related to the North Atlantic Oscillation (Section 4.5). Around the early 1990s, the NAO index was positive (Figures 4.40 and 4.41), and the Labrador Sea was subjected to unusually cold winds and winter storms, with the result that mixing there became deeper, and large amounts of Labrador Sea Water were formed (cf. Figure 6.17); at the same time, the Norwegian and Greenland Seas became warmer and less stormy, and convection became less intense and less deep, and the rate of formation of deep water there declined. Not surprisingly, there was a decline in the rate of formation of Labrador Sea Water in the early 1970s, when the Great Salinity Anomaly passed through the north-west Atlantic (Figure 5.28).

As will be discussed in Section 6.6, North Atlantic Deep Water plays a crucial role in the global climate system, so it is important to know how it is affected by changes in the rates of production of deep water in the Greenland and Labrador Seas. Oceanographers have been investigating the variation in the volume of dense water overflowing the Greenland–Scotland Ridge, and although (as mentioned earlier) individual overflow events seem to be triggered by meteorological changes over the Arctic region, there is evidence that during the course of the 1960s to 1990s, the rate at which dense water was spilling over into the deep Atlantic approximately doubled, and then returned to its previous value.

Antarctic Bottom Water

Antarctic Bottom Water (AABW) is the most widespread water mass in the world and is found in all three ocean basins, particularly in their southern parts (Figure 6.22). It seems to have two separate source areas – the continental shelf around the continent of Antarctica and deep levels in the Antarctic Circumpolar Current (cf. Figure 5.30).

Before we discuss how Antarctic Bottom Water forms, we should emphasize an important aspect of the stratification of polar waters, in the Arctic as well as the Antarctic. Except where ice is actually forming, surface water is relatively fresh, because of the combined effects of excess precipitation and the production of meltwater; this surface water is also very cold. Below the surface layer there is generally a layer of relatively warm water which is nevertheless denser because of its relatively high salinity. The stability of this layering may be destroyed through turbulent mixing caused by wind, combined with an increase in density of the surface waters. In northern polar latitudes, the density of surface waters is increased

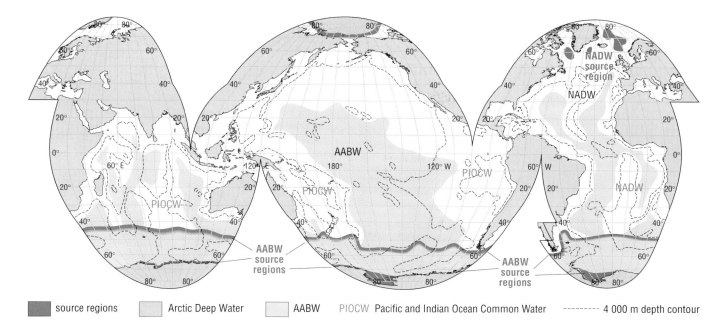

source regions   Arctic Deep Water   AABW   PIOCW   Pacific and Indian Ocean Common Water   --------- 4 000 m depth contour

Figure 6.22   The global distribution of deep and bottom water masses (between a depth of about 1500 m and the sea-floor). At high southern latitudes, the distribution of Antarctic Bottom Water is shown schematically; for a more realistic picture, see Figure 6.25. The source regions are shown by dark blue tone. The fine dashed line is the 4000 m isobath. (The unlabelled regions are to a large extent occupied by Pacific and Indian Ocean Common Water – see text.)

through winter cooling by cold winds (cf. Figure 6.7(b)) and, particularly around the Arctic basin, through ice-formation. In the Southern Ocean, where the seasonal production of ice is more extensive (Figure 5.27), the interaction between ice and surface waters plays an even greater role in the formation of dense water.

Ice–water interaction is perhaps best illustrated through discussion of **polynyas**, extensive areas of ice-free water within the winter ice cover. In the Antarctic region, there are two types of polynya: coastal polynyas and open ocean polynyas. *Coastal polynyas* develop when strong winds blowing off the Antarctic continent drive newly formed ice away from the shoreline, exposing a zone of open sea that might be 50–100 km wide. *Open-ocean polynyas* develop far from the coast within the pack ice (both over the continental shelf and in deeper water), and include the largest and most long-lived polynyas. The Weddell Polynya was an enormous area of open ocean that appeared in the Weddell Sea during three consecutive winters (1974–76); at its largest it measured about 1000 km by 350 km. The Weddell Polynya reappeared in approximately the same position each year (Figure 6.23) above a sea-bed topographic high, known as the Maud Rise. This implies that its formation was related to the upward deflection of relatively warm subsurface water, and in fact this is a possible explanation for a number of other open-ocean polynyas.

Recurrent polynyas are also found in the Arctic ice cover, but do not seem to be associated with sinking of dense surface water. In the Antarctic, both coastal and open-ocean polynyas are sites where surface water may be sufficiently cooled for deep convection to occur.

We stated earlier that water in contact with ice is cooled by conduction, until a new heat balance is attained. Why, then, are polynyas – regions with no ice cover – characterized by significant heat loss?

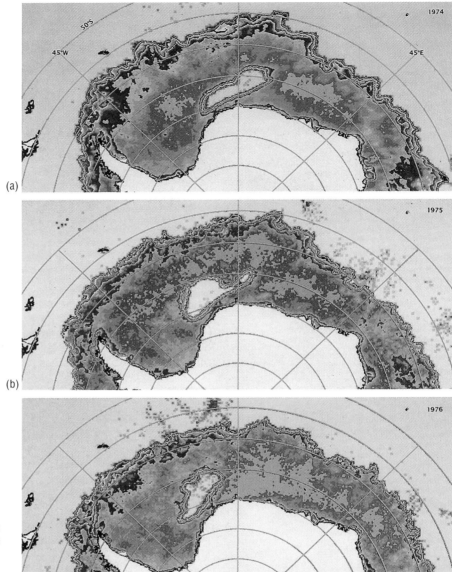

Figure 6.23 *Nimbus* satellite images showing the position of the Weddell Polynya during the southern winters of (a) 1974, (b) 1975 and (c) 1976. The westward movement of the polynya has been attributed to the generally westward current flow in the region (cf. Figure 5.30). The light blue tone corresponds to regions that are ice-free, while the purplish tone corresponds to regions that are more or less completely covered with ice. The Antarctic continent has been superimposed in white.

A layer of ice 'insulates' the ocean from the atmosphere and significantly reduces heat loss through conduction/convection, Q_h (as well as preventing heat loss through evaporation, Q_e). When there is no ice cover, heat losses to the atmosphere are greatly increased, especially when cold winds blow over the sea-surface.

In fact, the main mechanisms for heat loss to the atmosphere are different for the two types of polynya. Coastal polynyas have been described as 'sea-ice factories': the wind drives sea-ice away from the continent as soon as it freezes, re-exposing the sea-surface to the atmosphere so that more ice can form (Figure 6.24). The continued production of ice removes large amounts of heat from surface water, mainly in the form of *latent heat of freezing*. This is the heat lost to the atmosphere by water while it remains at freezing point (−1.9 °C for seawater) as ice crystals are in the process of forming, and is analogous to the latent heat of condensation released when water

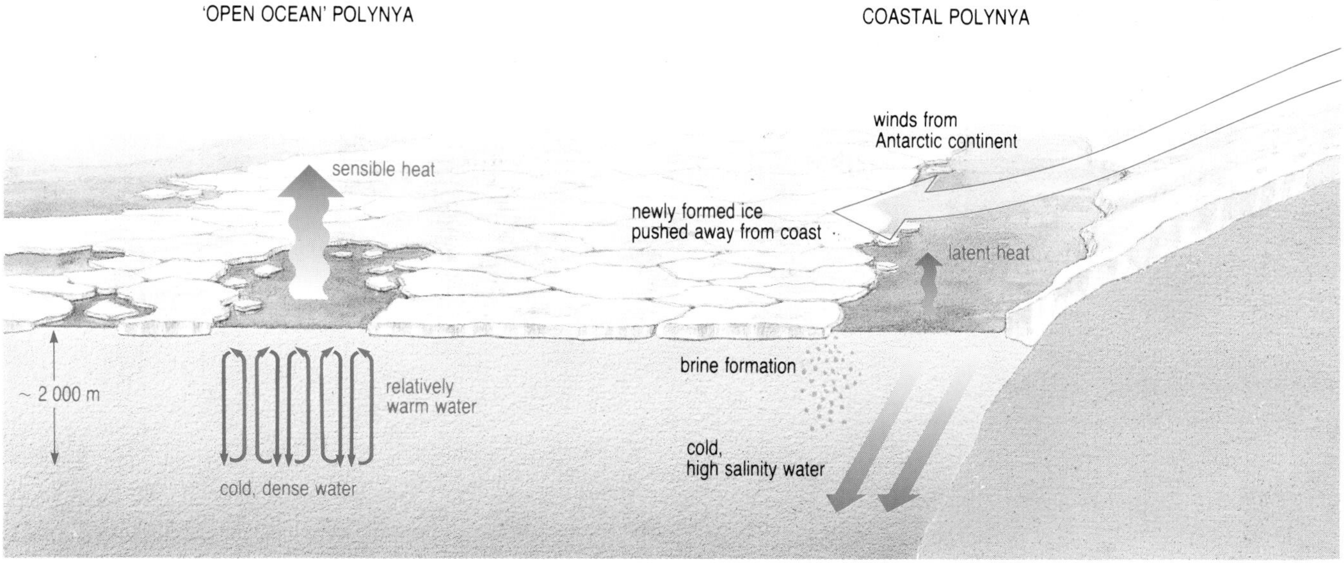

'OPEN OCEAN' POLYNYA   COASTAL POLYNYA

sensible heat

winds from
Antarctic continent

newly formed ice
pushed away from coast

latent heat

~ 2 000 m

relatively
warm water

brine formation

cold, dense water

cold,
high salinity water

Figure 6.24 The different roles played by coastal and 'open ocean' polynyas in the production of Antarctic Bottom Water.

vapour condenses to form liquid water. Coastal polynyas 'manufacture' ice on an enormous scale, perhaps producing much of the ice in the adjacent ocean. It has been calculated that the heat flux to the atmosphere from a coastal polynya is more than $300\,W\,m^{-2}$, enough to supply a ten-centimetre-thick layer of ice to the adjacent sea each day.

What effect will the continued production of sea-ice have on the salinity of surface water?

As discussed in connection with deep water in the Greenland and Norwegian Seas, the formation of ice results in an increase in the salinity of surface water. When ice forms, some salt is trapped amongst the ice crystals, but sea-ice is generally less saline than the water from which it forms and so the remaining water is correspondingly more saline. Sea-ice formation results in 'brine rejection' (as it is known) all round the Antarctic continent, but the effect is especially marked in coastal polynyas where ice is continually forming and being removed from the area.

In 'open-ocean' polynyas, heat is lost from the sea-surface mainly by conduction/convection (Q_h). Water cooled at the surface sinks and is replaced by warmer subsurface water which in turn is cooled and sinks, forming deep convection cells (Figure 6.24). Measurements made after the Weddell Polynya had formed showed that the temperature of deep water had changed dramatically, decreasing by $0.8\,°C$ all the way down to $2500\,m$ depth. It is reasonable to assume that the 'missing heat' had been carried to the surface by convection, and it has been estimated that the rate of overturning of water necessary to transport this much heat may have been as much as $6 \times 10^6\,m^3\,s^{-1}$ during the winter months, when the polynya was active.

QUESTION 6.8 It is thought that one of the factors that determine whether an open-ocean polynya will persist or freeze over is the production of meltwater from the surrounding ice. Why might the production of meltwater cause an open-ocean polynya to 'die', i.e. stop convecting?

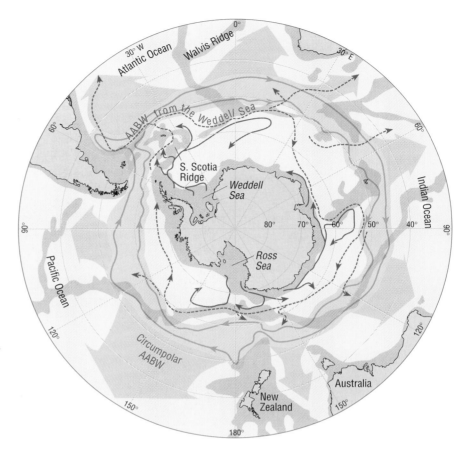

Figure 6.25 Schematic map to show the flow paths of the different varieties of Antarctic Bottom Water. Dark blue arrows show different components of the densest AABW, which originates on the shelf. The broken arrows show the AABW formed in the Weddell Sea, some of which eventually flows north in the western Atlantic and, to some extent, in the Indian Ocean. The thick blue–green arrows indicate production of the less dense component of AABW which forms from Antarctic Circumpolar Deep Water and flows north from the northern edge of the Antarctic Circumpolar Current. Blue–grey shading indicates water depths less than 3000 m.

As mentioned earlier, there seem to be two main types of Antarctic Bottom Water. The densest varieties originate at various locations over the Antarctic continental shelf, where water becomes sufficiently dense to sink as a result of winter ice formation and cooling, particularly in coastal polynyas. Having sunk from the surface, this water circulates for some time over the shelf. As a result, shelf waters have salinities of 34.4–34.8 and can be as cold as ~ −2 °C – this exceptionally low temperature can be attained because the freezing point of seawater decreases with increasing pressure. While flowing near the shelf break, these dense shelf waters mix with tongues of water from the Antarctic Circumpolar Current (Circumpolar Deep Water, discussed shortly) which have been carried southwards in cyclonic subpolar flows. The resulting mixtures flow westwards down the continental slope into the deep ocean (dark blue arrows on Figure 6.25).

These extremely cold varieties of Antarctic Bottom Water are all very dense but their temperature–salinity characteristics vary greatly, depending where they originate – for example, the dense water that forms in the south-western Weddell Sea is the coldest and the freshest, while that which forms in the north-western Ross Sea is the warmest and saltiest. Most of these water masses are so dense that they never escape from the Antarctic region, because having flowed down into one of the deep basins to the north of Antarctica, they remain trapped there (see unbroken dark blue arrows in Figure 6.25). However, some does manage to flow north into subtropical regions. This is the mixture of shelf water and Circumpolar Deep Water that, having circulated at depth around the Weddell Gyre for some time, overflows into the southern Scotia Sea, mostly through deep channels in the

South Scotia Ridge, and then flows northwards in the Atlantic western boundary current along the continental slope of eastern South America (broken blue arrows in Figure 6.25), beneath the Antarctic Circumpolar Current. Some of this water flows westward and then north into the southern basins of the Indian Ocean, and the remainder travels around the Antarctic continent until blocked at the Drake Passage, where its remnants fill the South Shetland Trench. The total transport of this 'true' Antarctic Bottom Water, with temperature–salinity characteristics of about −0.4 °C and 34.66, is probably about $10 \times 10^6 \, \mathrm{m^3 \, s^{-1}}$.

Most of the cold, dense Antarctic water that flows north to occupy the deepest parts of the three main ocean basins (cf. Figure 6.22) is exported from lower levels in the Antarctic Circumpolar Current. Water in the Antarctic Circumpolar Current is well mixed by wind and turbulence, and consists of North Atlantic Deep Water to which has been added Antarctic Bottom Water formed to the south, over the continental shelf (discussed above). This relatively warm (~0.25 to 2.0 °C), relatively saline (>34.6–34.72) water, known as Antarctic Circumpolar Water, may be clearly seen on Figure 5.31.

Poleward of about 50° S, there is only a small difference in density between the surface layer of cold, fresh water and the underlying (warmer, more saline) Antarctic Circumpolar Water, and so the stratification is easily destabilized and the layers mixed together by strong winds aided by the turbulence generated around ice-floes extending down into the water. Winter cooling of surface water, especially in open ocean polynyas, produces dense water that sinks while circulating eastwards around the Antarctic continent, perhaps a number of times. The resulting water mass is the 'Circumpolar Deep Water' mentioned earlier. An enormous volume of well mixed and homogeneous Antarctic water eventually flows equatorwards at depth from the northern edge of the Antarctic Circumpolar Current. This is labelled 'Circumpolar AABW' on Figure 6.25, and has a temperature close to 0 °C and salinity of ~34.6–34.7.

Because of the Coriolis force, the flow of Antarctic Bottom Water is concentrated along the western boundaries of the basins, but it is found in all the deepest parts of the basins of the Southern Hemisphere. In the western trough of the Atlantic Ocean, it flows northwards below southward-flowing North Atlantic Deep Water (readily distinguishable by its higher salinity), but on the eastern side of the Mid-Atlantic Ridge its northward passage is restricted by the Walvis Ridge, which extends south-westwards from south-west Africa to the Mid-Atlantic Ridge at depths of less than 3500 m.

Antarctic Bottom Water is the densest water mass in the open ocean. The components of North Atlantic Deep Water – newly formed Deep Waters of the Norwegian and Greenland Seas, and those of the Mediterranean – are denser, but they entrain such large volumes of less dense water in their turbulent passage through to the open ocean, that NADW ends up less dense than even the 'Circumpolar' type of Antarctic Bottom Water. This is not quite the whole story, however, as will be discussed in the next Section.

Before moving on, we should briefly return to Antarctic Circumpolar Water/ Circumpolar Deep Water, from which the circumpolar type of Antarctic Bottom Water forms. As mentioned above, its combination of relatively high salinity and relatively high temperature is due to the presence of North Atlantic Deep Water, which in flowing southwards has entrained Central Waters (themselves warm and saline) formed in the subtropical gyres, and has been influenced by the warm saline outflow from the Mediterranean.

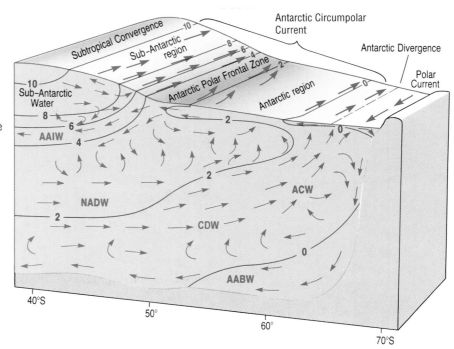

Figure 6.26 Block diagram showing the surface currents and vertical motion of water masses in the Atlantic poleward of about 40° S. North Atlantic Deep Water (NADW) becomes Antarctic Circumpolar Water (ACW) and rises to the surface at the Antarctic Divergence. Surface water flowing northwards from the Antarctic Divergence sinks at the Antarctic Polar Frontal Zone (as AAIW), while that flowing southwards may become AABW. CDW = Circumpolar Deep Water. Contours show isotherms in °C; this schematic diagram should be compared with Figure 5.31 which shows an actual temperature and salinity distribution measured in the Drake Passage between 56° and 62° S. Note: Here, we have not distinguished between the different types of AABW.

Figure 6.26 shows how the modified North Atlantic Deep Water/Antarctic Circumpolar Water/Circumpolar Deep Water flows up between Antarctic Intermediate Water and the very dense Antarctic Bottom Water, and rises towards the Antarctic Divergence (Section 5.5.2). Thus, warm water carried downwards in the subtropical gyres is transported polewards and then upwards until eventually it reaches the surface of the ocean around Antarctica, where a large amount of heat is given up to the atmosphere. Cooled water flowing northwards from the Antarctic Divergence will sink at convergences in the Antarctic Polar Frontal Zone in the form of Antarctic Intermediate Water, while that flowing polewards from the Antarctic Divergence may eventually be converted to Antarctic Bottom Water (Figure 6.26). The cycle of water mass formation is therefore an intrinsic part of the thermohaline circulation which, along with the surface current system, redistributes heat around the globe.

Pacific and Indian Ocean Common Water

The deep waters of the Pacific and Indian Oceans are very similar and are sometimes considered as making up one water mass, known as **Pacific and Indian Ocean Common Water**. Pacific and Indian Ocean Common Water is then the largest water mass in the ocean, accounting for about 40% of the total volume (cf. Figure 6.22).

There are no source regions of Deep Water in the northern Pacific, and Red Sea/Persian Gulf Water – a relatively warm, saline water mass – is the only dense water mass formed in the vicinity of the Indian Ocean (cf. Figure 6.16).

How, then, is the Pacific and Indian Ocean Common Water formed?

It must be formed through the mixing together of other water masses, specifically, Antarctic Intermediate Water, North Atlantic Deep Water and Antarctic Bottom Water. Of these, Antarctic Bottom Water is by far the most voluminous contributor to Pacific and Indian Ocean Common Water. It contributes about half the water, and North Atlantic Deep Water and Antarctic Intermediate Water together contribute the rest. As described earlier, North

Atlantic Deep Water flowing into the Southern Ocean mixes with Antarctic Intermediate Water above and Antarctic Bottom Water below. Some of the North Atlantic Deep Water which flowed south in the Atlantic continues its passage eastwards around the tip of South Africa, and – still mixing with the other water masses – flows northwards into the Indian basin and thence into the Pacific (the rest becomes subsumed into Antarctic Circumpolar Water). In Section 6.4, you will see how the observed temperature–salinity characteristics of Antarctic Intermediate Water, North Atlantic Deep Water and Antarctic Bottom Water can be used to quantify the contributions of these water masses to Pacific and Indian Ocean Common Water.

6.4 OCEANIC MIXING AND TEMPERATURE–SALINITY DIAGRAMS

By identifying the source regions of water masses, and following their subsequent passage through the oceans, we can build up a qualitative picture of the three-dimensional circulation of the oceans. As water masses move away from their source regions, their temperature–salinity characteristics are changed by mixing with adjacent water masses, and we can *use* these changes to assess *how much mixing* has occurred. Before describing how temperature–salinity characteristics may be used in this way, we should first say a little about how mixing occurs in the oceans.

6.4.1 MIXING IN THE OCEAN

As emphasized already, flow in the oceans is turbulent (Section 3.1.1), and mixing is predominantly the result of 'stirring' by turbulent eddies. Turbulent mixing is most pronounced along (gently sloping) isopycnic (density) surfaces, where it may occur with the least expenditure of energy, but mixing together of water with different densities also occurs. This is referred to as 'diapycnal mixing' (i.e. mixing *across* isopycnals).

In Section 6.3, we discussed the spread of water mass characteristics as if water masses always spread away from their source regions in an even manner. For example, Figure 6.14 suggests that Mediterranean Water spreads out gradually through the Atlantic in a continuous process, with the plume of Mediterranean water mixing at its edges with adjacent water masses and gradually losing its distinctive characteristics. This may be true for much of the Mediterranean Water entering the Atlantic but some, at least, is carried into the body of the Atlantic in spinning lenses of Mediterranean water known as 'meddies'. Meddies have been observed as close to the Straits of Gibraltar as the Canary Basin and as far away as the Bahamas, and it has even been suggested that the 'tongue' of Mediterranean Water shown in Figure 6.14 consists partly or wholly of 'decayed' meddies.

Meddies form from the Mediterranean outflow off south-west Portugal. They are typically 40–150 km across and may be as much as 1 km thick, and they can last as coherent bodies of identifiable Mediterranean origin for a considerable time. Figure 6.27 shows the tracks of three Sofar floats (Section 4.3.3) that were dropped into meddies in 1984 and 1985. Together, the tracks of the floats travelling in meddies 1 and 3 indicate that meddies may last at least two-and-a-half years. More recent observations indicate that meddies can last for up to five years, and that most eventually 'die' as a result of colliding with seamounts.

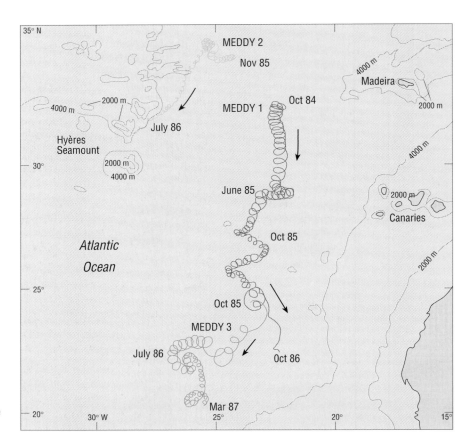

Figure 6.27 The paths of three meddies in the eastern Atlantic, as revealed by the tracks of Sofar floats, between October 1984 and March 1987. Remember that the floats were at about 1000 m depth, while the eddies themselves were occupying a depth of ~500–1500 m.

It is becoming clear that meddies and other mesoscale eddies play a significant part in the redistribution of heat and salt – exactly how much is not yet certain. On the one hand, their 'stirring' motion acts to even out inhomogeneities within water masses; on the other, individual eddies transport bodies of water with particular temperature and salinity characteristics from one part of the ocean to another (Sections 3.5.2, 4.3.5 and 5.5.2).

It was long thought that bulk mixing by turbulence was the only significant mechanism whereby water mass characteristics could be changed. According to this view, the behaviour of seawater is entirely determined by its *joint* temperature and salinity characteristics. However, if the discovery of mesoscale eddies shed new light on the processes contributing to oceanic mixing on a large scale, laboratory experiments demonstrated the existence of some very small-scale mixing processes. Over the past few decades, it has become clear that variations in temperature and salinity *of themselves* may cause mixing. Even in the absence of turbulent mixing, heat and salt diffuse through seawater as a result of processes occurring at the molecular level. Heat diffuses faster than salt, and the relatively fast transfer of heat from a layer of warmer, more saline water to a layer of colder, less saline water may be sufficient to cause small-scale instability. The best-known phenomenon associated with the double-diffusion of heat and salt is **salt fingering** (Figure 6.28), which was first observed below the outflow of Mediterranean Water into the Atlantic. It is now recognized that salt fingering and related processes can make a significant contribution to vertical mixing within the oceans generally, and that these very small-scale convective overturns affect the large-scale characteristics of water masses.

Figure 6.28 (a)   When warm saline water overlies cooler and less saline water, the more rapid diffusion of heat (red arrows) than salt (dots), leads to instability and the development of 'salt fingers', as shown in (b) and (c).

However, if we assume that mixing occurs through turbulent processes only, and ignore mixing through processes occurring at a molecular level, we may use joint temperature–salinity characteristics to study the global distribution of water masses and how different water masses mix together.

6.4.2 TEMPERATURE–SALINITY DIAGRAMS

In order to identify the water mass, or water masses, present at a given location in the ocean, a set of observations of temperature and salinity for successive depths at that location are plotted on a graph with temperature on the vertical axis and salinity on the horizontal axis, and the points are joined up in order of increasing depth. Today, this is generally achieved using data acquired by continuous profiling. The result is known as a **temperature–salinity (T–S) diagram**.

If a water mass is completely homogeneous, it will be represented on a T–S diagram by a single point, and will be described as a **water type**. Observations clustering around such a point indicate the presence of this water type. For example, a number of observations of newly formed Deep Water in the north-western Mediterranean would all plot at or around a point corresponding to $T = 12.8\,°C$ and $S \approx 38.4$.

Figure 6.29 shows what happens when two water types of differing temperature and salinity are mixed, and illustrates the following principles.

1   Whatever the relative proportions of the two water types, the point representing the temperature and salinity of any mixture must lie on the straight light joining the two water types on the T–S diagram.

2   Following on from 1, the actual position of the point representing the mixture will be determined by the relative proportions of the two water types. In the example shown in Figure 6.29, the point for the mixture, R, lies closer to type II than type I, so the mixture must contain a larger proportion of type II. The actual proportions of the two water types that are present in the mixture can be determined by measuring the lengths of the segments a and b, as shown in Figure 6.29. (Note that we have been using the term 'water type' rather than 'water mass', because we are assuming that mixing is occurring between two bodies of water each represented by a single temperature and salinity value, rather than by a range of values.)

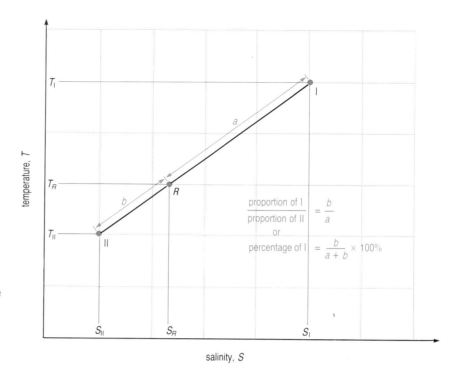

Figure 6.29 A temperature–salinity diagram showing the effect of mixing water type I having T_I and S_I with water type II having T_{II} and S_{II}. The resulting mixture R (having T_R and S_R) will be represented by a point on the line between I and II, the position of which will be determined by the relative proportions of the two water types in the mixture.

Although T–S diagrams can be used to predict the temperatures and salinities that would result from water masses mixing together, the usual application of the method is to determine the relative proportions of different (known) water masses contributing to the water we are interested in, and for which we know the temperature and salinity (from measurements). The following question illustrates how this can be done.

QUESTION 6.9   If water type I, with a temperature of 5 °C and a salinity of 35.5 mixes with water type II with a temperature of 2 °C and a salinity of 34.5, to give a mixture with T–S characteristics of 3 °C and 34.85, what are the proportions of water types I and II in the mixture? (You can use Figure 6.29 by lightly adding scales of your own choosing and then plotting the data directly onto it.)

Before moving on to consider how T–S diagrams can be used to interpret more complex situations, we should note that not all straight segments of T–S curves reflect mixing between water masses – they may indicate variations *within* one water mass. Such variations might result from waters of slightly different T–S characteristics forming at different times of year and sinking to different depths, according to their densities.

Alternatively, surface conditions may vary within the source region during the period (usually winter) when the water mass forms. Water mixing down along sloping isopycnic surfaces will eventually become *vertically* stratified (Figure 6.30(a)) and at a given hydrographic station, the water mass will be represented by a more or less straight line on a T–S diagram (Figure 6.30(b)). The water mass used as an example in Figure 6.30 is North Atlantic Central Water (Section 6.3.1), and because of the way that they form, Central Waters are generally recognizable as segments of T–S plots with this characteristic slope.

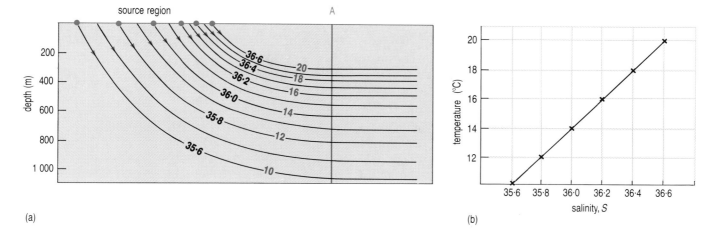

(a)

(b)

Figure 6.30 Highly schematic diagram showing how a water mass formed over an area with a range of surface conditions will be represented by a segment of a T–S diagram. (a) Water with a range of T–S values is formed at the surface and sinks/mixes down isopycnic surfaces, with the result that data collected at hydrographic station A plot onto a T–S diagram (b) to produce a line as shown. Temperatures are given in °C.

Returning to mixing *between* water masses, the procedure described above in relation to Figure 6.29 can be extended to the more complex situation where three water types, I, II and III, are mixing together. In this case, the mixture R must lie inside the triangle formed by joining points I, II and III together on the T–S diagram (as in Figure 6.31). If we know (from measurement) the temperature and salinity of R (T_R, S_R), the relative proportions of the intermixing water types contributing to R can be determined by a simple (if slightly tedious) graphical procedure. This is shown in Figure 6.31, which we can use to determine the relative proportions of water types I, II and III. Using a ruler to measure the segments a–f, and inserting values (here rounded off to the nearest 0.05) in the formula shown on the diagram, we get

$$\text{I} : \text{II} : \text{III} = 0.40 : 0.45 : 0.15 \text{ or } 40\%, 45\%, 15\%.$$

In other words, water type R on Figure 6.31 is the result of the mixing together of approximately 40% of water type I, 45% of water type II and 15% of water type III.

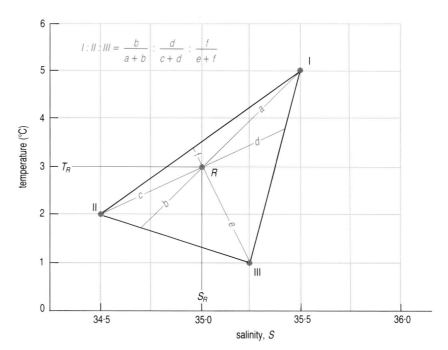

Figure 6.31 A temperature–salinity diagram showing the effect of mixing three water types (I, II and III) to give a mixture R with $T = 3\,°C$ and $S = 35$. The method used to determine the relative proportions of I, II and III in the mixture is shown here graphically and described in the text.

Of course, mixing occurs between adjacent water masses (e.g. the various upper water masses shown in Figure 6.15), but the mixing that is manifested in a *T–S* diagram plotted from data collected at a single hydrographic station is more often primarily from water masses which are one above the other. Let us consider how the distributions of temperature and salinity change as three such water masses mix together in the oceans, and how the *T–S* curve changes as a result.

Imagine three homogeneous water masses as follows: one at fairly shallow depth (200–600 m), one at intermediate depth (600–1000 m) and one fairly deep (1000–1400 m). We will assume that the intermediate and deep water masses have the same temperature but different salinities – this approximates to what is often found in the oceans. Temperature and salinity profiles are shown in Figure 6.32(a) and (b), while the *T–S* relationships are shown in part (c). The diagrams in 1 (at the top) illustrate the situation before any mixing has occurred, while 2 and 3 show subsequent stages as mixing progresses. As mixing occurs between *moving* water masses, the diagrams in 1, 2 and 3 should be seen as representing the situation at three different locations, although from the point of view of the moving water masses they correspond to three different stages in time.

Initially (stage 1), the three water masses are homogeneous and may be represented on the *T–S* diagram by three points (i.e. they are water types). As mixing progresses (stage 2), the sharp interfaces between the water masses become transition zones, so that the 'corners' of the temperature and salinity profiles become rounded off and water with characteristics between those at 400 m and 800 m, and between those at 800 m and 1200 m, appears on the *T–S* diagram. A layer of the intermediate water with its original temperature and salinity is still discernible at 800 m. This is known as **core water** and it shows up on the *T–S* diagram as a sharp point. As the core water continues to be affected by mixing both above and below, the sharp angle on the *T–S* diagram is eroded away, and the final *T–S* plot has the curved shape shown in stage 3. The changing shape of the *T–S* plot, from a sharp point to a curve, thus corresponds to the modification of the 'least-mixed layer' of the water mass as it spreads away from its site of formation (as shown by Figures 6.14 and 6.18 for Mediterranean Water and Antarctic Intermediate Water, respectively).

Stage 3 of the hypothetical scenario we have been considering is similar in many ways to what was actually observed at *Meteor* Station 200, at 9° S in the Atlantic Ocean. We can use the *T–S* curve for this station, shown in Figure 6.33, to apply the principles illustrated in Figures 6.29 and 6.31.

QUESTION 6.10   First locate the position of *Meteor* Station 200 on Figure 6.19. Now assume that at this station, the water between 400 m and 1800 m depth is the result of the mixing of Antarctic Intermediate Water with North Atlantic Deep Water (below it) and with water that has the *T–S* characteristics represented by the 400 m point on the *T–S* curve (above it).

(a) Is it possible to identify an eroded core of Antarctic Intermediate Water in Figure 6.33?

(b) What is the percentage of Antarctic Intermediate Water present in the water at 800 m depth (to the nearest 10%)? (Assume that the characteristics of Antarctic Intermediate Water and North Atlantic Deep Water correspond to the *centres* of the blue rectangles.)

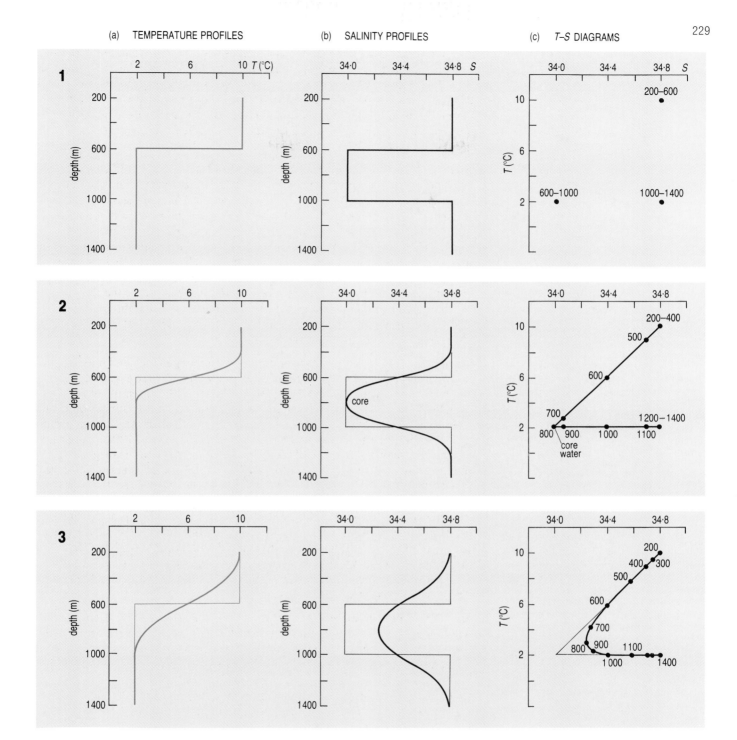

Figure 6.32 Profiles of (a) temperature and (b) salinity, along with (c) the corresponding *T–S* diagrams, to illustrate the mixing of three homogeneous water masses (water types). Stage 1 (top) represents the situations before any mixing has taken place; stage 2 (middle) shows an early stage of mixing when the core of intermediate water is very prominent; by stage 3 (bottom), the core has been eroded.

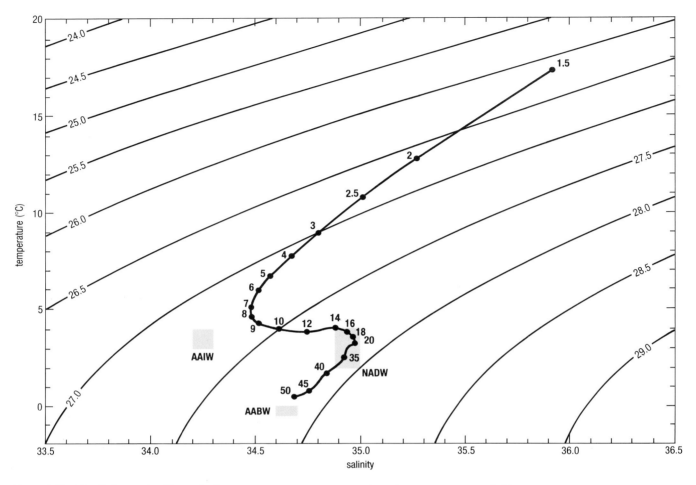

Figure 6.33 The *T–S* curve for *Meteor* Station 200, at 9° S in the Atlantic Ocean (for use in Question 6.10). Numbers on the curve give depths in hundreds of metres. The contours are lines of equal density, expressed as σ_t (see text). Observations were made at intervals from 150 m down to 5000 m. The *T–S* characteristics of Antarctic Bottom Water (AABW), North Atlantic Deep Water (NADW) and Antarctic Intermediate Water (AAIW) are also shown. Note that the AABW characteristics shown here are those for true AABW, and do not include that formed within the Antarctic Circumpolar Current; this is also true of Figure 6.19.

Temperature–salinity diagrams and stability

The density (ρ) of seawater is a function of temperature, salinity and pressure. By assuming constant (atmospheric) pressure, it is therefore possible to draw lines of equal density onto *T–S* diagrams. Variations in the density of seawater are very small – in the oceans as a whole, density at atmospheric pressure ranges between about 1025 and 1028 kg m^{-3}. For convenience, oceanographers generally write density in terms of σ (sigma), where: $\sigma = \rho - 1000$.

Using this system, densities of 1026.51 and 1026.94 kg m^{-3} are written as 26.51 and 26.94, and may more easily be compared with one another. These σ values were traditionally written without units, but are increasingly given units of kg m^{-3} (e.g. 26.94 kg m^{-3}).

Until the 1980s the temperature value used to calculate ρ and σ for a sample of water was generally the temperature at depth, i.e. its *in situ* temperature. The σ value calculated on the basis of *in situ* temperature, *in situ* salinity and atmospheric pressure is known as σ_t (**sigma-t**). In Figure 6.33, the equal-density lines used are lines of equal σ_t.

However, the situation is complicated by the fact that seawater is compressible, albeit only slightly.

Thinking back to Section 2.2.2 (concerning convection in the atmosphere), what implication does compressibility of seawater have for the temperature of seawater at depth in the ocean?

The temperature of deep water will be raised through adiabatic compression. As a result, the *in situ* temperature of a sample of water is higher than the temperature that would be recorded for the same sample of seawater at the surface, under atmospheric pressure. Adiabatic heating of seawater has two important consequences. The first is that because *in situ* temperature (T) is changed by pressure, it is not strictly a conservative property; however, **potential temperature** (θ) – i.e. *in situ* temperature corrected for compression and heating – *is* a conservative property.

The second consequence of adiabatic heating is that a T–S curve may give a misleading impression of the degree of stability of the water column. For a water column to be stable, its density must increase downwards, so that the T–S curve representing it would cross σ contours corresponding to successively increasing density values, in the direction of increasing depth (and the greater the rate of increase of density with depth, the more stable the water column). If you look at Figure 6.33, you will see that the part of the T–S curve corresponding to the very deepest water, between 4500 m and 5000 m, curves upwards, apparently indicating that at these depths density *decreases* with depth. This effect is spurious, and is partly the result of the *in situ* temperatures at depth being increased through adiabatic heating. If we were to plot θ against S, and then compare the trend of the resulting curve with contours of **sigma-theta** (σ_θ) – the σ values corresponding to the **potential density**, i.e. appropriate to the potential temperature, θ, salinity, and atmospheric pressure – we might well find that this water column appears to be stable all the way to the sea-floor (which is what we would expect, as an unstable water column could not persist for long).

However, this is not quite the whole story, because while using θ and σ_θ takes account of the effect of pressure on temperatures, and hence on calculated densities, it does *not* compensate for the direct effect of pressure on seawater density through compression. In other words, because σ_t and σ_θ correspond to density at atmospheric pressure, the shape of a temperature–salinity curve in relation to σ contours does not give a completely accurate impression of the stability of very deep water.

Under certain circumstances – including those represented by Figure 6.33 – the effect of compression can be significant. At atmospheric pressure, water with the temperature and salinity characteristics of Antarctic Bottom Water could well be slightly *less dense* than water with the characteristics of North Atlantic Deep Water – you can see this by comparing the density ranges represented by the blue NADW and AABW rectangles in Figure 6.33. However, because Antarctic Bottom Water is both colder and fresher than North Atlantic Deep Water, it is also more compressible, so at depth in the ocean it becomes sufficiently dense to flow beneath it.*

*Because of the complications caused by compressibility, in studies of deep and bottom water the σ_θ values quoted now often correspond to densities at a reference level other than the surface. In these circumstances, σ_θ values corresponding to densities at atmospheric pressure are referred to as σ_0, while values corresponding to densities at (say) 2 km, 3 km and 4 km, are referred to as σ_2, σ_3 and σ_4.

Water temperatures are increasingly recorded in terms of θ, and water masses are generally defined in terms of their θ–S characteristics, rather than their T–S characteristics. Likewise, σ_θ is often used instead of σ_t. More accurate methods of temperature measurement, combined with continuous profiling techniques, have meant that it is worth correcting for the effect of compression on temperature (and even, in some circumstances, its direct effect on density) in order to reveal interesting aspects of the temperature–salinity curve and/or deep-water flow paths. Nevertheless, for many purposes, measurements of *in situ* temperature are adequate, and both T and θ are widely used.

To consolidate your understanding of this Section, study Figure 6.34 – a θ–S curve for a station to the east of the Azores – and attempt Question 6.11.

QUESTION 6.11

(a) The water mass represented by the top 600 m or so of the θ–S curve is (Eastern) North Atlantic Central Water.
(i) What is the water mass below that, between about 800 m and 1200 m depth, and what is unusual about it at this location?
(ii) What is the deepest water mass represented by the θ–S curve?

(b) Imagine that two water masses with $\theta = 2\,°C$, $S = 35.04$ and $\theta = 8.5\,°C$, $S = 36.00$ mix together in approximately equal proportions. What is particularly interesting about the resulting mixture? (Begin by plotting the data points lightly onto Figure 6.34.)

The phenomenon identified in Question 6.11(b) – i.e. two water masses mixing together to form water with a higher density than either of the original contributions – is known as *cabelling*. It may be important in the production of deep and bottom water masses.

Water mass analysis has been invaluable to oceanographers attempting to build up a three-dimensional picture of large-scale flow within the oceans. Currents at depth are often too slow and/or too variable for their average motions to be easily determined directly. However, although temperature–salinity diagrams enable us to identify both the depth of the least-mixed (core) layer of a water mass and the direction in which it is spreading, they tell us nothing about the *rate* at which the water is moving. For this, we need to track water masses using a 'time-coded' tracer, or a *non-conservative* property of the water mass. How this may be done is discussed in Section 6.5.

There is, however, another *conservative* property of ocean water that we have not so far mentioned in connection with the tracking of water masses: its potential vorticity. As discussed in Section 4.2.1, away from regions of strong current shear, planetary vorticity f is very much greater than relative vorticity ζ and so, to a first approximation, potential vorticity is given by f/D where D is the thickness of the layer under consideration. In regions where water masses of significant vertical extent are forming, the water column is well mixed, i.e. there is a pycnostad in which isopycnic surfaces are widely spaced (Section 6.3.1). If we take D to be the distance between two selected isopycnic surfaces, then within the water mass f/D is relatively small (Figure 6.35); furthermore, it will remain so as the water mass spreads away from the source region, enabling it to be tracked. In Figure 6.36, the depth at which the potential vorticity within a particular water mass is at its lowest has been identified, and the salinity of the water at that depth has been contoured.

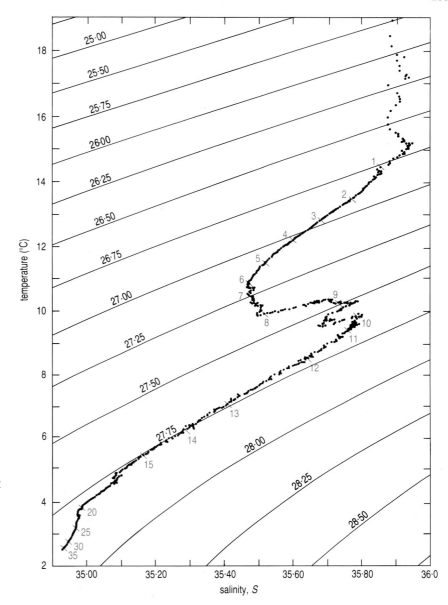

Figure 6.34   θ–S curve for station *Suroit* 1070, to the east of the Azores. The numbers on the curve are in hundreds of decibars, and so approximate to hundreds of metres. The topmost part of the curve, above 100 dbar, corresponds to the thermocline. Equal-density lines are lines of equal σ_θ. (Note the detail visible on this curve, which was computer-generated using θ and S data obtained by continuous profiling techniques. By contrast, the curve in Figure 6.33 was obtained from measurements made mostly at intervals of about 100 m or more.)

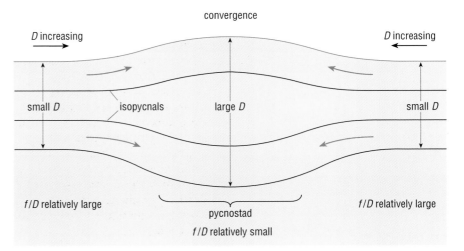

Figure 6.35   Schematic diagram to illustrate why potential vorticity is low in a homogeneous water mass; the example shown here is a central water mass. In the open ocean away from regions of strong current shear, $f \gg \zeta$ and so potential vorticity = f/D. If D is large, then f/D is small.

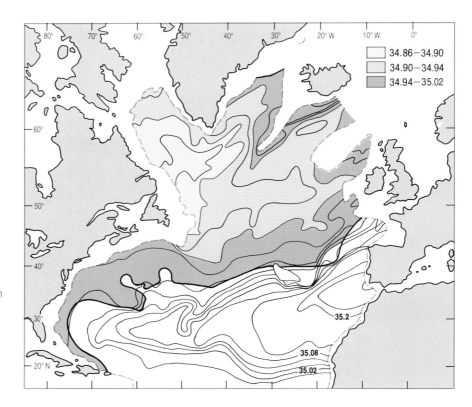

Figure 6.36 Distribution of salinity at the depth of the potential vorticity minimum (largest separation of isopycnic surfaces) associated with a particular water mass (see text). The depth of the potential vorticity minimum varies, but it is about 1500 m. The heavy line indicates the furthest limit at which the potential vorticity minimum can be identified.

The spread of which water mass is shown by Figure 6.36?

The region of origin of the water mass is the Labrador Sea, and this, combined with the low salinities associated with the water mass, identify it as Labrador Sea Water (cf. Figures 6.17, 6.18 and 6.20).

6.5 NON-CONSERVATIVE AND ARTIFICIAL TRACERS

The non-conservative property of ocean water traditionally most used as a tracer is the concentration of dissolved oxygen. Surface waters are supersaturated with dissolved oxygen because it goes into solution when air is churned down into the water by breaking waves and, in addition, is produced by phytoplankton photosynthesis. However, throughout the water column oxygen is used up in the bacterial oxidation (decomposition) of decaying material and in the respiration of organisms in general. Thus, the longer a water mass has been isolated from the atmosphere, the lower will be its content of dissolved oxygen. The oxygen content of surface waters is particularly high in polar regions, because cold water can contain more dissolved gas than can warm water. When water sinks from the surface in deep convection events, it carries with it high concentrations of dissolved oxygen. This can be seen clearly in Figure 6.17(c), illustrating the formation of Labrador Sea Water. On a larger scale, the oxygen maxima on Figure 6.19 correspond to the least-mixed, 'core' regions of cold water masses sinking down into the ocean, while the oxygen minima correspond to those 'older' parts of the water mass that are farthest from this least-mixed core. Thus oxygen minima correspond very roughly to boundaries *between* water masses.

Silica is also used successfully as a tracer of certain water masses. The bottom waters of the Southern Ocean contain high concentrations of silica because the diatoms that flourish in the nutrient-rich waters that have upwelled at the Antarctic Divergence use silica to build their skeletons, and most of this silica dissolves when they sink to the sea-floor on death. Figure 6.37 shows the distribution of dissolved oxygen and of silica along about 30° S, between South America and the Mid-Atlantic Ridge.

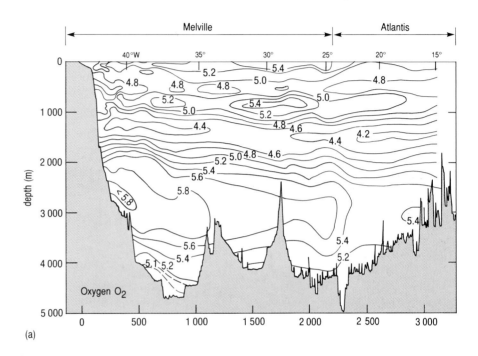

(a)

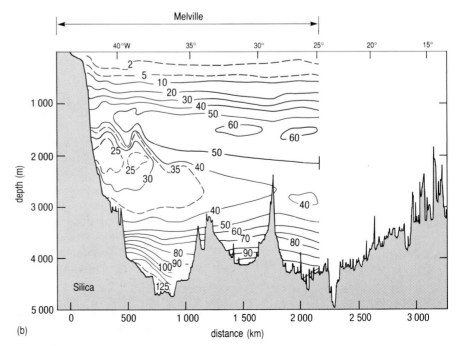

(b)

Figure 6.37   Sections of (a) dissolved oxygen concentration (ml l^{-1}) and (b) silica concentration (μmol l^{-1}) between South America and the Mid-Atlantic Ridge. The sections were made along 30° S (approximately) by the research vessels *Melville* and *Atlantis* in November 1976 and May 1959 respectively.

QUESTION 6.12   Given that North Atlantic Deep Water is characterized by low silica concentrations and Antarctic Bottom Water by high silica concentrations, can you distinguish these two water masses in Figure 6.37(b)? What striking aspect of their flow patterns may be inferred from Figure 6.37?

Silica concentration is particularly useful for studying the mixing of North Atlantic Deep Water with Antarctic waters. This is because the silica concentration of North Atlantic Deep Water is very uniform, in contrast to its temperature–salinity characteristics which, because it is the sum of various contributions, are fairly broad.

Dissolved oxygen and silica are of limited value as tracers, because without further information about their production and/or consumption in the water column we cannot use them to get an indication of when the water mass concerned was formed at the surface – that is, of its 'age'. However, during the latter part of the twentieth century, oceanographers were supplied with a number of useful tracers for which they *know* the exact time of entry (or at least the earliest possible time).

One group of substances that began to enter the oceans relatively recently is the chlorofluorocarbons (CFCs), which can be used in refrigeration systems and as propellants in aerosols. Chlorofluorocarbons, sometimes referred to as 'Freons' (their US trade name), are particularly useful to the oceanographer because they are relatively easy and cheap to measure at sea, in relatively small amounts of seawater. Some tracers – radioactive ^{39}Ar and ^{85}Kr, for example – may only be accurately measured by sampling hundreds of litres of water.

There is no natural source of chlorofluorocarbons. They were first manufactured in the 1930s, and up until the 1990s their concentrations in the atmosphere were increasing almost exponentially. Different chlorofluorocarbons have increased in concentration at different rates: for example, the atmospheric ratio of CCl_3F (known as CFC-11) to CCl_2F_2 (CFC-12) rose from near-zero in the mid-1940s to between 0.5 and 0.6 in the 1970s (Figure 6.38(a) and (b)). The CFC-11 : CFC-12 ratio of water now at depth in the ocean is a measure of the relative concentrations of the two CFCs in the atmosphere when the water was last at the surface (after allowance is made for their different solubilities). It is thus possible to 'age' the water; that is, to work out when it was last in contact with the atmosphere.

The CFC-11 : CFC-12 ratio stopped increasing in about 1977, but significant production of a third CFC – CFC-113 – started in 1975 (Figure 6.38(a) and (b)). Since then, its atmospheric concentration has increased so rapidly that it can be used to pinpoint the actual *year* that a water mass was last at the surface (providing it was post-1975). CFC-113 has proved particularly useful for the study of water masses that sink and spread very rapidly – North Atlantic Deep Water, for example.

The concentrations of chlorofluorocarbons may also be used simply as a 'fingerprint'. Figure 6.38(c), plotted from data collected in the late 1980s, shows concentration profiles for CFC-11 and CFC-12 (and, for comparison, dissolved oxygen) for the South Georgia Basin in the South Atlantic. The increased concentrations below 4000 m show the presence of Antarctic Bottom Water which had only recently sunk down away from the atmosphere. By tracking these elevated CFC (and oxygen) concentrations, it is possible to observe the northward spread of Antarctic Bottom Water. (The inert chemical sulphur hexachloride (SF_6) is often used similarly in short-term experiments involving the tracking of bodies of water; in this case, the marker chemical is added directly to the water to be tracked.)

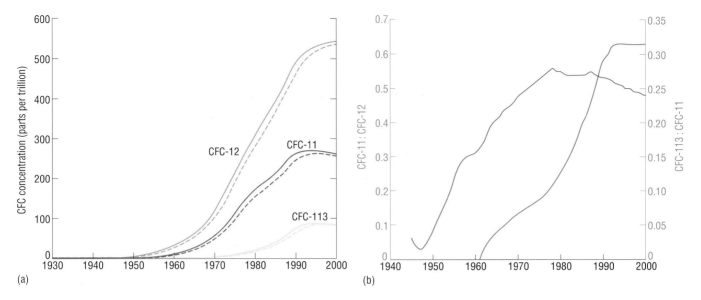

(a)

(b)

Figure 6.38 (a)   Changes in the atmospheric concentrations of CFC-11, CFC-12 and CFC-113 between 1930 and 2000.
(b)   Changes in the CFC-11 : CFC-12 ratio and the CFC-113 : CFC-11 ratio, up until 2000. As CFC-11 : CFC-12 ratios have remained more or less constant since 1977, they are not very useful for 'ageing' recently formed water masses.
(c)   Concentration profiles of CFC-11, CFC-12 and dissolved oxygen, in the South Atlantic at about 55° S. (1 pmol (= picomol) is 10^{-12} mol.) Note that despite its lower concentrations in the atmosphere, CFC-11 has higher concentrations in seawater than does CFC-12 because it is more soluble. These observations were made in the late 1980s.

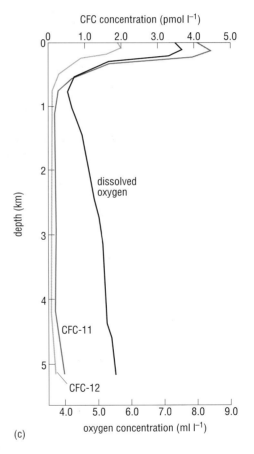

(c)

In Figure 6.39 (overleaf), the two sections of CFC-12 concentration, made 10 years apart, show vividly how surface water has been mixed downwards to varying extents at different locations in the North Atlantic. The high concentrations at 50° N correspond to Labrador Sea Water which, having been carried down to 1000–2000 m depth by winter convection, has then spread southwards and eastwards across the ocean.

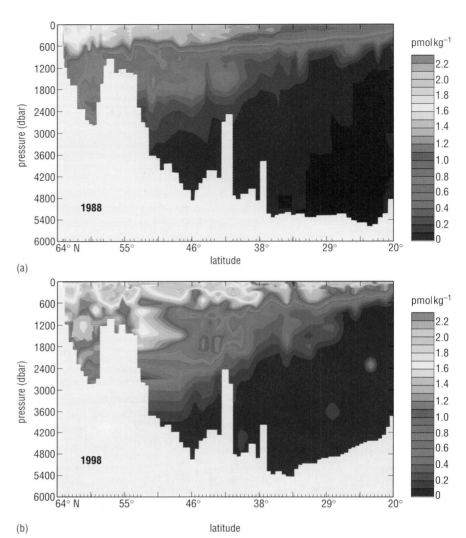

Figure 6.39 Sections of CFC-12 concentration, made along ~ 20° W, from Iceland to 20° N, (a) in 1988, from RV *Oceanus*, and (b) in 1998, from RV *Discovery*. During the intervening ten years, surface water has mixed down from the surface to varying extents, depending on the latitude. In (b), the high concentrations at 1000–2000 m depth at ~ 50°–55° N correspond to CFC-12 which dissolved in the surface waters of the Labrador Sea and was carried down by deep convection, before spreading across to intersect the transect along 20° W. (In both sections, the ticks along the bottom indicate the locations of the hydrographic stations.)

Two other tracers that have been used extensively are ^{14}C (carbon-14 or radiocarbon) and ^{3}H, tritium. Both are radioactive and are introduced into the atmosphere by nuclear explosions, whence they are carried into the ocean via precipitation and run-off. Both are also produced naturally in the atmosphere by cosmic-ray bombardment. ^{3}H has a relatively short half-life of about 12 years; ^{14}C, on the other hand, has a half-life of 5700 years, which makes it very suitable for studying the time-scales of oceanic mixing (though corrections have to be made to compensate for 'bomb' carbon produced in the 1950s and 1960s). Indeed, naturally occurring ^{14}C is perhaps the tracer most used to calculate ages and mixing times in the oceans.

Chemically, ^{14}C behaves exactly like the dominant isotope of carbon, ^{12}C. It therefore occurs in atmospheric CO_2, and is found in the oceans in dissolved and particulate form in both the hard and soft tissues of marine organisms. The highest ratio of ^{14}C to ^{12}C atoms is found in the atmosphere, and in surface waters in equilibrium with the atmosphere. When the carbon is removed from the air–sea interface – either carried down in solution by sinking water or through the sedimentation of organic particles (dead plankton and other organic debris) – the $^{14}C : ^{12}C$ ratio begins to decrease as the radiocarbon decays back to ^{14}N. Therefore, the older the seawater – i.e. the longer it has been away from the surface – the lower its ^{14}C : ^{12}C ratio will be. The ^{14}C : ^{12}C ratio of a sample of seawater can be measured using a mass spectrometer and the result used to calculate its age.

Carbon isotope data have been used to estimate a value for the residence time for water in the deep ocean of about 1500 years. However, it is now known that carbon in particulate form sinks from surface to deep waters very much faster than was originally thought, and as a result this estimate has been considerably revised. For example, in the North Atlantic, the mean residence time of water colder than 4 °C is probably closer to 200 years. This average conceals large geographical variations: for example, the residence time for water in the European Basin (the north-eastern North Atlantic) is about 13 years, while that for the Guinea Basin (off equatorial Africa) is more than 938 years. Calculations based on the ^{14}C distribution in the abyssal waters of the oceans indicate that average residence (or replacement) times for water below 1500 m depth are approximately 510, 250 and 275 years in the Pacific, Indian and Atlantic Oceans, respectively. By taking into account the volumes of the water masses involved, these residence or replacement times may be used to estimate the rates at which the various water masses form.

The implication of these finite residence times for the deep water masses is that, ultimately, the water must recirculate to the surface. In summary, the deep and bottom waters spread from the polar regions along the western sides of the ocean basins (Section 6.3.2, Figures 6.20 and 6.37). They spread eastwards from these deep western boundary currents (which flow counter to the surface western boundary currents) and well up through the ocean at speeds of perhaps $3–4 \times 10^{-7}$ m s^{-1} (~ 3 cm day^{-1}).This upwelling is a more or less uniform upward diffusion throughout most of the ocean, but is to some extent concentrated in the northern North Pacific. In certain well-defined regions, such as the Equatorial and Antarctic Divergences, or where there is coastal upwelling (Section 4.4), rates of upward movement of water are enhanced, and higher in the water column velocities may be several orders of magnitude greater (upward velocities of 2 m day^{-1} have been recorded in the vicinity of the thermocline). On returning to the upper layers of the oceans, the water rejoins the wind-driven circulation; eventually, it returns to polar regions where it enters the cycle again.

Figure 6.40 shows the simplified model of the deep circulation, which was derived theoretically by Stommel before any deep boundary currents had actually been observed. You may notice that according to this model the predominant flow of deep water in the Atlantic is from the northern source regions, so that the deep western boundary current flows southwards in both the North and South Atlantic. As discussed in Section 4.3.2, in the 1950s and 1960s, the southward-flowing deep western boundary current in the North Atlantic was observed directly using Swallow floats (Figure 4.24(b)), and indirectly by means of geostrophic calculations (Figure 4.22); it is shown schematically in Figure 6.20, flowing south from the Labrador Sea (having originated as the overflows to the east of Greenland). In the South Atlantic, however, the flow along the bottom is actually the northward flow of Antarctic Bottom Water, which is concentrated on the western side of the ocean (Figure 6.25) and over-ridden by the southward-flowing western boundary current of North Atlantic Deep Water. Figure 6.37 showed both these deep western boundary currents most clearly (cf. Question 6.12).

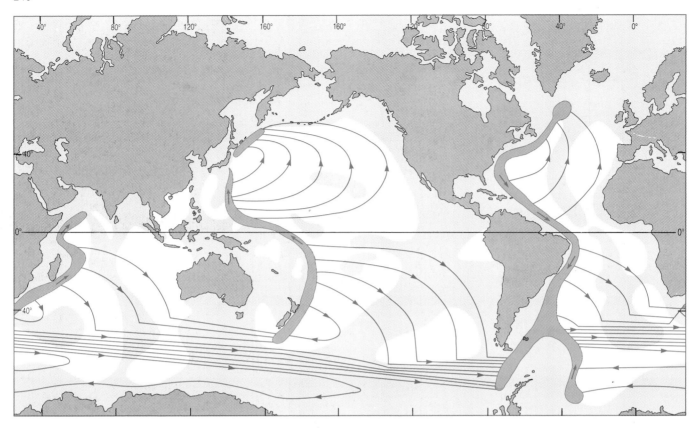

Figure 6.40 Stommel's simplified model of the deep circulation of the world ocean, with source regions of Deep and Bottom Water in the North Atlantic and the Weddell Sea. According to this model, in the Atlantic, flow in the deep western boundary current is southwards (apart from the flow from the Weddell Sea, which travels north and then eastward with the Antarctic Circumpolar Current); in the Indian Ocean and most of the Pacific, flow in the deep western boundary current is northwards.

6.6 GLOBAL FLUXES OF HEAT AND FRESHWATER

In this Volume, we have seen how heat is redistributed over the surface of the globe by winds in the atmosphere and by wind-driven surface currents and deep density-driven flow within the ocean. The simple plots showing heat transport in the atmosphere and ocean (Figure 1.5) show the end-result of very complicated processes, involving three dimensional flow of bodies of air and water of differing temperatures.

Fluxes of heat and salt are intimately inter-related (Section 6.3), and the global redistribution of heat and salt may also be considered in terms of the global redistribution of heat and freshwater. Of course, these fluxes are not independent, as the evaporation–precipitation cycle which results in the transport of freshwater from one part of the globe to another also transfers latent heat.

6.6.1 THE GLOBAL THERMOHALINE CONVEYOR

In the 1980s, Wallace Broecker suggested that the fluxes of heat and freshwater around the globe in ocean currents and water masses could be viewed as a kind of 'thermohaline conveyor belt' (Figure 6.41). This was not intended to be a realistic picture of warm and cold currents (although it is sometimes wrongly interpreted as such), but is a representation of the *overall effect* of warm and cold currents on the vertical circulation within the ocean. Its usefulness lies in the fact that because it reduces the oceanic circulation to its essentials, it allows us to think more easily about oceanographic and climatic problems. In particular, viewing the

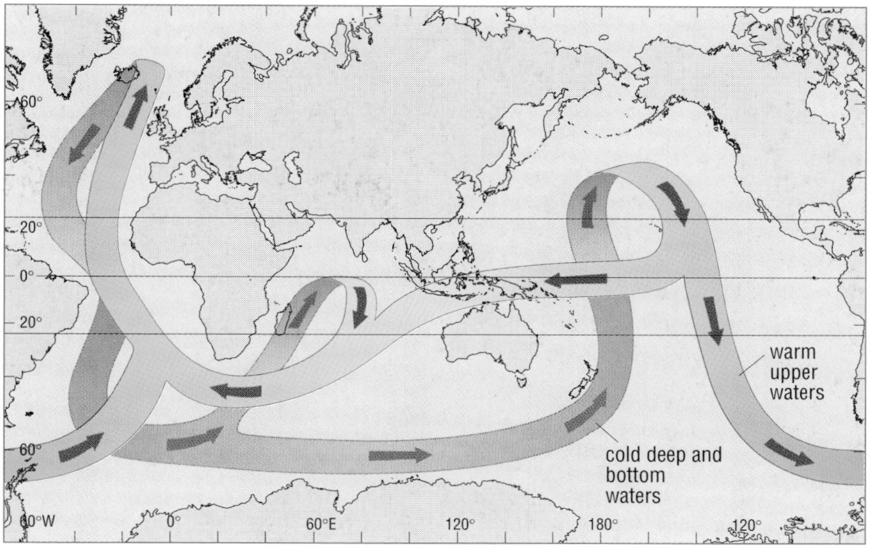

Figure 6.41 The thermohaline conveyor, first envisaged by Wallace Broecker. The red/orange part of the conveyor represents the net transport of warm water in the uppermost 1000 m or so, the blue part the net transport of cold water below the permanent thermocline. You may see other versions of this diagram elsewhere, because the current pattern in the Indian Ocean and the Pacific Ocean, and the seas in between, is not as well known as that in the Atlantic. This version assumes that there is a strong throughflow of warm water westward between the islands of the Indonesian archipelago.

thermohaline circulation as a 'conveyor' has attracted attention to the possible consequences of global warming in response to increased concentrations of atmospheric CO_2 produced by the burning of fossil fuels and deforestation – the so-called enhanced **greenhouse effect**.

Before addressing this question in particular, let's look at another related matter. The conveyor 'cartoon' graphically illustrates the idea that the driving mechanism for the global thermohaline circulation – sometimes referred to as the *meridional overturning circulation* – is the formation of North Atlantic Deep Water.

So, why is no Deep Water formed in the North Pacific?

It is sometimes said that the reason for there being no 'North Pacific Deep Water' is that the topography of the Pacific basin at northern high latitudes is so different from that of the Atlantic. It is true that there are no semi-enclosed seas like the Norwegian and Greenland Seas, where water that has acquired characteristic temperature and salinity values can accumulate at depth behind a sill. However, there *is* a well-developed subpolar gyral circulation (Figure 3.1). Might we not therefore expect some cold deep water masses to form and then spread out at depth, rather in the manner of Labrador Sea Water/North West Atlantic Deep Water (cf. Figures 6.20 and 6.36)?

The formation of deep water masses depends on the production of relatively dense surface water, through cooling and/or increase in salinity: North Atlantic Deep Water is both cold and relatively saline ($S \sim 35.0$). We have already noted that surface salinities in the Pacific are significantly lower than those in the Atlantic, particularly in the northernmost part of the basin, where values may be 32.0 or less (Figure 6.11(a)). The input of freshwater to the North Atlantic, through precipitation, rivers and melting ice, is about 104 cm yr^{-1}, while that to the North Pacific is about 91 cm yr^{-1}. Evaporation rates are about 103 cm yr^{-1} for the Atlantic and 55 cm yr^{-1} for the Pacific.

What does this suggest about the cause of the relatively low surface salinities in the North Pacific, in comparison with those in the North Atlantic – is it low evaporation or high precipitation?

The important difference is in the *evaporation* rates – compare the differing values of Q_e in the two areas (Figure 6.8). The resulting $E-P$ values ($-36 \, \text{cm yr}^{-1}$ for the North Pacific and $-1 \, \text{cm yr}^{-1}$ for the North Atlantic) reflect a net transfer of freshwater from the sea-surface of the Atlantic to the sea-surface of the Pacific, in the form of water vapour carried in the atmosphere.

But why the low rates of evaporation in the North Pacific? As graphically illustrated by the 'conveyor belt' image in Figure 6.41, in contrast to the North Atlantic which is supplied with warm water from the South Atlantic, and is a region where cooled surface water sinks, the North Pacific is continually supplied by cool water from below (cf. end of Section 6.5) and, as a result, has relatively low sea-surface temperatures (Figure 6.5). As discussed in Section 6.1.2, a cool sea-surface cools the overlying air, thereby reducing its ability to hold moisture and so reducing the evaporation rate. What is more, the effect of a low evaporation rate is to limit the extent to which the density of surface water may be increased through increase in salinity. It has been calculated that even if the surface waters of the North Pacific were cooled to freezing point, because of their low salinity they would still not be dense enough to sink and initiate deep convection. Paradoxically, therefore, in the North Pacific, a cool sea-surface prevents the surface layers from becoming sufficiently dense to sink, and there can be no North Pacific Deep Water.

Concerns about the effect of global warming on the thermohaline circulation are focused on whether such warming could result in North Atlantic Deep Water no longer being formed. The observed fluctuations in the rates of production of deep water in the Greenland Sea and the Labrador Sea (Section 6.3.2) seem to confirm that whether or not deep water production occurs may be very finely balanced, and it is possible that an increased production of fresh meltwater from glaciers and sea-ice will in effect 'turn off' the production of deep water in the Greenland Sea, by preventing surface layers from becoming sufficiently dense to sink, and hence inhibiting convection (cf. Question 6.8). Many (but not all) climate researchers believe that the episode of cooling known as the Younger Dryas, which occurred after the end of the last glacial period, was caused by a layer of fresh meltwater (from the large North American ice-sheet) spreading across the North Atlantic, so preventing the production of North Atlantic Deep Water.

QUESTION 6.13   What effect might a reduction in the rate of formation of deep water in the Greenland and Norwegian Seas have on the surface circulation of the North Atlantic in general (see Section 4.3.1), and why would this have consequences for the climate of north-west Europe?

On a more general note, at present, any increase in heat content caused by global warming is being distributed through the atmosphere *and the body of the ocean*. As a result, any rise in temperature at the surface of the sea (and the land) will be relatively slow. If formation of deep water masses ceased completely, and there were no sinking of surface water, any increase in the heat content of the atmosphere and ocean would be shared between the atmosphere and the uppermost layers of the ocean, and the rise in sea-surface temperature, and in the temperature of the atmosphere, would be relatively fast.

6.6.2 THE WORLD OCEAN CIRCULATION EXPERIMENT

At the end of the 1980s, in an effort to address the lack of knowledge about the role of the ocean in the global climate system, the international oceanographic community launched the **World Ocean Circulation Experiment** (**WOCE**). In summary, the primary aim of WOCE was to develop the means to predict climate change. To this end, the following aspects of the world oceanic circulation were investigated more thoroughly than ever before:

1   The large-scale fluxes of heat and freshwater, their redistribution within the ocean, and their annual and interannual variability.

2   The dynamic balance of the global circulation, and its response to changes in surface fluxes of heat and freshwater.

3   The components of ocean variability on time-scales of months to years, and space-scales of thousands of kilometres and upwards (and the statistics of shorter/smaller scale variability).

4   The mechanisms of formation of the water masses that influence the climate system on time-scales from 10 to 100 years, and their subsequent circulation patterns at depth.

In practical terms, the goal was to develop computer models to predict climate change, and to collect the enormous amount of data necessary to test them. The red lines in Figure 6.42 are the positions of the hydrographic sections – the traverses made by oceanographic research vessels (in some cases a number of times) – as part of WOCE during the 1990s. Previously, there had been a limited number of 'basin-scale' sections; the traverse of the Atlantic at 30° S by the *Melville* and *Atlantis* (Figure 6.37) was one of only a few such sections in the South Atlantic. In the Pacific, where the distances involved are enormous, the situation was even worse.

Figure 6.42   The hydrographic sections – the traverses made by oceanographic research vessels – as part of WOCE during the 1990s (red lines).

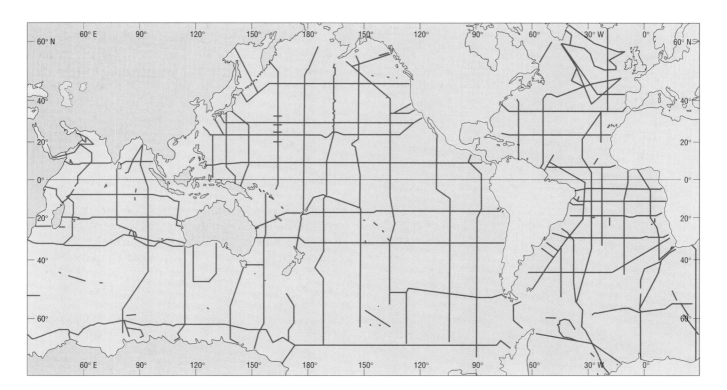

The transports of heat, salt and water across a section may be calculated on the basis of *T–S* data and information about current velocities. Where temperature and salinity data are available, it is possible to estimate the velocity across a section using the geostrophic method (Section 3.3.3). However, if no direct current measurements are available, such estimates are likely to be too low.

By reference to Figure 3.18, can you remember the reason for this?

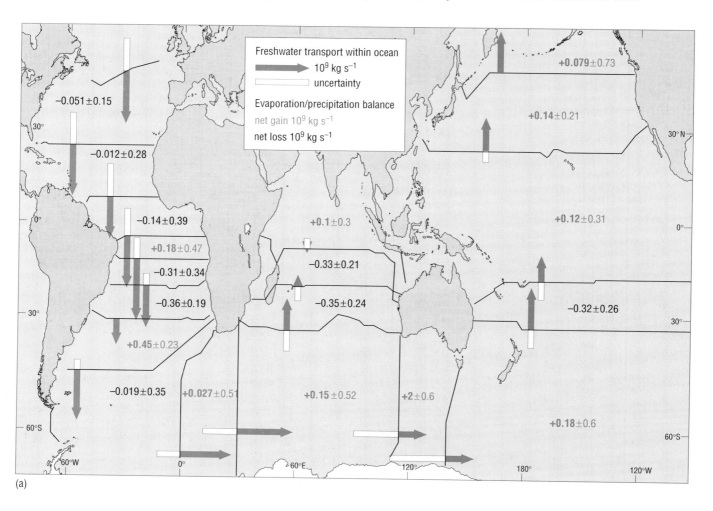

(a)

Figure 6.43   Estimated transports and air–sea fluxes of (a) freshwater and (b) heat around the globe. These values were obtained using hydrographic data collected during WOCE and (for the section between Australia and Indonesia) the Franco–Indonesian JADE programme. In both maps, lengths of arrows indicate transport within the ocean, the open part of the arrow corresponding to uncertainty in the value. In (a), the arrow in the key corresponds to a transport of 10^6 m^3 s^{-1}, or 1 sverdrup. As far as air–sea fluxes are concerned, in (a), positive numbers indicate net gain of freshwater by the sea, negative numbers net loss, and uncertainties by ±; in (b), red ovals indicate net gain of heat by the sea, blue ovals a loss of heat by the sea, with the uncertainty as an open oval. (1 pW = 1 petawatt = 10^{15} W.)

(*Further information:* The northernmost Atlantic and the Arctic are treated as one region in both maps. In (a), transport of freshwater between the northernmost Pacific and northernmost Atlantic, via the Bering Straits, is represented by a net northward transport of 0.73×10^6 m^3 s^{-1} at 47° N in the Pacific and a southward transport of 0.95×10^6 m^3 s^{-1} at ~48° N in the Atlantic (both on the basis of previous estimates). Net transport of freshwater across the southern boundary of the North Atlantic 'subtropical region' has been calculated using this value of 0.95×10^6 m^3 s^{-1} plus data on *E – P* for the region; transport across the southern boundary of the 'equatorial region' has been calculated using the transport from the north and *E – P* data for this equatorial region; and so on. Net freshwater fluxes through the Drake Passage and through the Indonesian Islands have arbitrarily been assumed to be zero.)

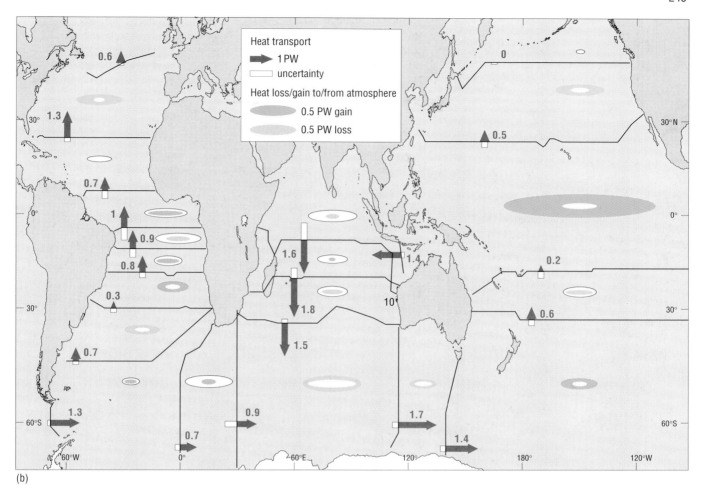

(b)

Geostrophic calculations only provide information about *relative* current velocities and so, without direct current measurements at depth, cannot reveal any 'barotropic' or depth-independent current flow. WOCE cruises resulted in a large number of direct current measurements, and the deployment of large numbers of surface and subsurface floats (e.g. ALACE floats, Section 4.3.4 and Figures 5.34 and 6.45).

An intrinsic part of the WOCE programme was the development of powerful computer models which could use the data collected to produce realistic estimates of global transports and fluxes of heat and of freshwater/salt (Section 4.2.4). Figures 6.43 and 6.44 are two examples of important results obtained by combining sophisticated modelling techniques with a very large database. In calculating the transports shown in these Figures, transports in the surface layer were assumed to be directly wind-driven (producing Ekman transport) and geostrophic velocities across the sections were calculated on the basis of large numbers of temperature and salinity measurements (cf. Section 3.3), and then adjusted until mass and conservative tracers were 'conserved' (i.e. fulfilled the condition: total inflow to the volume between two sections = total outflow).

Figure 6.43 shows net global transports, and fluxes across the air–sea interface, of freshwater and heat. Not surprisingly, for the North Pacific, Figure 6.43(a) shows a net gain of freshwater across the air–sea interface, while the North Atlantic shows a net loss. Note also, however, the large amount of freshwater carried within the ocean from the North Pacific into the Arctic through the Bering Straits – although the inflow through the Bering Straits is small (~ 1 m³ s⁻¹), as we have seen, it is of very low salinity – and the large amount of freshwater carried into the North Atlantic from the Arctic (much in the form of ice and meltwater).

Figure 6.43(b) shows a large northward heat transport in the Atlantic (chiefly a result of the fact that the South Equatorial Current crosses the Equator, cf. Figure 3.1), a small northward heat transport in the Pacific, and a large southward heat transport in the Indian Ocean.

In what important respect does the heat transport into the Atlantic presented in Figure 6.43(b) differ from the heat transport carried by the 'conveyor belt', as shown in Figure 6.41?

According to Figure 6.43(b), there is net transport of heat into the (South) Atlantic as a result of flow from the Pacific through the Drake Passage, but a net *export* of heat from the (South) Atlantic into the Indian Ocean, to the south of Africa. This result does not support the idea of heat transport *from* the Indian Ocean into the Atlantic, shown in Figure 6.41. On the other hand, the net transport of heat from the Pacific into the Indian Ocean through the Indonesian seas, is supported by Figure 6.43(b). If enough warm 'Indonesian throughflow' water eventually enters the Agulhas Current system, it is possible that a significant volume could make it into the South Atlantic.

Thinking back to Section 5.2.2, can you suggest how might this happen?

By means of Agulhas Current rings, formed from the Agulhas retroflection (Figure 5.14). This example of how mesoscale eddies might be playing an important role in the global thermohaline conveyor illustrates why computer models developed for climate studies need to have a sufficiently fine spatial resolution to be able to reproduce such eddies. The model used to produce Figure 6.43(a) and (b) did not take account of mesoscale eddies, except in the estimates of the uncertainties. The present lack of knowledge about how much heat and salt is carried in eddies, as opposed to transported by the mean flow, is largely responsible for the big uncertainties in both transports and fluxes, which are significant when compared with the estimated values themselves, particularly in the Indian Ocean.

Figure 6.44 is a best-estimate, coast-to-coast integrated (i.e. total net) transport of water, calculated using essentially the same model as used for Figure 6.43(a) and (b), again incorporating data collected largely during WOCE cruises. The arrows represent flow in three density classes, which we can take to correspond more or less to warm surface water, including upper and intermediate water masses (red arrows), deep water masses (blue arrows) and water masses flowing along the bottom (dark blue–green arrows). Note that, as in Figure 6.43, the arrows represent net transports. For example, the value for net northward transport of warm, upper layer water in the North Atlantic is given as about 16×10^6 m³ s⁻¹. This is because, although the northward transport in the Gulf Stream attains a maximum of about 150×10^6 m³ s⁻¹ (cf. end of Section 4.3.1), most of this

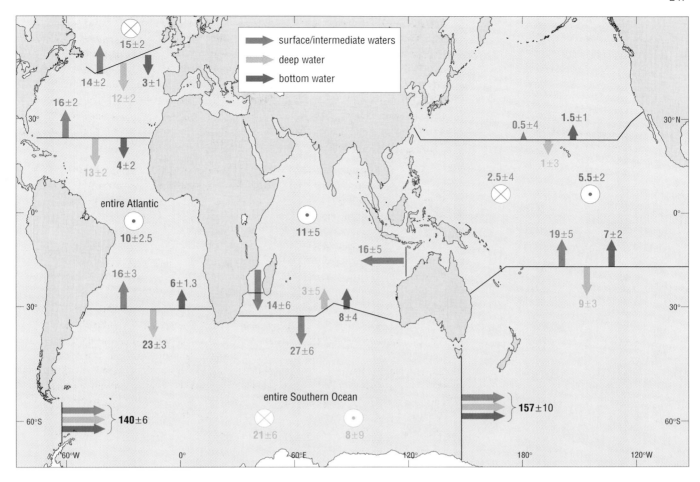

Figure 6.44   Transports of water across selected sections, in $10^6 \, m^3 \, s^{-1}$. Estimates of transports have been made for three different density classes, which for our purposes can be taken to represent warm surface water and upper and intermediate water masses (red arrows), deep water masses (blue arrows) and water masses flowing along the bottom (dark blue–green arrows). (In terms of σ_θ, the three density classes correspond approximately to $\sigma_\theta < 27.6$, $\sigma_\theta = 27.4$–27.6, and $\sigma_\theta = 27.8$–28.05.) The calculations used to produce this diagram make the assumption that the throughflow from the Pacific to the Atlantic, via the Bering Strait, is $0.8 \times 10^6 \, m^3 \, s^{-1}$. Also, according to the model, most of the Indonesian throughflow water flows south and joins the Antarctic Circumpolar Current, so increasing its transport south of Australia.
The colours of the 'upwelling' and 'downwelling' symbols ($\odot$ and $\otimes$) indicate the layer from which the water is coming.

(i.e. about $134 \times 10^6 \, m^3 \, s^{-1}$) recirculates and flows south in the eastern limb of the subtropical gyre. The important point is that for the circulation in a vertical plane, the net effect is an 'overturning circulation', corresponding to the 'loop' of the conveyor belt in the North Atlantic (cf. Figure 6.41), of $12 \times 10^6 \, m \, s^{-1}$ (see blue arrow in Figure 6.44).

In the North Atlantic, we can take the blue and blue–green arrows to correspond to North Atlantic Deep Water. The northerly blue–green arrow crossing the 30° S section in the South Atlantic (and similar arrows in the South Indian Ocean and South Pacific Ocean) correspond to the 'Circumpolar Antarctic Bottom Water' exported from lower levels in the Antarctic Circumpolar Current (Figure 6.25).

According to Figure 6.44, what is the southward transport of North Atlantic Deep Water at about 50° N? What is it at 30° S? Can you explain why the volume of the water mass has changed during its southward passage through the Atlantic?

At around 50° N, the southward passage of North Atlantic Deep Water is about $(3 + 12) \times 10^6 \, m \, s^{-1}$, i.e. $15 \times 10^6 \, m \, s^{-1}$. By 30° S, it has increased to about $23 \times 10^6 \, m \, s^{-1}$. This is because of entrainment of Antarctic Intermediate Water (included in the transport represented by the red arrow) flowing northwards above it, and of Circumpolar Antarctic Bottom Water flowing northwards below.

Estimates of transports and fluxes like those in Figures 6.43 and 6.44 – i.e. estimates made on the basis of a large amount of high quality data such as that collected during WOCE, with clearly defined assumptions and carefully calculated uncertainties – are essential as a reference for climate studies. If we do not have accurate data on the current state of the ocean, we will not be able to measure or predict future climatic change with a useful degree of certainty.

6.6.3 OCEANOGRAPHY IN THE 21st CENTURY: PREDICTING CLIMATIC CHANGE

In addition to the group of aims listed earlier, WOCE had a second main goal, which was to find ways of predicting climatic change over decadal time-scales. Building on work done in WOCE to determine the representativeness of the WOCE dataset, and to identify which specific aspects of the ocean and atmosphere will need to be continuously monitored if prediction of decadal climatic change is to be possible, the international research programme CLIVAR (*Cli*mate *Var*iability and Predictability) is now investigating variability and predictability on time-scales from months to hundreds of years, and the response of the climate system to global warming resulting from the greenhouse effect.

Hydrographic data collected over relatively long time-scales at fixed locations in the ocean indicate clearly that temperatures (and salinities) within the ocean are continually varying, typically over a ten-year time-scale. These changes are very small – only a few tenths of a degree Celsius in the uppermost part of the water column, decreasing with depth – but they are observed across the width of ocean basins and so represent large changes in oceanic heat content. We now know that much of this decadal variability may be explained in terms of climatic fluctuations such as the North Atlantic Oscillation, ENSO, the Pacific Decadal Oscillation and the Antarctic Circumpolar Wave (and other oscillations not mentioned in this Volume), which means that, in theory at least, future climatic fluctuations may be predicted. However, we do not yet fully understand how the observed decadal changes in the ocean relate to changes in the atmosphere. In particular, there is much to learn about the role of the ocean in driving, sustaining and modulating variability in the atmosphere.

Since the 1960s, it has been appreciated that El Niño events are phenomena resulting from the interaction of the atmosphere and ocean, in which the effect of a warm sea-surface feeds back into the atmosphere. However, until relatively recently it was not thought likely that the ocean could significantly influence the atmosphere *outside* tropical regions. In other words, it was assumed that the atmosphere–ocean system is driven primarily by the atmosphere, and so computer models designed to simulate changes in the atmosphere and ocean were not realistically coupled together. Progress is now being made in producing weather forecasts using coupled atmosphere–ocean models; however, although the models used for routine European weather forecasting 'update' the conditions at the sea-surface at intervals, they do not allow the atmosphere and ocean to evolve together. (Because the atmosphere changes faster than the ocean, this usually does not matter – but if, as in October 1987, the sea-surface is unusually warm, there can be unexpected consequences for the weather!) Coupled ocean–atmosphere climate models, in which changes in the ocean feed back into the atmosphere (and *vice versa*) do already exist, but are not generally linked to sophisticated general circulation models, such as that which produced the transports and fluxes shown in Figures 6.43 and 6.44.

One of the greatest advances in modelling has been the development of **data assimilation**, whereby real data are incorporated into models, which then adjust themselves (e.g. by changing the values chosen for the eddy viscosities) until the best possible match is achieved between the modelled distributions of (say) temperature, salinity and current velocity, and the distributions according to observations. Thus, not only does assimilation of data improve our understanding of the dynamics of the ocean, but it also effectively produces 'new data' in regions of the ocean where none were collected. This is clearly a very powerful tool, and the huge numbers of observations collected during WOCE, both within the ocean and at the sea-surface, thanks to a new generation of oceanographic satellites, have allowed it to become a much more profitable approach to modelling than was previously possible.

As described in Section 6.1, satellite-borne instruments have already greatly improved the supply of information about surface wind fields, and allowed better estimates of fluxes of heat and water across the air–sea interface, atmospheric water vapour content and sea-surface temperature. Arguably the most successful satellite launched during WOCE was *TOPEX–Poseidon*; since 1992, its two radar altimeters have been continually monitoring the topography of the sea-surface. The mean sea-surface topography obtained from *TOPEX–Poseidon* data (see Frontispiece) has provided information about the average surface current system (cf. Figure 3.21), while observed changes in the sea-surface topography have recorded fluctuations such as the development and decline of El Niño and La Niña conditions in the Pacific (cf. Figure 5.25). During the 21st century, our capability to observe the ocean from space will improve still further with the launch of new satellites.

In the future, the rate of acquisition of data via instruments *within* the ocean will also increase dramatically. The Argo Project involves 'seeding' the ocean globally with of the order of 3000 floats, similar in type to those shown in Figure 4.26(b). Argo floats will sink to 2000 m depth, and every ten days will rise to the surface to relay their position, along with data on temperature and salinity collected during the journey to the surface, via the *Jason* satellite. In addition, there will be an increase in the use of Autonomous Underwater Vehicles, or AUVs (one of the best known AUVs currently in use is *Autosub*). Together, freely drifting floats and directable AUVs should provide a continuous stream of valuable information about conditions in the upper ocean, so that changing oceanic conditions may not only be monitored almost in 'real time' but also, eventually, be predicted, as the severe 1997–98 El Niño event was predicted by the ENSO Observing System (Section 5.4).

The ultimate aim is to develop a cost-effective ongoing 'climate observation and prediction' system, analogous to the observation system which provides meteorological information worldwide. This system has been given the name **GOOS (the Global Ocean Observing System)**. When fully operational (from about 2010 onwards), GOOS will involve data-collecting, modelling and forecasting, and unlike WOCE, will be explicitly multidisciplinary. It is intended that there will be continual collection and dissemination of useful practical information relating to natural marine resources (e.g. fisheries), pollution and other natural hazards that might threaten ecosystems (and/or human health), plus other information needed for safe and efficient marine operations. There will also, of course, be an improved supply of data to the scientific community. If we cannot prevent global climate change and continually rising sea-levels, we should at least be able to make sensible plans to cope with them, based on accurate data and reliable predictions.

6.7 SUMMARY OF CHAPTER 6

1   The temperature and salinity of water in the ocean are determined while that water is at the surface. The temperature of surface water is determined by the relative sizes of the different terms in the oceanic heat budget equation; the salinity is determined by the balance between evaporation and precipitation $(E-P)$ and, at high latitudes, by the freezing and melting of ice.

2   The heat-budget equation for a part of the ocean is:

$$Q_s + Q_v = Q_b + Q_h + Q_e + Q_t$$

where Q_s is the amount of heat reaching the sea-surface as incoming short-wave radiation, Q_v is heat advected into the region in currents, Q_b is the heat lost from the sea-surface by long-wave (back) radiation, Q_h is the heat lost from the sea-surface by conduction/convection, Q_e is the net amount of heat lost from the sea-surface by evaporation, and Q_t is the net amount of heat available to raise the temperature of the water.

3   The net radiation balance $(Q_s - Q_b)$ is largely controlled by variations in Q_s; these depend partly on the latitudinal variation in incoming solar radiation but also on the amount of cloud and water vapour in the atmosphere. The amount of long-wave radiation emitted from the sea-surface depends on its temperature. However, Q_b is the *net* loss of heat from the sea-surface, and so is also affected by cloud cover, and by the water vapour content, etc., of the overlying air. $Q_s - Q_b$ is generally positive.

4   Q_h and Q_e depend upon the gradients of temperature and water content, respectively, of the air above the sea-surface. Both Q_h and Q_e generally represent a loss of heat from the sea, and both are greatly enhanced by increased atmospheric turbulence above the sea-surface.

5   Within 10°–15° of the Equator, there is a net gain of heat by the sea all year, but outside these latitudes, the winter hemisphere experiences a net heat loss. The patterns of heat loss in the two hemispheres are very different, with Q_h playing a greater role in winter cooling in the Northern Hemisphere. This is largely a result of cold, dry continental air moving over the warm western boundary currents, which also increases evaporative heat losses, Q_e.

6   The formation of ice at the sea-surface greatly influences the local heat budget; in particular, it leads to an increase in the albedo and a substantial decrease in Q_s, while Q_b is not much affected. Thus, once formed, ice tends to be maintained.

7   The principle of conservation of salt, combined with the principle of continuity, may be used to make deductions about the volume transports into and out of semi-enclosed bodies of water or, alternatively (if these are known), about the evaporation–precipitation balance in the region concerned.

8   Water masses are bodies of water that are identifiable because they have certain combinations of physical and chemical characteristics. The properties most used to identify water masses are temperature (strictly potential temperature) and salinity, because away from the sea-surface they may only be changed through mixing, i.e. they are conservative properties. Deep and bottom water masses, and 'thick' upper water masses, are formed in regions of convergence, and where deep convection results from the destabilization of

surface waters through cooling and/or increase in salinity. Whether a water mass can form in this way depends not on the absolute density of surface waters but on their density relative to that of underlying water.

9   Central water masses are 'thick' upper water masses that form in winter in the subtropical gyres. They are characterized by relatively high temperatures and relatively high salinities. The water mass that forms in the Sargasso Sea has a remarkably uniform temperature of about 18 °C. This '18 °C water' is an example of a mode water.

10   The most extensive intermediate water mass is Antarctic Intermediate Water (AAIW), which forms in the Antarctic Polar Frontal Zone. Like other intermediate water masses formed in subpolar regions (e.g. Labrador Sea Water), AAIW is characterized by relatively low temperature and relatively low salinity. Although Labrador Sea Water is referred to as an intermediate water mass, in some years it may be formed down to depths of 2000 m or more, and will contribute significantly to North Atlantic Deep Water (see 11). In contrast to AAIW and Labrador Sea Water, Mediterranean Water is characterized by relatively high temperature and relatively high salinity.

11   North Atlantic Deep Water is formed in winter, mainly through cooling of surface waters and ice-formation in the Greenland Sea. As in the Labrador Sea and the Mediterranean, the near-surface waters are more easily destabilized because isopycnals bow upwards as a result of cyclonic circulation. Deep convection seems to occur in small, well-defined regions ('chimneys') and produces a dense water mass that mixes at depth with a cold, highly saline outflow from the Arctic Sea. The resulting water mass circulates and accumulates in the deep basins of the Greenland and Norwegian Sea. Intermittently, it overflows the Greenland–Scotland ridge, and cascades down into the deep Atlantic, mixing with the overlying water masses. The densest overflow water eventually reaches the Labrador Sea where it mixes with overlying Labrador Sea Water to produce a slightly less dense variety of North Atlantic Deep Water than that which flows south at depth in the eastern basin of the Atlantic. There seems to be a 'sea-saw' relationship between the intensity of convection in the Labrador Sea and that in the Greenland Sea; this may be related to the North Atlantic Oscillation.

12   Antarctic Bottom Water is the most widespread water mass in the oceans. There are two types, the much more voluminous 'Circumpolar AABW', which flows north from lower levels in the Antarctic Circumpolar Current, and the extremely cold 'true' AABW, which forms around the Antarctic continent at various locations on the shelf. Ice-formation plays an important part in its production, particularly in coastal polynyas; cooling of surface waters by winds is also important, especially in 'open ocean' polynyas. Most of this extremely dense water never escapes the Antarctic region, but some formed in the Weddell Gyre spills over the Scotia Ridge and flows north in the western Atlantic and in the Indian Ocean. Like other deep water masses, it is affected by the Coriolis force and so flows as deep western boundary currents (as predicted by Stommel).

13   The deep water mass with the largest volume is Pacific and Indian Ocean Common Water. It is a mixture, being about half Antarctic Bottom Water, and half Antarctic Intermediate Water plus North Atlantic Deep Water.

14   Dense water tends to sink along isopycnic surfaces, and mixing along isopycnic surfaces occurs with a minimum expenditure of energy. However, mixing does occur between water of different densities. In certain circumstances, mixing also occurs as a result of double-diffusive processes occurring on a molecular scale (e.g. salt fingering). Generally, different water masses are envisaged as mixing at their boundaries, but 'meddies' are an example of eddies of one water mass being carried into the body of an adjacent water mass; eddies of Indian Ocean water, formed at the Agulhas retroflection, are another example.

15   Temperature–salinity diagrams may be constructed for different locations in the ocean. For a given location, the shape of the temperature–salinity plot is determined by the different water masses present and the extent to which they have mixed together; also (as in the case of central water masses) by variations within a particular water mass. Assuming that mixing occurs through turbulent processes only, temperature–salinity diagrams may be used to determine the proportions of different water masses contributing to the water at a given depth. Sequences of temperature–salinity diagrams may be used to trace the least-mixed (or core) layer of a water mass, as it spreads through the ocean.

16   For convenience, density (ρ) is usually written in terms of σ (sigma), where $\sigma = \rho - 1000$; temperature–salinity diagrams show equal-σ contours. Temperature–salinity diagrams now generally use potential temperature (θ) rather than *in situ* temperature (T) (hence also σ_θ rather than σ_t), because θ is corrected for adiabatic heating as a result of compression and so is more truly a conservative property than is T. Comparison of the trend of a θ–S curve with contours of σ_θ may be used to evaluate the degree of stability of a water column, but using θ instead of T does not compensate for the direct effect of pressure on seawater density. This effect may be significant, especially as compressibility is affected by temperature and salinity.

17   Because it is a conservative property, potential vorticity may be used to track water masses. Dissolved oxygen and silica concentrations, which are non-conservative properties, may be used to identify as well as track water masses. The 'age' of a water mass (the time since leaving the sea-surface) is indicated by its concentration of dissolved oxygen, and may be determined with some accuracy from the concentrations of certain other substances that enter the ocean at the surface. These include radioactive isotopes such as ^{14}C (radiocarbon) and ^{3}H (tritium) – which are formed both naturally and in nuclear tests – and chlorofluorocarbons, i.e. CFCs. Using radiocarbon data, residence times for water in the deep oceans have been estimated to be of the order of 250–500 years.

18   The 'global thermohaline conveyor' is a schematic representation of the *vertical* circulation (or 'meridional overturning circulation') of the global ocean. It encapsulates the idea of global convection being driven by the sinking of dense (cold, high salinity) water in the northern North Atlantic, with net heat transport in the topmost 1000 m or so of the ocean being southward in the Pacific and Indian Oceans and northward in the Atlantic Ocean. The overall net transport of freshwater within the oceans is believed to be in the opposite direction, i.e. southward in the Atlantic and northward in the Indian and Pacific Oceans.

19   The observational phase of the World Ocean Circulation Experiment (WOCE) took place during the 1990s. WOCE was designed to investigate large-scale fluxes of heat and freshwater, ocean variability and climate (both now the subject of the CLIVAR programme), and mechanisms of water-mass formation. The enormous amount of data collected during WOCE, combined with improvements in modelling techniques, have added greatly to our understanding of the ocean and its role in the global climate system. In the future, the rate of supply of oceanographic data will increase still further, through (for example) Argo floats, underwater autonomous vehicles and new satellites. This improved supply of data will enable assimilation of data into models to be an even more useful tool than it is at present. Plans for a multidisciplinary, Global Ocean Observing System (GOOS) are well advanced, and the programme may become operational around 2010.

Now try the following questions to consolidate your understanding of this and earlier Chapters.

QUESTION 6.14   Figure 6.45 shows the tracks of ALACE floats deployed at about 1000 m during WOCE, as recorded up until 1999. Bearing in mind that the floats were initially launched at various locations along the transects shown in Figure 6.42, outline briefly what the tracks indicate about the characteristics of current flow in (1) the equatorial region, (2) mid-latitudes in the ocean gyres.

Figure 6.45   Tracks of ALACE floats deployed in the Pacific during WOCE.

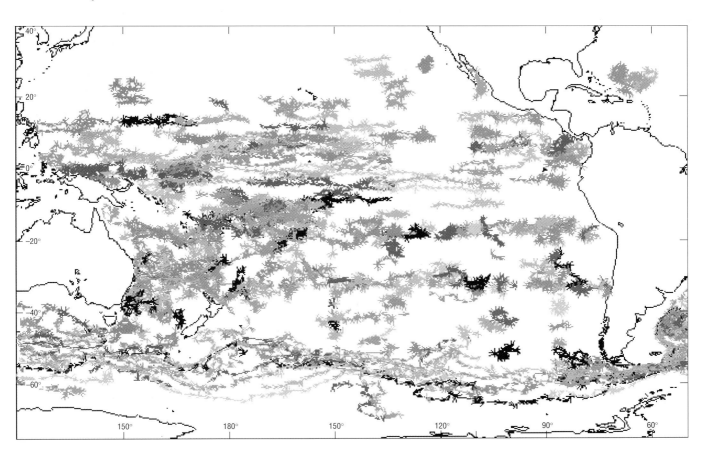

254

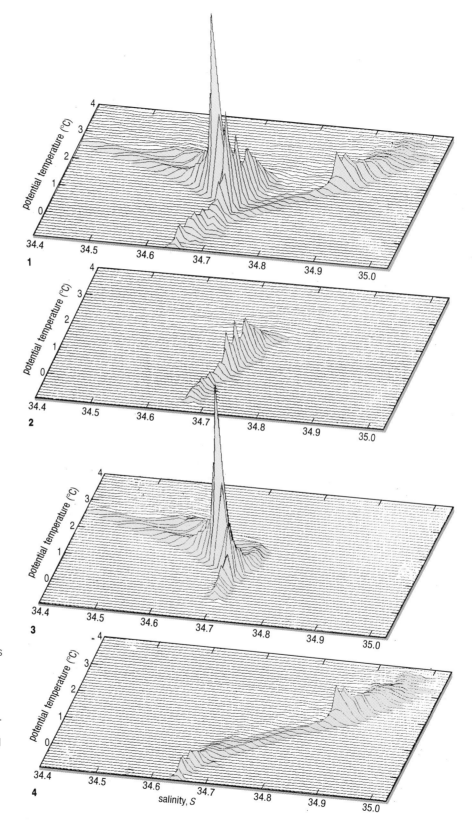

Figure 6.46 Computer-generated θ–S diagrams for water colder than 4 °C (for use with Question 6.17). The elevation of the surface corresponds to the volume of water with given combinations of θ and S, the volumes having been calculated for θ–S classes of 0.1 °C × 0.01. Diagram **1** is for the world ocean, **2** is for the Indian Ocean, and **3** and **4** are for the Pacific and Atlantic Oceans (not necessarily in that order). In **1** and **3**, the elevation of the highest peak corresponds to $26 \times 10^6 \text{ km}^3$ of water, in **2** it corresponds to $6.0 \times 10^6 \text{ km}^3$ of water, and in **4** to $4.7 \times 10^6 \text{ km}^3$.

QUESTION 6.15 As mentioned in Section 6.3.2, there was a decline in the rate of formation of Labrador Sea Water in the early 1970s, when the Great Salinity Anomaly passed through the north-west Atlantic. Can you suggest why?

QUESTION 6.16 As mentioned in the text, the Antarctic Convergence was renamed the Antarctic Polar Frontal Zone. What features of the zone mean that the use of the term 'frontal' is appropriate?

QUESTION 6.17 In Figure 6.46, diagram 1 is a computer-generated three-dimensional θ–S diagram for all ocean water with θ less than 4 °C, and diagrams **2–4** show similar information for the three individual ocean basins. In each diagram, the elevation of the surface corresponds to the volume of water with given θ–S characteristics.

(a) Diagram **2** corresponds to the Indian Ocean. Given the characteristics of North Atlantic Deep Water, can you say which of the remaining diagrams (**3** and **4**) corresponds to the Atlantic and which to the Pacific?

(b) What is the water mass, distinguishable in all the diagrams, which has potential temperatures less than 0 °C?

(c) To what is the peak (visible in diagrams **1**, **3** and, to some extent, **2**) attributable?

QUESTION 6.18 In Section 6.3.2, we stated that Pacific and Indian Ocean Common Water is approximately 50% Antarctic Bottom Water, with North Atlantic Deep Water and Antarctic Intermediate Water together making up the rest. Use Figure 6.47, along with the principles outlined in Section 6.4.2, to make your own estimates of the proportions of each of these water masses in Pacific and Indian Ocean Common Water.

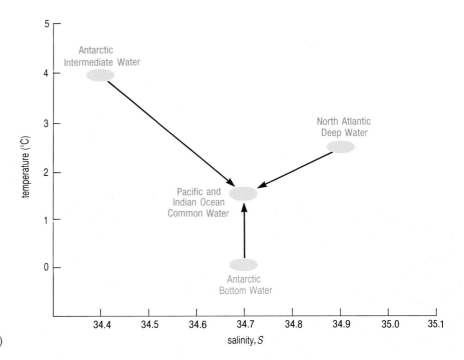

Figure 6.47   *T–S* diagram showing the characteristics of the water masses that mix together to form Pacific and Indian Ocean Common Water. (For use with Question 6.18.)

QUESTION 6.19 Briefly explain the greenhouse effect by reference to one of the terms in the heat-budget equation (Equation 6.1).

SUGGESTED FURTHER READING

GLANTZ, M.H. (2001) *Currents of Change: Impacts of El Niño and La Niña on Climate and Society* (2nd edn), Cambridge University Press. A readable discussion of the El Niño phenomenon in general (i.e. without heavy emphasis on the details of the science).

GORDON, A.L. AND COMISO, J.C. (1988) Polynyas in the Southern Ocean, in *Scientific American*, **257**, pp. 70–77. An informative article discussing not only how polynyas form and are maintained, but also their importance for the global climate.

KNAUSS, J.A. (1997) (2nd edn) *Introduction to Physical Oceanography*, Prentice-Hall. A concise overview of the physics of ocean circulation (with more mathematics than in this Volume), tides and surface waves, sound, optics and the oceanic heat budget.

McGREGOR, G.R. AND NIEUWOLT, S. (1998) *Tropical Climatology: An Introduction to the Climates of the Low Latitudes* (2nd edn) John Wiley and Sons. Although written from a meteorological viewpoint, this book illustrates the importance of the ocean for the tropical climate.

PICKARD, G.L. AND EMERY, W.H. (1982) *Descriptive Physical Oceanography: An Introduction* (4th (SI) enlarged edn), Pergamon Press. An introduction to the physical properties of the ocean and to oceanographic techniques (current measurement, etc.) along with descriptions of hydrography and circulation, region by region.

POND, S. AND PICKARD, G.L. (1983) *Introductory Dynamic Oceanography*, Pergamon Press. A discussion of the mathematical relationships used in physical oceanography, with explanations of their physical bases and descriptions of how they may be manipulated.

STOMMEL, H. (1965) *The Gulf Stream: A Physical and Dynamical Description* (2nd edn), University of California Press/Cambridge University Press. A classic text, which describes and explains ideas and theories about the Gulf Stream and hence about ocean circulation in general; contains a fair amount of mathematics.

SUMMERHAYES, C. AND THORPE, S. (1996) *Oceanography: An Illustrated Guide*, Manson Publishing. This well illustrated book is a collection of chapters on different aspects and disciplines of oceanography, with emphasis on recent scientific and technological breakthroughs.

TOMCZAK, M. AND GODFREY, J.S. (1994) *Regional Oceanography: An Introduction*, Pergamon Press. Descriptions of the current systems and water masses of the oceans of the world (and their adjacent seas), with some theoretical discussion.

WELLS, N. (1995) (2nd edn) The *Atmosphere and Ocean: A Physical Introduction*, John Wiley & Sons. This text covers a wide range of atmospheric and oceanic phenomena at a fairly advanced level.

WEIBE, P.H. (1982) Rings of the Gulf Stream, in *Scientific American*, **246** (March), pp. 50–60. A well-illustrated article, describing the formation and subsequent history of a number of warm-core and cold-core rings, along with general information about the impact of such rings on the oceanic environment.

WHITWORTH III, T. (1988) *The Antarctic Circumpolar Current*, in *Oceanus*, **31** (Summer) pp. 53–58. A lively discussion of the Antarctic Circumpolar Current in the context of the Antarctic Polar Frontal Zone and water masses. (This particular issue of *Oceanus* concentrates on legal, scientific and political problems concerning Antarctica.)

ANSWERS AND COMMENTS TO QUESTIONS

CHAPTER 1

Question 1.1   On the real Earth, the eastward velocity of the surface is greatest at the Equator and zero at the poles. For the hypothetical cylindrical Earth, the eastward velocity of the curved surface would be the same whatever the distance from the poles, and so a body moving above the surface would not appear to be deflected, and there would be no Coriolis force.

Question 1.2   (a) According to Figure 1.4(a), there must be a radiation surplus all year within about 10–13° of the Equator. Poleward of about 75° of latitude, there is always a deficit.

(b) (i) According to Figure 1.4(b), surface waters within 10° of the Equator average about 27 °C, while those poleward of 70° average about −1 °C. Thus, equatorial surface waters are about 28 °C warmer than surface waters poleward of 70°. In answering this question, you will no doubt have noticed that the temperature curve is asymmetrical, with its peak shifted northwards by about 10°. The reason for this will become clear in Chapter 2. (The curve in Figure 1.4(b) was included for comparison with those in (a). However, sea-surface temperatures are determined by gains and losses of heat *at the sea-surface* and, as will be discussed in Chapter 6, these are not just radiative gains and losses.)

(ii) False. In the vicinity of the Equator, day and night are of comparable length, being equal at the equinoxes. The high equatorial sea-surface temperatures are the result of relatively large amounts of radiation being received all year – i.e. a high *average* insolation. (As shown in Figure 1.4(a), the largest amounts of incoming solar radiation are received by *mid-latitudes* in summer.)

Question 1.3   (a) The area under the solid blue curve in Figure 1.5 is very slightly greater than that under the dashed blue curve, indicating that the ocean transports polewards fractionally more heat than does the atmosphere.

(b) The ocean contributes more to poleward heat transport in the tropics (between about 20° S and 30° N) and the atmosphere contributes more in mid-latitudes (~30°–55° S and 40°–65° N).

Question 1.4   (a) In addition to its southward firing velocity, the missile has the eastward velocity of the surface of the Earth at the Equator. As the missile travels southwards, the eastward velocity of the Earth beneath it becomes less and less, so that *in relation to the Earth* the missile is moving not only southwards but eastwards. Thus, a missile fired southwards from the Equator is deflected towards the east.

(b) (i) In the Northern Hemisphere, the Coriolis force deflects currents to the right, and so would tend to deflect southwards any current flowing initially eastwards.

(ii) On the Equator, the Coriolis force is zero and there is no deflection.

Question 1.5   Like that in Figure 1.4(a), the horizontal axis of (b) is scaled according to surface area, but in this case it is not the area of the Earth's surface that is relevant but the area of *ocean* surface. The proportion of the Earth's surface that is ocean is much smaller in the Northern Hemisphere than in the Southern Hemisphere, which between 30° and 65° of latitude is largely ocean. The horizontal axis is therefore even more compressed at northern high latitudes than southern high latitudes.

CHAPTER 2

Question 2.1   The answers to this question relate to the contrasting conditions associated with rising and sinking air. In the case of (1) subpolar lows, air spirals in cyclonically and rises (Figure 2.5(a)). The rising air expands and cools adiabatically, causing moisture in the air to condense, releasing heat to the atmosphere, and further encouraging convection. Condensation from the rising air produces clouds and rain.

In the case of (2) the subtropical high pressure regions, air is sinking (cf. Figure 2.5(b)). Sinking air becomes compressed and *warms* adiabatically, causing water droplets in it to evaporate. Thus subtropical highs are typically dry, with little cloud; land areas affected by subtropical highs tend to be arid, and include desert regions, such as the Sahara and Sahel.

Question 2.2   Easterly waves and tropical cyclones are generated in the vicinity of the equatorial low pressure zone/Intertropical Convergence Zone (zone of highest sea-surface temperature). Figure 2.3 shows that the ITCZ is generally displaced northwards with respect to the geographical Equator; in the Atlantic and eastern Pacific it generally does not lie more than 5° south of the Equator, even in the southern summer.

Question 2.3   (a) True. Moist air convects more readily than dry air, because when it rises, condensation releases latent heat of evaporation which partly offsets adiabatic cooling.

(b) False. *Rising* air cools adiabatically and cannot contain as much water vapour as air at ground/sea-level, so that clouds and precipitation result. The ITCZ is a global-scale example of rising air leading to high precipitation; the depressions of mid-latitudes are smaller-scale examples. *Sinking* air tends to be dry, as a result of warming adiabatically.

(c) True; see Figure 2.3(a) and (b).

(d) False. As discussed in the text in connection with monsoonal circulations, the flow of air over Eurasia is cyclonic (inwards and upwards) in the northern summer (Figure 2.3(a)); and anticyclonic (downwards and outwards) in the northern winter (Figure 2.3(b)).

(e) False. Figure 2.2(b) shows that in polar regions the tropopause (the top of the troposphere) is about 9–10 km above the surface of the Earth. It *is* about 15 km above the surface of the Earth in mid-latitudes.

(f) True. Undulations in the polar jet stream mean that the British Isles are sometimes beneath poleward-trending 'limbs' of the jet stream, and sometimes beneath equatorward-trending 'limbs', and so are sometimes affected by trains of lows/depressions (Figure 2.8), sometimes by highs.

Question 2.4 (a) Most of the clouds visible in the Hurricane 'Andrew' satellite image are at the *top* of the troposphere, where air which has spiralled cyclonically in and up, diverges and spirals outwards from the 'eye' anticyclonically (clockwise in the Northern Hemisphere) (cf. Figure 2.16). The inward spiralling cloud bands below are obscured. Animated satellite images of tropical cyclones, as sometimes seen on weather or news programmes, clearly show the outward anticyclonic flow.

(b) The 'fuel' that drives a tropical cyclone could be said to be water vapour, because it is the change in state from gas (water vapour) to liquid (cloud droplets) that provides the latent heat which enables the upward cyclonic movement of air around the central region of the cyclone to continue.

Question 2.5 See Figure A1. Note that the subtropical high pressure regions are dark in Figure 2.20 because they are regions where *dry* air sinks, in contrast to the ITCZ, which is a region where moist air rises.

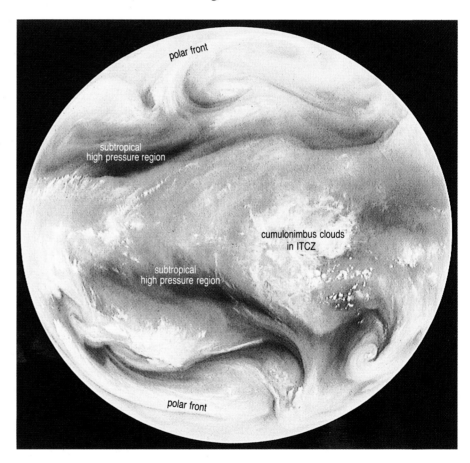

Figure A1   Answer to Question 2.5.

Question 2.6   The surface wind pattern shown in Figure 2.3(c) is for a day in the northern summer. It is quite hard to see the arrows on the flow lines, but there are three obvious clues which do not rely on your being able to see the wind direction. They are:

(1) in the eastern Pacific and Atlantic, the ITCZ is well north of the Equator (though not clearly defined on the western side of either ocean);

(2) a tropical cyclone is developing north-east of Taiwan (i.e. in the Northern Hemsiphere); and

(3) wind speeds are high, i.e. conditions are very stormy, in the Southern Ocean – compare the cyclonic subpolar storms with those in Figure 2.7(b).

In addition (though perhaps hard to see), there are strong anticyclones in the North Pacific and North Atlantic, a situation which corresponds to the typical pattern for the northern summer (cf. Figure 2.3(a)). In fact, the winds shown are those for 1 August 1999, and the tropical cyclone in the western North Pacific is typhoon 'Olga', *en route* to landfall in Korea.

Question 2.7   Examples given in this Chapter are the large-scale undulations of the jet stream in the upper westerlies (i.e. Rossby waves, Section 2.2.1) and the easterly waves in the Trade Wind belts (Section 2.3.1). You may also have thought of the waves that often occur in the polar front, leading to the formation of depressions and cyclonic storms (Figure 2.8).

CHAPTER 3

Question 3.1   (a) (i) The main cool surface currents in the subtropical Pacific are the California Current and the Peru or Humboldt Current, while (ii) in the subtropical Atlantic they are the Canary Current and the Benguela Current.

(b) All these currents flow equatorwards along the *eastern* boundaries of the oceans. (The characteristics of these 'eastern boundary currents' will be discussed in Chapter 4.)

Question 3.2   Using Equation 3.1 ($\tau = cW^2$), and substituting for the values of τ and W given, we have:

$$0.2 = c\,(10)^2$$

$$\therefore\ c = \frac{0.2}{200}$$

$$= 2 \times 10^{-3}$$

The units of c are $\dfrac{\mathrm{N\,m^{-2}}}{(\mathrm{m\,s^{-1}})^2} = \dfrac{(\mathrm{kg\,m\,s^{-2}})\,\mathrm{m^{-2}}}{\mathrm{m^2\,s^{-2}}} = \mathrm{kg\,m^{-3}}$

i.e. the units of density. Indeed, c is usually regarded as being composed of ρ_a, the density of the overlying air (about $1.3\,\mathrm{kg\,m^{-3}}$) multiplied by C_D, a dimensionless drag coefficient (of the order of 1.5×10^{-3}).

Question 3.3   In the thermocline, temperature decreases markedly with depth and, as a result, density *increases* markedly with depth. The water is well stratified and very stable and so mixing is inhibited. This is in contrast to the situation near the sea-surface where water is well mixed by wind and waves. The degree of turbulence, and hence the eddy viscosity, is therefore much greater in the mixed surface layer than in the thermocline.

Question 3.4 (a) The Coriolis parameter $f = 2\Omega \sin\phi$.

At 40° of latitude this is equal to: $2 \times (7.29 \times 10^{-5}) \times \sin 40° \, \text{s}^{-1}$

$$= 2 \times 7.29 \times 10^{-5} \times 0.643 \, \text{s}^{-1}$$

$$\approx 9.4 \times 10^{-5} \, \text{s}^{-1}$$

(b) (i) According to Ekman's theory, the surface current speed is given by:

$$u_0 = \frac{\tau}{\rho\sqrt{A_z f}}$$

Using the values given, and putting f equal to $10^{-4}\,\text{s}^{-1}$, to make the sums easier:

$$\mu_0 = \frac{0.1}{10^3\sqrt{10^{-1} \times 10^{-4}}}$$

$$= \frac{0.1}{10^3 \times 10^{-2}\sqrt{0.1}} = \frac{0.1}{10 \times 3.162}$$

$$= 0.036 \approx 0.03\,\text{ms}^{-1}$$

(ii) According to Ekman's theory, the surface current should be 45° *cum sole* of the wind direction. The situation under consideration is in the Southern Hemisphere, and so 45° *cum sole* is 45° to the left. The wind is westerly (towards the east) and so the surface current should be towards the north-east.

(c) According to the 'rule of thumb', the surface current is typically about 3% of the wind speed. In this case, the wind speed is $5\,\text{m s}^{-1}$ and so the surface current might be expected to be about $(3/100) \times 5 = 0.15\,\text{m s}^{-1}$. This is about five times the value we calculated in (b); however, in that calculation, we used a value for A_z towards the upper end of the likely range of values (10^{-5} to $10^{-1}\,\text{m}^2\,\text{s}^{-1}$), and a smaller value for A_z would have resulted in a larger value for u_0.

Question 3.5 The period of an inertia current is given by Equation 3.7:

$$T = \frac{2\pi}{f} \text{ where } f = 2\Omega \sin\phi$$

Now, $\Omega = 2\pi/24\,\text{hr}^{-1}$, and at the poles $\phi = 90°$ and $\sin\phi = 1$.

Therefore, $T = \dfrac{2\pi}{2 \times (2\pi/24) \times 1} = \dfrac{24}{2} = 12$ hours.

At a latitude of 30°, $\sin\phi = \frac{1}{2}$ and so

$$T = \frac{2\pi}{2(2\pi/24) \times \frac{1}{2}} = 24 \text{ hours.}$$

Question 3.6 (a) Pressure gradients act from areas of higher pressure to areas of lower pressure. In this case, the slope of the isobars tells us that at any given depth, the pressure is greater on the right than on the left. The horizontal pressure gradient force must therefore be represented by the arrow pointing to the left, and the Coriolis force by the arrow pointing towards the right (Figure A2).

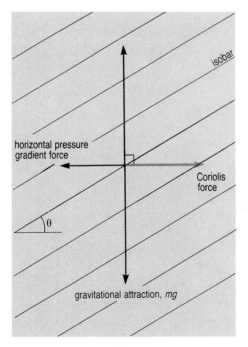

Figure A2 Answer to Question 3.6(a).

(b) We are told that the current is flowing into the page, and we know that the Coriolis force acts *cum sole* of the direction of motion, i.e. to the right in the Northern Hemisphere and to the left in the Southern Hemisphere. In this case, the Coriolis force is acting to the right of the direction of motion, and so the situation illustrated must be in the Northern Hemisphere.

Question 3.7 (a) The first step is to calculate f. At $30°$ of latitude, $\sin \phi = \frac{1}{2}$, so $f = 2\Omega \times 0.5 = 2 \times 7.29 \times 10^{-5} \times 0.5 = 7.29 \times 10^{-5}\,\mathrm{s}^{-1}$. In this case, $z_1 - z_0$ is $2000 - 1000 = 1000\,\mathrm{m}$, and we are assuming that h_B is the same:

$$\frac{\rho_B}{\rho_A} = \frac{1.0262 \times 10^3}{1.0265 \times 10^3}$$

$$= \frac{1.0262}{1.0265} = 0.9997$$

$L = 100 \times 10^3\,\mathrm{m}$ and $g = 9.8\,\mathrm{m\,s}^{-2}$.

Therefore, substituting into Equation 3.13:

$$u = \frac{gh_B}{fL}\left(1 - \frac{\rho_B}{\rho_A}\right)$$

$$= \frac{9.8 \times 1000\,(1 - 0.9997)}{7.29 \times 10^{-5} \times 10^5} = \frac{9800 \times 0.0003}{7.29}$$

$$= \frac{2.94}{7.29}$$

$$\approx 0.40\,\mathrm{m\,s}^{-1}$$

(b) (i) According to Equation 3.12:

$$u = \frac{g}{f}\left(\frac{h_B - h_A}{L}\right)$$

so $$\frac{fu}{g} = \frac{h_B - h_A}{L}$$

and $h_B - h_A = \dfrac{Lfu}{g}$

$$= \frac{10^5 \times 7.29 \times 10^{-5} \times 0.4}{9.8} = \frac{2.92}{9.8} = 0.29$$

$$\approx 0.3\,\mathrm{m}$$

The distance $h_B - h_A$ (which is the same as the difference between h_B and $z_0 - z_1$) is therefore about 0.03% of $z_0 - z_1$, so the assumption you made in part (a), that h_B and $z_0 - z_1$ were equal, was a reasonable one.

(ii) If $h_B - h_A \approx 0.3\,\mathrm{m}$, $\tan \theta = 0.3/L = 0.3/100 \times 10^3 = 0.000\,003$, which (typically) is a very small gradient indeed. The angle θ is correspondingly small, at $\sim 0.000\,17° \approx 0.000\,2°$.

(c) In the situation described here, as in the example given in the text, $\rho_A > \rho_B$ so the isobar p_1 slopes up towards B and the horizontal pressure gradient force acts from B to A. In the Southern Hemisphere, the Coriolis force acts to the *left* of the direction of current flow, so in conditions of geostrophic equilibrium the horizontal pressure gradient force must be acting to the *right* of the direction of flow. In this case, therefore, the current must be flowing out of the page, i.e. towards the south (as station B is due east of station A). For the diagram showing this, see Figure 3.17(a). Note that the arrow representing the pressure gradient force is *horizontal*.

Question 3.8 (a) The contours in Figure 3.21 show the dynamic topography of the sea-surface, assuming that the 1500 dbar isobaric surface is horizontal. For a more accurate picture, any topography in this 1500 isobaric surface would have to be 'added on' to that shown.

(b) You should have had no difficulty in identifying the Gulf Stream and the Antarctic Circumpolar Current.

(c) The stronger the geostrophic current, the closer together are the contours of dynamic height. Therefore, according to Figure 3.21 the fastest part of the Antarctic Circumpolar Current is to the east of the southern tip of Africa, at about 60° E.

(d) By inspecting the contour values to the north of the 4.4 dynamic metre closed contour, we can see that it represents a depression of the sea-surface, rather than a hill. In the Southern Hemisphere, isobaric surfaces slope up to the *left* of the current direction and so, given that the current is flowing clockwise, we would indeed expect the sea-surface to slope down towards the middle of the region defined by the closed contour.

Question 3.9 The conditions that would lead to convergence and sinking of surface water in the Southern Hemisphere are illustrated in Figure A3(a) and (b).

Question 3.10 (a) Kinetic energy is proportional to the *square* of current speed. So, if current speeds associated with eddies are 10 times those associated with the mean, the energy must be $10^2 = 100$ times bigger.

(b) According to Figure 3.32, a severe thunderstorm is the atmospheric phenomenon which has a kinetic energy closest to that of a mesoscale eddy.

(c) The Antarctic Circumpolar Current has the greatest kinetic energy. This is not surprising as, acted upon by prevailing westerlies, this broad current can flow unimpeded around the globe.

Question 3.11 (a) 1 Wind stress is an external force (i), as it acts on the upper surface of the ocean.

2 Viscous forces are internal frictional forces (ii). However, frictional forces act to oppose motion and so you would also have been justified in choosing (iii).

3 The tides are produced by the gravitational attraction of the Sun and the Moon, and so the tide-producing forces are external forces (i).

4 Horizontal pressure gradient forces that result from sea-surface slopes caused by variations in atmospheric pressure (i.e. when conditions are barotropic) could be argued to be external forces (i). However, it could also be argued that in baroclinic conditions, the horizontal pressure gradient force is partly an internal force (ii), as with increasing depth it is determined more and more by the distribution of mass (density) within the body of ocean water.

(It is also possible to argue that as *all* horizontal pressure gradient forces result from lateral differences in the pressure of overlying water, and these pressures derive ultimately from the gravitational attraction of the water to the mass of the Earth, horizontal pressure gradient forces are themselves external forces. Do not worry if you did not think of this.)

5 The Coriolis force is a secondary force (iii), as it only acts on water that is already moving. Note also that the Coriolis force is an *apparent* secondary force. In effect, it had to be 'invented' to explain movement relative to the Earth that results from the rotation of the Earth itself.

ANTICYCLONIC WINDS
IN SOUTHERN HEMISPHERE

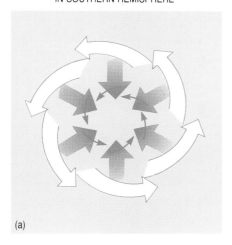

(a)

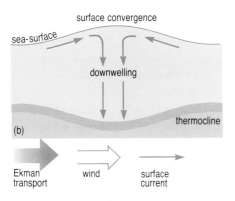

surface convergence

sea-surface

downwelling

thermocline

(b)

Ekman transport   wind   surface current

Figure A3   Answer to Question 3.9.

(b) (i) In geostrophic flow, the horizontal pressure gradient force is balancing the Coriolis force.

(ii) The mean flow of the Ekman layer at right angles to the wind is the result of the wind stress balancing the Coriolis force.

Question 3.12 At the end of Section 3.1.2, we stated that the Ekman transport – the total volume transport in the wind-driven layer – may be calculated by multiplying $\bar{u}$, the depth mean current, by the thickness of the wind-driven layer, D. From Equation 3.4, $\bar{u}$ is given by $\tau/D\rho f$. Multiplying this by D gives $\tau/\rho f$. A_z does not appear in this expression, and so we can say that the Ekman transport resulting from a given wind stress is independent of the eddy viscosity. This is an important result because it means that we can calculate Ekman transports without needing to know an appropriate value for eddy viscosity which, as you might imagine, is very hard to estimate.

Question 3.13 Inertia currents are indeed manifestations of the Coriolis force in action, because they are what results when the Coriolis force acts unopposed on already-moving water (ignoring frictional forces, which will eventually cause an inertia current to die away). (This obviously contrasts with the situation in a geostrophic current, where the Coriolis force is balanced by a horizontal pressure gradient force, and motion is at right angles to both.)

Question 3.14 (a) Because the water is well mixed, there will be no lateral variations in density, and so conditions will be barotropic.

(b) Flow is to the east, so the Coriolis force to the right of the flow, i.e. to the south, will be balanced by the horizontal pressure gradient force acting towards the left, i.e. to the north. In other words, the mean sea-level must be higher on the *southern* (French) side of the Straits.

Your sketch should therefore look something like Figure A4, with isobars parallel to the sea-surface, and horizontal arrows showing the Coriolis force balanced by the horizontal pressure gradient force. (If isopycnals had been drawn, they would have been parallel to the isobars.)

(c) If the slope of the sea-surface, and all other isobars, is given by tan θ, from the gradient equation (Equation 3.11):

$$\tan \theta = \frac{fu}{g}$$

$$= \frac{2\Omega \sin \phi \times 0.2}{9.8}$$

Taking Ω as $7.29 \times 10^{-5}\,\text{s}^{-1}$, and ϕ as $51°\,\text{N}$ (so that $\sin \phi = 0.778$),

$$\tan \theta = \frac{2 \times 7.29 \times 10^{-5} \times 0.778 \times 0.2}{9.8}$$

$$= 0.23 \times 10^{-5}$$

As the Straits of Dover are 35 km wide, this means that if the difference in height between the two sides is d m,

$$\tan \theta = 0.23 \times 10^{-5} = d/35\,000$$

$$\text{i.e. } d = 35\,000 \times 0.23 \times 10^{-5}\,\text{m}$$

$$= 0.08\,\text{m}$$

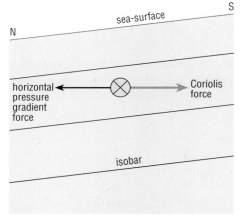

Figure A4   Answer to Question 3.14(b).

So, on average, the sea-level must be 0.08 m (8 cm) higher on the French side of the Straits.

Question 3.15   As described in Section 3.3.3, the geostrophic method can only determine large-scale flows because the hydrographic stations at which profiles of temperature and salinity are taken are typically many tens of kilometres apart and the measurements can take an appreciable time to make. This was particularly true before electronic equipment made it possible to obtain continuous vertical profiles of water properties – until relatively recently, researchers had to depend on individual temperature and salinity measurements made at specific depths. In short, the geostrophic method tends to conceal any small-scale, intermediate- or short-term phenomena, and provide information only about average conditions which approximate to the mean flow.

Question 3.16   As discussed in Section 3.1.2, wind-driven surface currents are deflected *cum sole* of the wind direction, i.e. to the right in the Northern Hemisphere, to the left in the Southern Hemisphere. As a result, cyclonic winds cause surface waters to move away from the centre of the low pressure region. This divergent flow of surface waters results in colder water upwelling from below (Section 3.4) – an effect that would be very marked if the winds were the powerful winds of a tropical cyclone, although as the cyclone would be moving the upwelling would be somewhat to the rear of the storm. The situation illustrated in Figure 3.24(a) would probably not arise, as a full Ekman spiral current pattern, and Ekman transport at right angles to the wind-direction, is not likely to have time to develop in such a fast-moving system.

Note: In answering this question, it may have struck you that there appears to be a contradiction between information in Section 2.3.1 and Section 3.4. In the former, we stated that the low atmospheric pressure associated with tropical cyclones leads to a *rise* in sea-level, while in the latter we stated that cyclonic winds lead to a *lowering* of sea-level. There is no conflict here: both mechanisms operate together. Whether the sea-level at a given location is raised or lowered will depend on a number of factors. In particular, cyclonic winds tend to lead to large rises in sea-level (storm surges) when they blow over shallow, semi-enclosed areas of sea where water can pile up (e.g. the southern North Sea, or the Ganges–Brahmaputra Delta).

Question 3.17   (a) The Antarctic Circumpolar Current flows through the Drake Passage (Figure 3.1). The contribution of mesoscale eddies to the spectrum is likely to be above average because such eddies are known to form from meanders in the Antarctic Circumpolar Current, which is a powerful frontal current.

(b) Yes it does. The broad 'hump' between about 30 days and 200 days must correspond to mesoscale eddies, which have periods of one to a few months (the beginnings of a similar hump may be seen in Figure 3.30).

(c) There was no need to do any calculations. All the peaks shown in Figure 3.33 have periods of longer than a day. The latitude of the Drake Passage is about 60° S, and the inertial period T must therefore be close to that of the inertia current in the Baltic at 57° N (Figure 3.8), i.e. about 14 hours. (In fact, for 60° of latitude, $T = 2\pi/f$ is $(2 \times 3.142)/(2 \times 7.29 \times 10^{-5} \times \sin 60°) = 0.498 \times 10^5$ s = 13.8 hours.)

Question 3.18 (a) The regions of the ocean that show the greatest variation in sea-surface height are regions with fast currents: the Gulf Stream, the Kuroshio, the Agulhas Current and the Antarctic Circumpolar Current (Figure 3.1), which are in fact all frontal currents. Such intense currents are associated with slopes in the sea-surface – the faster the current, the greater the slope and the more elevated some areas of the sea-surface become (Figure 3.21 and Frontispiece). However, Figure 3.34 shows *variability* in sea-surface height: as discussed in Section 3.5 (and Question 3.17), the pattern seen in Figure 3.34 is a result of the fact that these fast currents are not large steady 'rivers' of water but are continually shifting and changing, producing meanders and eddies. (Variability also occurs in regions of slow surface currents but in these cases the sea-surface slopes will be small, and the variations in the height at a given point, associated with variations in current flow, will also be small.)

(b) The equatorial current systems do not show up clearly on Figure 3.34 mainly because the slope of the sea-surface associated with geostrophic currents is proportional to the Coriolis parameter, f, and hence to the sine of the latitude. The Coriolis force is therefore zero on the Equator and very small close to it, and sea-surface slopes are correspondingly small.

CHAPTER 4

Question 4.1 (a) Both Franklin and Rennell believed that the Gulf Stream was the result of water being piled up in the Gulf of Mexico, so that the Stream was driven by a head of water and flowed in such a way as to tend to even out the horizontal pressure gradient. Maury clearly did not understand that in this case it would be the variation of pressure *at a given horizontal level* that would be important; in other words, he did not understand about horizontal pressure gradients.

(b) Figure 3.1 shows that the volume of water transported in the Gulf Stream must *increase* between the Straits of Florida and Cape Hatteras because of the addition of water that has recirculated in the subtropical gyre. (The volume transport is also increased as a result of mixing and entrainment of water adjacent to the Gulf Stream.)

Question 4.2 At that time, the idea of eddy viscosity – i.e. of frictional coupling resulting from turbulence – had not been formulated. If the effect of the wind were transmitted by molecular viscosity alone, it would indeed take a long time for major ocean currents to be generated. Once the role of turbulence was appreciated, it became easier to understand how the wind could drive currents (Section 3.1.1, especially Figure 3.5). Incidentally, the breakthrough came in 1883 when Reynolds published his now famous paper on the nature of turbulent flow in pipes; his influential ideas on the generation of frictional forces by turbulence were published in 1894.

Question 4.3 (a) The component of the Earth's rotation about a vertical axis at latitude ϕ is $\Omega \sin \phi$. Vorticity is $2 \times$ angular velocity, and so the planetary vorticity possessed by a parcel of water on the surface of the Earth at latitude ϕ is $2\Omega \sin \phi$. This is the expression for the Coriolis parameter, f (Section 3.1.2.).

(b) (i) and (ii) At the North and South Poles, the planetary vorticity will be equal to $2\Omega \sin 90°$, i.e. 2Ω, because $\sin 90° = 1$. As illustrated in Figure 4.7(a), the surface of the Earth rotates anticlockwise in the Northern Hemisphere and clockwise in the Southern Hemisphere. If we use the convention, given in the text, that clockwise rotation corresponds to negative vorticity, the planetary vorticity possessed by a parcel of fluid at the South Pole will be written as -2Ω.

(iii) At the Equator, $\phi = 0$, and so planetary vorticity $= 2\Omega \sin 0 = 0$.

Question 4.4 (a) (i) A body of water carried southwards from the Equator is moving into regions of increasingly negative planetary vorticity (Figure 4.7(a)). (ii) Its absolute vorticity $f + \zeta$ must remain constant, so to compensate for the decrease in f, its relative vorticity ζ must *increase*. In other words, the water must increasingly acquire a tendency to rotate with positive vorticity (i.e. anticlockwise) in relation to the Earth. (If the force originally driving the water ceases to act, we will have a simple inertia current, anticlockwise in the Southern Hemisphere – Figure 3.7(a).)

(b) Negative relative vorticity will be acquired by a body of water from (i) winds blowing in a clockwise direction, *and* (ii) from cyclonic winds in the Southern Hemisphere, because these are also clockwise (Figure 4.5).

Question 4.5 (a) and (b) In both cases, if D is reduced then for f/D to remain constant f must be reduced. While over the shallow bank, the current must therefore be diverted towards the Equator; this involves it curving southwards in case (a) and northwards in case (b). Note that it does not make any difference what the direction of current flow is initially.

Question 4.6 (a) The resemblance is fairly close. In both cases the overall motion is clockwise (cf. Figure 4.5): the part of the wind field between 15° and 30° N corresponds to the more southerly part of the subtropical anticyclonic gyre, especially the Trade Winds, which have a strong easterly component; the part between 30° and 45° N corresponds to the westerlies forming the higher-latitude part of the subtropical anticyclones.

(b) Ekman transport is to the right of the wind in the Northern Hemisphere and so will be in the directions shown by the wide arrows in Figure A5.

(c) The Ekman transports towards the centre of the ocean will converge and lead to a raised sea-surface (cf. Figure 3.24(c) and (d)). As a result, there will be horizontal pressure gradient forces acting northwards and southwards away from the centre (black arrows on Figure A5), so that geostrophic currents will flow in the directions shown by the thin blue arrows (cf. Figure 3.24).

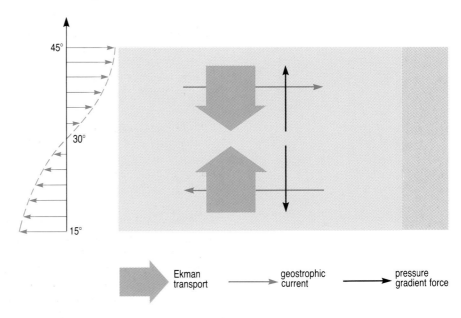

Figure A5   Answer to Question 4.6(b) and (c).

Question 4.7   The flow pattern in (2) is symmetrical, like that in (1), and so the existence of the Coriolis force *per se* cannot be the reason for the intensification of western boundary currents. However, when the Coriolis force is made to increase with latitude – albeit linearly – the flow pattern does show the asymmetry observed in the real oceans. Figure 4.12 therefore strongly suggests that the intensification of western boundary currents is a result of the fact that the Coriolis parameter (and hence the Coriolis force) increases with latitude.

Question 4.8   (a)   If $dw/dt = 0$, and there are no forces acting vertically other than the weight of the water and the vertical pressure gradient (i.e. $F_z = 0$), Equation 4.3c becomes:

$$0 = -\frac{1}{\rho}\frac{dp}{dz} - g$$

$$+\frac{1}{\rho}\frac{dp}{dz} = -g$$

$$\text{or } dp = -\rho g dz$$

This is the hydrostatic equation (3.8), which relates the pressure at a given depth to the weight of the overlying seawater.

Question 4.9   The western boundary currents are fast and deep, while eastern boundary currents are slow and shallow. This allows the rate at which water is transported polewards in the western boundary currents to be equal to the rate at which water is carried equatorwards in the eastern boundary currents, despite the fact that western boundary currents are narrow, and flow in the eastern boundary currents occupies much of the rest of the ocean. (Here we are ignoring currents flowing into and out of the gyres – in the case of the North Atlantic gyre, the Labrador Current and the northern branches of the North Atlantic Current.)

Question 4.10   (a)   The sections indicate that flow through the Florida Straits is baroclinic. Figure 4.21(a) and (b) show that there are strong lateral variations in *T*, *S* and hence density, and the slopes of the isotherms indicate that the isopycnals slope up to the left so as to intersect the sea-surface (the topmost isobaric surface). Furthermore, the velocity sections (c) and (d) show that the current changes with depth, decreasing to zero above the sea-bed, consistent with baroclinic conditions (cf. Figure 3.15(b)).

(b)   As this is in the Northern Hemisphere, the sea-surface must slope up to the right and the current must flow 'into the page' so that Florida is on the left-hand (west) side of the section and Bimini on the right-hand (east) side.

(c)   The isotherms and isohalines (Figure 4.21(a) and (b)) indicate that, in the upper 100 m, the least dense water (warmest *and* lowest salinity) is in the middle of the section. As a result, although the sea-surface will generally slope up to the right, there will be a slight 'hump' in the middle (see Figure A6).

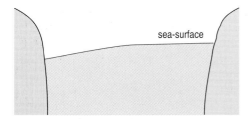

sea-surface

Figure A6   Answer to Question 4.10(c).

Question 4.11   In geostrophic flow (to which flow in mesoscale eddies approximates), motion is along contours of dynamic height, in such a direction that (in the Northern Hemisphere) the 'highs' are on the right and the 'lows' are on the left. Thus, the positive numbers correspond to anticyclonic (here clockwise) flow, and the negative numbers to cyclonic

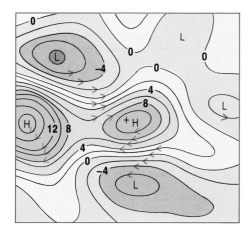

Figure A7   Answer to Question 4.11.

(anticlockwise) flow. As shown in Figure A7, flow in mesoscale eddies resembles that in the atmosphere – cyclonic around 'lows' and anticyclonic around 'highs'.

Question 4.12   Comparison of Figure 4.29(a) and (b) suggests that results obtained by Lagrangian methods may be more difficult to interpret than those obtained by Eulerian methods. On the other hand, they do provide an idea of current flow patterns, and give information about a large area. Eulerian measurements provide information about how a current changes with time at a given place, which Lagrangian methods do not, but a very large number of fixed current meters would be needed to give equivalent spatial coverage. In practice, experiments using drifters do involve a large number of them – in the MODE experiment, 20 Sofar floats were required to monitor adequately a region about 300 km wide – but they are very much cheaper than moored current meters.

Question 4.13   (i) The blue area on Figure 4.31(a) must be the cold water of the Labrador Current; and (ii) the yellow area must be warm Sargasso Sea and Gulf Stream water, with the reddest region corresponding to the warm core of the Stream. In image (b) these warm waters are shown blue, as they have low primary (phytoplankton) productivity.

Question 4.14   False. Both Franklin and Rennell believed that the Gulf Stream is driven by the 'head' of water in the Gulf of Mexico, i.e. by the horizontal pressure gradient *in the direction of flow*. They were not aware of the idea of geostrophic balance between the Coriolis force and a horizontal pressure gradient at *right angles* to the flow.

Question 4.15   As it is a frontal region, the edge of the Gulf Stream is also a *convergence*, where surface debris will tend to accumulate (cf. Section 3.4).

Question 4.16   The changes in depth associated with the seamounts mean that D in $(f + \zeta)/D$, the potential vorticity of water in the current, is changing. If $(f + \zeta)/D$ is to remain constant, a reduction in D must be accompanied by a decrease in the value of f and/or ζ – i.e. by equatorward meanders (cf. Question 4.5) and/or a tendency to clockwise rotation. (Note that as the Gulf Stream is a region of strong velocity shear, we cannot assume that $f \gg \zeta$.)

Question 4.17   Flow is in geostrophic equilibrium when the only significant forces acting in a horizontal direction are the horizontal pressure gradient force and the Coriolis force, acting in opposite directions and at right angles to the flow. In such circumstances, the flow is steady and non-accelerating. Water circulating in a relatively small-scale meander or eddy experiences a centripetal force toward the centre of the eddy (as given in Equation 3.5), and its rate of change of speed and direction (i.e. its acceleration) becomes significant.

Question 4.18   (a) This is true. The average velocity will be calculated from the estimates of the distance travelled and time taken. As the actual path taken by the drifter is likely to be complicated, the distance it has travelled may well be underestimated, especially if its position has been fixed infrequently or not at all (as in the case of cheap plastic drifters). Also, if a drifter is washed ashore and not found for a long time, the travel time will be overestimated.

(b) This is not strictly true. Some Lagrangian techniques rely on people voluntarily returning drifters, stating when and where they were found. However, any oceanographer will tell you that moored current meters are at the mercy of other users of the sea lanes, and may be easily damaged accidentally.

(c) False. The Doppler methods (ADCPs and OSCR) are Eulerian methods as they measure the velocity of patches of water at given positions. They provide no information about the paths taken by specific parcels of water.

(d) False. The calculation of surface currents using ship's drift is a Lagrangian technique because it relies on the deduction of the path taken by the ship under the influence of the current. It is the average current velocity over some distance that is calculated, rather than the current speed at any particular location. A significant proportion of all current measurements obtained by Lagrangian methods have been calculated from ship's drift.

CHAPTER 5

Question 5.1 (1) The sea-surface slopes down from the north towards the divergence at 10° N, leading to a southward horizontal pressure gradient force; if there is to be geostrophic equilibrium, this must be balanced by the Coriolis force acting northwards. The Coriolis force acts to the right of the current in the Northern Hemisphere, and so geostrophic flow must be westward.

(2) The same argument can be used here.

(3) In the Southern Hemisphere, on the other hand, the Coriolis force acts to the left of the current. A sea-surface slope down from the south will lead to a northwards horizontal pressure gradient force which, in geostrophic flow, will be balanced by the Coriolis force acting to the south, i.e. to the left of a westward-flowing current.

Question 5.2 The position of the ITCZ, the zone in which the Trade Winds meet, does not simply run east–west across the Atlantic. It is distorted by the effect of continental masses, in particular the bulge of the African continent. The wind field that results is reflected in the pattern of surface currents.

Question 5.3 (a) (i) According to Figure 5.5(a), the Cromwell Current is about 200 m deep. It is 4° of latitude wide, which is $4 \times 110 \times 10^3 = 440\,000$ m = 440 km. It is therefore $440\,000/200 = 2200 \approx 2000$ times wider than it is deep. (Do not worry if your values were slightly different – the main purpose of this question was to give you practice in interpreting oceanographic data and making order-of-magnitude estimates.)

(b) (i) The cross-sectional area of the Cromwell Current is about $440\,000 \times 200 = 88 \times 10^6$ m^2.

(ii) Figure 5.5(a) suggests that the average velocity in the Cromwell Current is about 0.4 m s^{-1}. This implies a volume transport of about $88 \times 0.4 \times 10^6 \approx 35 \times 10^6$ m^3 s^{-1}.

This is quite close to 31×10^6 m^3 s^{-1}, the higher of the two estimates shown in Figure 5.2. However, it is important to remember that averages can conceal wide variations, and in the northern summer the volume transport of the Cromwell Current may be as high as 70×10^6 m^3 s^{-1}.

Question 5.4 (a) As discussed in connection with Figure 4.38, to the north of Cape Blanc the Trade Winds blow roughly parallel to the coast so as to cause offshore transport of surface water, and hence upwelling, more or less all the year round. The same is true to the south of Cape Frio. Between Cape Blanc and Cape Frio, wind directions vary markedly in response to the changing position of the ITCZ, and so upwelling occurs only seasonally.

(b) The upwelling in the region between the Equator and about 4° S is at least partly the result of the surface divergence caused by the South-East Trades crossing the Equator (as illustrated in Figure 5.1(a)).

Question 5.5 (a) Winds are cyclonic around regions of low pressure (see, for example, Figures 2.3 and 2.5).

(b) When the ITCZ is in its most northerly position, it passes over the Guinea Dome which is then affected by cyclonic winds. As illustrated in Figure 3.24(a) and (b), cyclonic winds lead to a raised thermocline, surface divergence and upwelling, so the effect of the ITCZ is to intensify the 'doming' of the isotherms in the near-surface waters.

Question 5.6 No, it is not possible. As the Coriolis force acts to the right of the direction of motion in the Northern Hemisphere and to the left in the Southern Hemisphere, the opposing pressure gradient, resulting from the sea-surface slope, must act to the left in the Northern Hemisphere and to the right in the Southern Hemisphere. This condition is satisfied by Kelvin waves travelling with the coast on the right in the Northern Hemisphere, and on the left in the Southern Hemisphere. (This is not the whole story, however, because the coastal current and the Kelvin wave can travel in opposite directions. How this comes about is beyond the scope of this Volume.)

Question 5.7 The Coriolis parameter, f, is equal to $2\Omega \sin \phi$. At 10° N, $\sin \phi = 0.174$, so $f = 2 \times (7.29 \times 10^{-5}) \times 0.174 = 2.5 \times 10^{-5}\,\text{s}^{-1}$. Hence, the Rossby radius of deformation, $L = c/f = 2/(2.5 \times 10^{-5})\,\text{m} = 0.8 \times 10^5\,\text{m} = 80\,\text{km}$.

Question 5.8 Perhaps the most obvious features are:

The Agulhas retroflection (cf. Figure 5.14)

The Falklands Current (cf. Figure 3.1)

Flow along the northern limit of the Antarctic Circumpolar Current. The lengths of the arrows (i.e. the distances travelled between transmissions) indicate that – not surprisingly – this was where floats had been travelling the fastest.

You may also have spotted the float caught in the Benguela Current (Figure 3.1). Other noticeable features, not specifically discussed in the text, are an eddy or gyre to the north-east of the Falkland Islands, and what looks like a retroflection off the tip of South America.

Question 5.9 Yes, in principle, monsoon winds are the same as land and sea breezes. In both cases, the wind direction depends on differential heating of land and sea, being from land to sea in the winter/night when the atmospheric pressure over the land is greater than that over the sea, and from sea to land in the summer/day when the atmospheric pressure is higher over the sea. However, monsoon winds are intensified by the increased convection generated by the release of latent heat, when water condenses to form rain. Also, of course, they are on a much greater scale, so their flow paths are affected by the Coriolis force and they interact with winds in the upper troposphere.

Question 5.10 It is a western boundary current, in the form of the Somali Current, which in the northern summer is an intense poleward flow, similar to those found in the other oceans (and in the southern Indian Ocean).

Question 5.11 To estimate the Rossby radius of deformation we need to know where the Kelvin wave's amplitude has fallen to ~0.4 of its maximum amplitude, which is between 22.5 and 27.5 m. We can assume for convenience that the maximum amplitude is 24 m, and $0.4 \times 24\,m = 9.6\,m \approx 10\,m$. The contour corresponding to 10 m would be between those for 7.5 m and 12.5 m, i.e. at about 3–4° of latitude, or 330–440 km. (Baroclinic equatorial Kelvin waves usually have Rossby radii of 100–250 km, so this is rather large; see Section 5.3.1.)

Question 5.12 In the case of shelf waves, it is the depth D that varies, rather than latitude (i.e. planetary vorticity, f), as is the case in Rossby waves (cf. the discussion of topographic steering in Section 4.2.1). If the current flowing parallel to a particular depth contour is displaced either towards or away from the coast, D will decrease or increase. For potential vorticity $(f + \zeta)/D$ to remain constant (and assuming for convenience that latitude remains more or less constant) the relative vorticity ζ must change. The current will oscillate to and fro about the original depth contour, with alternate cyclonic and anticyclonic eddies forming along it (cf. Figure 5.19(b)).

Question 5.13 (a) The *Fram* must have travelled with the Transpolar Current.

(b) A distance of 4000 km in three years gives an average current speed of:

$$\frac{4000 \times 10^3}{3 \times 365 \times 24 \times 3600} \approx 0.04\,m\,s^{-1}$$

This is within the usual range of estimated mean current speeds obtained using ice-drift (0.01–0.04 m s⁻¹).

CHAPTER 6

Question 6.1 (a) The effect of water in the atmosphere is greater at low latitudes because evaporation occurs at a greater rate from a warmer sea-surface, and warm air can contain more water vapour than cold air. This is borne out by Figure 6.2, in which the distortion of contours over the oceans is greatest at very low latitudes. Away from tropical regions, the difference in insolation between ocean and land areas does not seem to show any obvious relation to latitude.

(b) (1) The relatively low values of incoming solar radiation over Indonesia are a result of the Indonesian Low, which is a region where moist air is drawn in, leading to vigorous convection and cumulonimbus formation, especially during monsoon rains (Figure 5.21).

(2) The south-easterly bulge in the 205 W m⁻² contour is a result of the South Pacific High which, like all regions affected by anticyclonic air flow, is characterized by dry subsiding air and clear skies.

Note that as Figure 6.2 shows annual averages, seasonal variations have been smoothed out, so the 'highs' and 'lows' in the contours are less marked than they would be in a similar map for a particular season.

Question 6.2   The heat-budget equation is:

$$
\begin{array}{ccccc}
\text{energy gained} & & \text{energy lost} & \text{energy lost} & \text{energy lost by} \\
\text{as short-wave} & = & \text{as long-wave} + & \text{as latent} + & \text{conduction to} \\
\text{radiation} & & \text{radiation} & \text{heat} & \text{atmosphere}
\end{array}
$$

i.e. $Q_s = Q_b + Q_e + Q_h$.

Question 6.3   (a) The areas in question are affected by the Kuroshio and the Gulf Stream. The water carried from low latitudes in these western boundary currents is significantly warmer than the overlying air, and so the transfer of heat from sea to air by conduction/convection is above average here. In addition, evaporation occurs readily from a warm sea-surface, and the overlying warmed air can contain more water vapour than cooler air before becoming saturated. Both Q_h and Q_e are increased by turbulent convection in the atmosphere, which is encouraged by the warming of the air in contact with the sea-surface.

(b) By contrast, in the eastern equatorial Pacific, sea-surface temperatures are anomalously low because of upwelling of cooler subsurface water (Section 5.1.2). When the air is warmer than the sea, the sea gains heat from the air by conduction and Q_h is negative. A cool sea-surface results in the overlying air being stable and convection being inhibited. This, along with the cool sea-surface and the relatively high humidity of air cooled by contact with the sea-surface, leads to low values of Q_e.

Question 6.4   (a) In general, surface salinities are higher in the Atlantic than in the Pacific. This is particularly true in the Northern Hemisphere where, at a given latitude, values are about 2 (parts per thousand) higher in the Atlantic than in the Pacific.

(b) A notable feature of Figure 6.11(b) is the large negative $E - P$ value and low salinity just north of the Equator. This is the result of the high precipitation characteristic of the ITCZ.

(c) The northern summer is the time of the South-West Monsoon, when the rivers entering the Bay of Bengal – the Ganges and the Brahmaputra – will be swollen with rainwater (adding to meltwater from Tibet and the Himalayas). We would therefore expect the surface waters of the Bay of Bengal to have a region of very low salinity, corresponding to the outflow from the great rivers.

Question 6.5   (a) From the information given, $F = -7 \times 10^4\,\text{m}^3\,\text{s}^{-1}$. From the estimated values of S_1 and S_2,

$$
\frac{S_1}{S_2} = \frac{36.3}{37.8} = 0.9603
$$

$$
\frac{S_2}{S_1} = \frac{37.8}{36.3} = 1.041
$$

Substituting in Equations 6.5a and 6.5b, we get

$$
V_2 = \frac{-7 \times 10^4}{1 - 1.041} = \frac{-7 \times 10^4}{-0.041} = 1.71 \times 10^6\,\text{m}^3\,\text{s}^{-1}
$$

and $V_1 = \dfrac{-7 \times 10^4}{0.9603 - 1} = \dfrac{-7 \times 10^4}{-0.0397} = 1.76 \times 10^6\,\text{m}^3\,\text{s}^{-1}$

Thus, according to our calculations, the rate of inflow through the Straits of Gibraltar is about $1.76 \times 10^6\,\mathrm{m^3\,s^{-1}}$, while the rate of outflow is $1.71 \times 10^6\,\mathrm{m^3\,s^{-1}}$.

An alternative method would have been to use Equations 6.2 and 6.3 and hence obtain two equations relating V_1 and V_2 which could then be solved.

(b) Dividing the total volume by V_1, the rate of inflow, gives:

$$\frac{3.8 \times 10^6 \times 10^9\,\mathrm{m^3}}{1.76 \times 10^6\,\mathrm{m^3\,s^{-1}}} = 2.16 \times 10^9\,\mathrm{s}$$

$$\text{or } \frac{2.16 \times 10^9}{365 \times 24 \times 3600} = 68\ \text{years}$$

That is, it would take about 70 years for all the water in the Mediterranean to be replaced once.

Question 6.6 Important regions of convergence in mid-latitudes are the centres of the subtropical gyres, where anticyclonic winds lead to Ekman transport towards the centres of the gyres (Figure 3.24(c) and (d)). The high-latitude region of convergence that should have come to mind is, of course, the Antarctic Polar Frontal Zone, previously known as the Antarctic Convergence (Figure 5.31).

Question 6.7 (a) The cross-sections in Figure 6.17 indicate the presence of a large volume of Labrador Sea Water because they show water with very similar temperature and salinity characteristics occupying a region from about 100 m depth down to about 2000 m, in the central part of the gyre. (As discussed in the text following this question, mode waters such as Labrador Sea Water are by definition characterized by such pycnostads, but this is an unusually thick one.)

(b) The contour patterns in Figure 6.17 are consistent with a doming of isotherms (isopycnals), divergence of surface water and upwelling, under the influence of cyclonic winds (as shown in Figure 3.24). The fact that relatively dense water has been brought close to the surface means that the uppermost layers are easily destabilized by extra cooling (or an increase in surface salinity).

(The distribution in (c) indicates oxygen dissolving in surface water which is being mixed down into upwelled water which is 'older' and so somewhat depleted in oxygen. That is why concentrations in the centre of the gyre are not as high as around the edge.)

Question 6.8 Meltwater is fresh and therefore of low density. The formation of a layer of low-density water at the surface *increases* the stability of the upper water column and so tends to 'damp out' vertical convection.

Question 6.9 See Figure A8 for the result. Measurement of segments a and b shows that they are in the proportion of $1:2$, which means that the mixture contains twice as much of water type II as of water type I (i.e. water type I makes up 33% of the mixture). You could have obtained the same result mathematically by using values of T (or S). By proportion,

$$\frac{b}{a} = \frac{3-2}{5-3}\left(\text{or } \frac{34.85 - 34.5}{35.5 - 34.85}\right) = \frac{1}{2}$$

However, the graphical method is useful because it helps you to estimate proportions of contributing water masses 'by eye': the closer the position of a mixture to a particular water mass, the more of that water mass must be in the mixture.

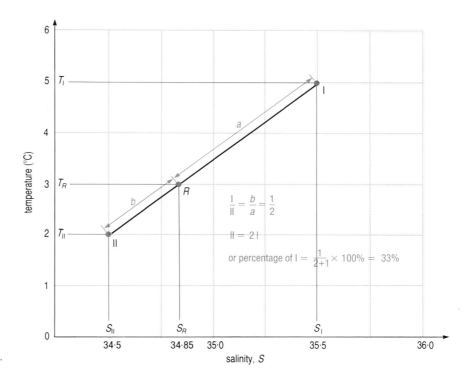

Figure A8   The completed T–S diagram for Question 6.9, showing water types I and II mixed in proportions of 1 : 2, to give a resultant mixture with a temperature of 3 °C and a salinity of 34.85.

Question 6.10   If you find *Meteor* Station 200 on Figure 6.19, you will see that at this location (9° S) warm, saline upper water flows above cooler, fresher AAIW, below which flows relatively saline NADW and finally, at depth, AABW. All these features may be identified in the T–S curve in Figure 6.33. (Remember that in Figure 6.19 the 10 °C and 0 °C isotherms and the 34.8 isohaline are *not* actually boundaries between water masses, although they give fairly good indications of the relative positions of the different water masses.)

(a) Yes, the water at 800 m is the eroded core of Antarctic Intermediate Water (see Figure A9).

(b) If you draw construction lines as shown in Figure A9 (cf. Figure 6.31) and measure the segments a and b, you get:

$$\frac{b}{a + b} \times 100 = 55\%$$

So, the percentage of Antarctic Intermediate Water at 800 m depth is about 55%.

The graphical method enables you to find out the proportion of one water mass without having to find it for all three. However, for completeness, the proportions of the other water masses contributing to the mixed water at 800 m are about 25% for North Atlantic Deep Water and about 20% for the water at 400 m depth.

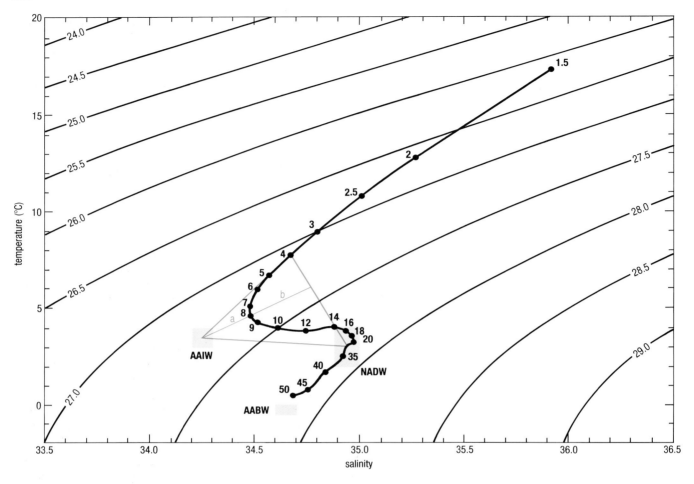

Figure A9 The completed *T–S* diagram for Question 6.10. The ratio *a*:*b* indicates that the proportion of Antarctic Intermediate Water in the core is about 55%.

Question 6.11 (a)(i) The water mass below North Atlantic Central Water is of high salinity (reaching nearly 35.8) and is warm (9–10 °C). This, combined with the latitude and the depth of the water mass, indicates clearly that it is Mediterranean Water. The trend of the θ–S curve in this region is more or less parallel to the σθ contours, reflecting the spread of Mediterranean Water at depths where it is neutrally buoyant. Interestingly, there seems to be a 'double core' of Mediterranean Water. This could be because Mediterranean Water of slightly different densities is spreading out along two different isopycnic surfaces (corresponding to σθ ~ 27.6 and ~ 27.55), allowing a local intrusion of slightly fresher, cooler water between 900 and 1000 m depth.

(ii) The deepest water mass represented by the θ–S curve is North Atlantic Deep Water, clearly identified by its temperature–salinity combination of ~2–4 °C and ~35.0 (as well as by the geographical location of the station). If the θ–S curve had been plotted for sufficiently great depths, some influence of Antarctic Bottom Water would have been detected.

(b) The two water masses are of approximately the same density. However, if you plotted them on Figure 6.34, you would have seen that when they mix together, the resulting water mass (on the line joining the two points) is *denser* than either of the original contributions (with σθ ~ 28.05 as opposed to 28.00).

Question 6.12   Yes, the two water masses can be distinguished. The lowest silica concentrations ($< 40\,\mu\text{mol}\,l^{-1}$), corresponding to North Atlantic Deep Water, are seen between 1500 m and 3000 m. Below that, concentrations increase again, reaching $125\,\mu\text{mol}\,l^{-1}$, in Antarctic Bottom Water. The way the contours of silica concentration relate to the sea-bed topography strongly suggests that both North Atlantic Deep Water and Antarctic Bottom Water are flowing along the western boundary of the ocean (albeit in different directions), as deep western boundary currents (cf. Figure 6.21). The same can be said of the contours of oxygen concentration in Figure 6.37(a).

Question 6.13   If, as is now thought, the Gulf Stream is driven partly by the sinking of water in the subpolar seas to the north-east of the Atlantic (Section 4.3.1), a reduction in the rate of formation of deep water would cause the Gulf Stream to slow down (and/or penetrate less far north). Currently, the waters of the Gulf Stream/North Atlantic Current give up large amounts of heat (and moisture) to the atmosphere over the northern Atlantic, and this is carried to north-west Europe in the prevailing westerlies. If this heat-supply declined markedly, the climate of north-west Europe would become much colder (in particular, winters would become longer and harsher).

Question 6.14   (1) In equatorial regions, flow is predominantly zonal and fast (though not as fast as that in the Antarctic Circumpolar Current, cf. Question 5.8). (Remember these floats are at about 1000 m depth.)

(2) Within the subtropical gyres, flow is much slower and apparently more random – some floats in the eastern Pacific have hardly moved from their launch sites, which can still be distinguished because they are along a line of longitude (cf. Figure 6.42).

You may also have spotted some locations where floats seem to have been somehow 'trapped', for example those apparently in a gyre just to the north of the Falklands Plateau.

Question 6.15   As discussed in Section 6.3.1, Labrador Sea Water is formed when the density of surface water in the Labrador Sea is increased through cooling and an increase in salinity, as a result of cold, dry winds blowing off the adjacent continent in winter. A large volume of low salinity water moving into the area would reduce the density of the upper part of the water column and make it much less likely that it could be destabilized in this way. (We can assume that the low salinity water was not exceptionally cold.)

Question 6.16   Features you may have thought of are (i) the convergence of surface waters, (ii) sloping boundaries between denser and less dense water (cf. sloping isotherms/isohalines on Figures 5.31 and 6.26), (iii) swift currents flowing along these boundaries (Figure 5.32), and (iv) a tendency for meanders and eddies to form from these currents (Section 5.5.2).

Question 6.17   (a) The high-salinity ridge evident in diagram **4** corresponds to North Atlantic Deep Water. Diagram **4** is therefore for the Atlantic, while diagram **3**, with much lower salinities, is for the Pacific.

(b) The low-temperature water mass visible on all the diagrams (though clearest in **4**, for the Atlantic) is Antarctic Bottom Water.

(c) The peaks in diagrams **1**, **2** and **3** correspond to the huge volume of water making up Pacific and Indian Ocean Common Water.

Question 6.18 Our answer is illustrated in Figure A10. The results obtained here differ from those of the researchers who first published estimates of these proportions. They obtained the following values: Antarctic Bottom Water, 60%; North Atlantic Deep Water, 24%; Antarctic Intermediate Water, 16%. The discrepancy probably arises because these researchers allowed for the fact that the temperature of Antarctic Bottom Water is raised by heat flow through the sea-bed.

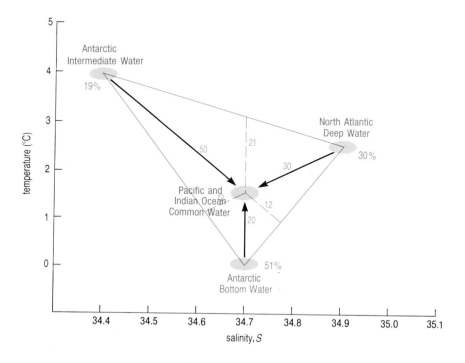

Figure A10 Completed *T–S* diagram for the answer to Question 6.18. The lengths of the segments *a–f* are given in millimetres.

Question 6.19 As discussed in Section 6.1.1, the surfaces of the oceans and continents emit long-wave (back-) radiation which is absorbed by clouds and water vapour and other gases including carbon dioxide. As a result of this warming, these all emit radiation of a longer wavelength in all directions, including back to the Earth's surface. This reduces the *net* amount of long-wave radiation emitted by the sea-surface, and hence reduces the heat-loss term Q_b. Because the concentration of carbon dioxide in the atmosphere is increasing, the amount of heat absorbed in the atmosphere and re-emitted back towards the surface of the Earth is also increasing, so the net heat loss from the sea-surface as Q_b is probably *decreasing* (along with heat lost from the continents as long-wave radiation).

ACKNOWLEDGEMENTS

The Course Team wishes to thank the following: Neil Wells and Martin Angel (Volume Assessor and Course Assessor), Harry Bryden, Alexandre Ganachaud, Simon Josey, Mike McCullough, Adrian New and Roger Proctor. The Course Team would also like to thank the many students, tutors and other readers who provided useful feedback on the first edition. This second edition builds on the first, which benefited from a advice from Martin Angel, Mike Davey, John Harvey, Malcolm Howe, Mike Hosken, Peter Killworth, Mary Llewellyn, Denise Smythe-Wright, Phillip Woodworth and Chris Vincent; John Harvey also contributed to Chapter 6.

The structure and content of the Series as a whole owes much to the experience of producing and presenting the first Open University course in Oceanography (S334) from 1976 to 1987. We are grateful to those people who prepared and maintained that Course, to the tutors and students who provided valuable feedback and advice and to Unesco for supporting its use overseas.

Grateful acknowledgement is also made to the following for material used in this Volume:

Table 2.1 C.J. Neumann (1993) in *Global Guide to Tropical Cyclone Forecasting*, WMO/TC-No. 560, World Meteorological Organization; *Frontispiece, Figures 3.34, 5.25* CLS, France; *Figures 1.1, 2.15, 3.22, 3.28(d), 3.31(a), 4.31, 6.23* NASA; *Figure 1.4(a,b)* G.R. McGregor and S. Nieuwolt (1998) *Tropical Climatology: An Introduction to the Climates of Low Latitudes*, (2nd edn), Wiley; *Figure 1.4(c)* J.W. Hedgpeth (1957) *Treatise on Marine Ecology and Palaeoecology*, Vol. 1, *Memoirs of Geol. Soc. of America*, **67**; *Figure 1.5* B.C. Carrissimo *et al.* (1985) in *Journal of Physical Oceanography*, **15**, American Meteorological Society; *Figures 2.1, 2.20* European Space Agency; *Figures 2.2(b), 2.3(a,b)* A.H. Perry and J.M. Walker (1977) *The Ocean–Atmosphere System*, Longman; *Figure 2.3(c)* NASA/NOAA-sponsored data system, Seaflux, JPL, courtesy W. Timothy Liu and Wenquing Tang; *Figure 2.7(b)* NOAA/NESDIS/NCDC/SDSD; *Figure 2.8(b)* Meteorological Office; *Figure 2.11(a)* L.R.A. Melton; *Figure 2.11(b)* H.G. Mullett; *Figure 2.13* Institute for Computational Earth System Science, University of California, Santa Barbara; *Figure 2.18* J.G. Harvey (1976) *Atmosphere and Ocean*, Artemis Press; *Figure 2.19* J.H. Golden, courtesy Meteorological Office; *Figures 3.2, 5.11* J. Crease *et al.* (eds.) (1985) *Essays on Oceanography, Progress in Oceanography*, **14**, Pergamon; *Figure 3.8* G. Neumann (1968) *Ocean Currents*, Elsevier; *Figures 3.19(a), 5.26, 6.14* G.L. Pickard and W.J. Emery (1982) *Descriptive Physical Oceanography*, Pergamon; *Figures 3.21, 3.33, 6.46* B.A. Warren and C. Wunsch (eds) (1981) *Evolution of Physical Oceanography*, MIT Press; *Figure 3.26* NSF/NASA-sponsored US Global Flux Study Office, Woods Hole, with Goddard Space Flight Center, University of Miami, University of Rhode Island; *Figure 3.28(a)* M. Hosken; *Figure 3.28(b)* J.B. Wright; *Figure 3.30* F. Webster (1969) in *Deep-Sea Research*, **16**, Pergamon; *Figure 3.31(b)* Steve Groom, NERC (BOFS Project); *Figure 3.32* N. Wells (1986) *The Atmosphere and Ocean*, Taylor and Francis; *Figure 4.1(a)* L. de Vorsy Jr. (1974) *The Atlantic Pilot* (facsimile of De Brahm (1772)) Univ. of Florida Press; *Figure 4.4* A.M. Colling; *Figures 4.11, 4.12, 4.14, 4.30* H. Stommel (1965) *The Gulf Stream*, University of California Press; *Figure 4.19(b,c)* R. Proctor and J. Wolf, Proudman Oceanographic Laboratory; *Figure 4.20(b) (base map)* Marie Tharp, 1 Washington Ave., NY 10960; *Figures 4.21, 6.6, 6.26* H.U. Sverdrup *et al.* (1942) *The Oceans*, Prentice-Hall; *Figure 4.22* C.A.M. King (1975) *Introduction to Physical and Biological Oceanography*, Arnold; *Figure 4.23* W.S. von Arx (1962) *An Introduction to Physical Oceanography*, Addison-Wesley; *Figure 4.24(a)* The MODE Group (1978) in *Deep-Sea Research*, **25**, Elsevier; *Figure 4.24(b)* P. Rhines (1976) in *Oceanus*, **19**, Woods Hole;

Figure 4.25 J.C. McWilliams (1976) in *Journal of Physical Oceanography*, **6**, American Meteorological Society; *Figures 4.26(a), 4.27* Ian Waddington, George Deacon Division, Southampton Oceanography Centre; *Figure 4.28(a,b)* Wimpol; *Figure 4.29* A.R. Robinson (ed.) *Eddies in Marine Science*, Springer-Verlag; *Figure 4.32* D. Tolmazin (1985) *Elements of Dynamic Oceanography*, Allen & Unwin; *Figure 4.33* R.E. Cheyney and J.G. Marsh (1981) *Journal of Geophysical Research*, **86**, Americal Geophysical Union; *Figure 4.35* DYNAMO (EU MAST project no. MAS2-CT93–0060) S. Barnard *et al.*; *Figure 4.37* J.A. Knauss (1997) *Introduction to Physical Oceanography*, Prentice-Hall/Pearson; *Figure 4.3* J.L. Fellous and A. Morel/ESA; *Figures 4.39, 6.18* A. Defant (1961) *Physical Oceanography*, Vol. 2, Pergamon; *Figure 4.41* Y. Kushnir (1999) Europe's winter prospects, *Nature*, **398**; *Figures 5.2, 5.6, 5.14, 6.17* M. Tomczak and J.S. Godfrey (1994) *Regional Oceanography: An Introduction*, Pergamon; *Figure 5.4(a,b)* J.A. Knauss (1978) *Introduction to Physical Oceanography*, Prentice-Hall; *Figure 5.4(c)* L. Lemasson and B. Piton, Institut Française de Recherche Scientifique pour le Developpement en Corporation; *Figure 5.5(a,b)* American Meteorological Society; *Figure 5.5(c,d)* P. Pullen and D. Paul/NOAA/ PMEL and RD Instruments; *Figures 5.7/5.8* B. Voituriez in *Oceanologica Acta*, **4**, Gauthiers Villars; *Figure 5.9* B. Voituriez and A. Herbland, Centre Océanologique de Bretagne; *Figures 5.10, 5.15(b)* J. Merle (1980) in *Journal of Physical Oceanography*, **10**, American Meteorological Society; *Figure 5.12* A.N. Cutler and J.C. Swallow (1984) *Surface Currents of the Indian Ocean, Institute of Oceanographic Sciences Report 187*, Southampton Oceanography Centre; *Figures 3.31(c), 5.13* CZCS/SeaWiFS data, courtesy NASA SeaWiFS Project and OrbImage Inc.; *Figure 5.20* D.B. Enfield (1987) in *Endeavour*, New Series, Pergamon; *Figure 5.24(a)* N. Nicholls (1993) in *South-East Asia's Environmental Future: The Search for Sustainability*, ed. H. Brookfield and Y. Byron, UN University Press/Oxford University Press; *Figure 5.24(b)* NOAA–CIRES Climate Diagnostics Center, University of Colorado; *Figure 5.27* NASA (1992) *Arctic and Antarctic Sea Ice, 1978–1987: Satellite Passive–Microwave Observations and Analysis* (NASA SP-511); *Figure 5.28* David Ellett (Scottish Marine Biological Association); *Figures 5.30, 5.31* T. Whitworth III (1988) in *Oceanus*, **31**, Woods Hole; *Figure 5.32* S. Cunningham, Southampton Oceanography Centre; *Figure 5.33* W.B. White and R.G. Peterson (1996) in *Nature*, **380**; *Figure 6.5* B.W. Meeson *et al.* (1995) *ISLSCP Initiative I — Global Data Sets for Land–Atmosphere Models, 1987–1988*, Vol. 1–5, published on CD by NASA; *Figures 6.7–6.10* S. Josey *et al.* (1998) *Southampton Oceanography Centre (SOC) Ocean–Atmosphere Heat, Momentum and Freshwater Flux Atlas*, SOC Report No. 6; *Figures 6.13, 6.30* P. Tchernia (1980) *Descriptive Regional Oceanography*, Pergamon; *Figures 6.15, 6.16* W.J. Emery and J. Meinke (1986) in *Oceanologica Acta*, **9**, Gauthier Villars; *Figure 6.21(a)* L.V. Worthington (1969) in *Deep Sea Research*, **16** (suppl.), Pergamon; *Figure 6.21(b)* R.R. Dickson *et al.* (1990) in *Nature*, **344**; *Figure 6.27* P.L. Richardson *et al.* (1989) *Journal of Physical Oceanography*, **19**, No. 3, American Meteorological Society; *Figure 6.28* M.C. Gregg (1973) in *Scientific American*, **228**, W.H. Freeman; *Figure 6.34* IFREMER, Brest, France; *Figure 6.36* L.D. Talley and M.S. McCartney (1982) in *Journal of Physical Oceanography*, **12**, American Meteorological Society; *Figure 6.37* J.L. Reid *et al.* (1977) in *Journal of Physical Oceanography*, **7**, American Meteorological Society; *Figures 6.38(c), 6.39* D. Smythe-Wright, George Deacon Division, Southampton Oceanography Centre; *Figure 6.40* H. Stommel (1958) in *Deep-Sea Research*, **5**, Pergamon; *Figures 5.34, 6.42, 6.45* WOCE International Project Office, Southampton Oceanography Centre; *Figure 6.43(a)* A. Ganachaud (2000) *Large-scale oceanic circulation and fluxes of freshwater, heat, nutrients and oxygen* (Ph.D Thesis, MIT–Woods Hole); *Figures 6.43(b), 6.44* A. Ganachaud and C. Wunsch (2000) in *Nature*, **408**; *Figure 6.47* B. Bolin and H. Stommel (1961) in *Deep-Sea Research*, **8**, Pergamon.

INDEX

283